PROGRESS IN

Molecular Biology
and Translational Science

Volume 103

PROGRESS IN

Molecular Biology and Translational Science

Molecular Assembly in Natural and Engineered Systems

edited by

Stefan Howorka

Department of Chemistry
Institute of Structural and Molecular Biology
University College London
London, United Kingdom

Volume 103

AMSTERDAM • BOSTON • HEIDELBERG • LONDON
NEW YORK • OXFORD • PARIS • SAN DIEGO
SAN FRANCISCO • SINGAPORE • SYDNEY • TOKYO

Academic Press is an imprint of Elsevier

Academic Press is an imprint of Elsevier
32 Jamestown Road, London, NW1 7BY, UK
Radarweg 29, PO Box 211, 1000 AE Amsterdam, The Netherlands
225 Wyman Street, Waltham, MA 02451, USA
525 B Street, Suite 1900, San Diego, CA 92101-4495, USA

This book is printed on acid-free paper. $\circledinf$

Library of Congress Cataloging-in-Publication Data
A catalog record for this book is available from the Library of Congress

British Library Cataloguing in Publication Data
A catalogue record for this book is available from the British Library

ISBN: 978-0-12-415906-8
ISSN: 1877-1173

For information on all Academic Press publications
visit our website at elsevierdirect.com

Printed and Bound in the USA
11   12   13   14   10   9   8   7   6   5   4   3   2   1

Working together to grow
libraries in developing countries

www.elsevier.com | www.bookaid.org | www.sabre.org

ELSEVIER   BOOK AID
              International   Sabre Foundation

Contents

Spider Silk: Understanding the Structure–Function Relationship of a Natural Fiber 131

Martin Humenik, Thomas Scheibel, and Andrew Smith

Protein Modifications Giving Rise to Homo-oligomers 187

Georg E. Schulz

Alpha-Helical Peptide Assemblies: Giving New Function to Designed Structures 231

Elizabeth H.C. Bromley and Kevin J. Channon

Nanobiotechnology with S-Layer Proteins as Building Blocks 277

Uwe B. Sleytr, Bernhard Schuster, Eva M. Egelseer, Dietmar Pum, Christine M. Horejs, Rupert Tscheliessnig, and Nicola Ilk

Viral Capsids as Self-Assembling Templates for New Materials 353

Michel T. Dedeo, Daniel T. Finley, and Matthew B. Francis

Contributors

Numbers in parentheses indicate the pages on which the authors' contributions begin.

Elizabeth H.C. Bromley, Department of Physics, Durham University, Durham, United Kingdom (231)

Kevin J. Channon, Department of Physics, Cavendish Laboratory, University of Cambridge, Cambridge, United Kingdom (231)

Michel T. Dedeo, Department of Chemistry, University of California; and Materials Sciences Division, Lawrence Berkeley National Laboratories, Berkeley, California, USA (353)

Eva M. Egelseer, Department of NanoBiotechnology, University of Natural Resources and Life Sciences, Vienna, Austria (277)

Daniel T. Finley, Department of Chemistry, University of California; and Materials Sciences Division, Lawrence Berkeley National Laboratories, Berkeley, California, USA (353)

Matthew B. Francis, Department of Chemistry, University of California; and Materials Sciences Division, Lawrence Berkeley National Laboratories, Berkeley, California, USA (353)

Christine M. Horejs, Department of NanoBiotechnology, University of Natural Resources and Life Sciences, Vienna, Austria (277)

Stefan Howorka, Department of Chemistry, Institute of Structural and Molecular Biology, University College London, London, United Kingdom (73)

Martin Humenik, Lehrstuhl Biomaterialien, Universität Bayreuth, Bayreuth, Germany (131)

Nicola Ilk, Department of NanoBiotechnology, University of Natural Resources and Life Sciences, Vienna, Austria (277)

Walter Keller, Institute of Molecular Biosciences, Structural Biology, University of Graz, Graz, Austria (73)

Tea Pavkov-Keller, Institute of Molecular Biosciences, Structural Biology, University of Graz, Graz, Austria (73)

Dietmar Pum, Department of NanoBiotechnology, University of Natural Resources and Life Sciences, Vienna, Austria (277)

Han Remaut, Structural & Molecular Microbiology, VIB/Vrije Universiteit Brussel, Brussels, Belgium (21)

Thomas Scheibel, Lehrstuhl Biomaterialien, Universität Bayreuth, Bayreuth, Germany (131)

Georg E. Schulz, Institut für Organische Chemie und Biochemie, Albert-Ludwigs-Universität, Freiburg im Breisgau, Germany (187)

Bernhard Schuster, Department of NanoBiotechnology, University of Natural Resources and Life Sciences, Vienna, Austria (277)

Uwe B. Sleytr, Department of NanoBiotechnology, University of Natural Resources and Life Sciences, Vienna, Austria (277)

Andrew Smith, Lehrstuhl Biomaterialien, Universität Bayreuth, Bayreuth, Germany (131)

Sophia J. Tsai, UCLA Department of Chemistry and Biochemistry, UCLA-DOE Institute for Genomics and Proteomics, Los Angeles, California, USA (1)

Rupert Tscheliessnig, Department of NanoBiotechnology, University of Natural Resources and Life Sciences, Vienna, Austria (277)

Nani Van Gerven, Structural & Molecular Microbiology, VIB/Vrije Universiteit Brussel, Brussels, Belgium (21)

Gabriel Waksman, Institute of Structural and Molecular Biology, University College London and Birkbeck College, London, United Kingdom (21)

Todd O. Yeates, UCLA Department of Chemistry and Biochemistry, UCLA-DOE Institute for Genomics and Proteomics, Los Angeles, California, USA (1)

Preface

The topic of this book—regular protein assemblies—is highly interdisciplinary, rapidly expanding, and of great interest for several reasons. Natural protein assemblies are essential components in every virus or biological cell. They act as cytoskeletal scaffold, storage containers, or molecular engines for directional transport, cell contraction, and locomotion. Protein assemblies are also of interest due to their sophisticated molecular architecture. While difficult to analyze, their structural complexity is increasingly elucidated in atomistic details. Moreover, the self-organization of individual molecular entities into higher-order systems is—from the biophysical perspective—an appealing yet not fully understood phenomenon. Further, the inspiration by Nature has led researchers to rationally design synthetic protein- and peptide-based higher-order structures that expand the scope of biology. Finally, natural and engineered protein assemblies can be exploited in biomedical or nonbiological applications. For example, natural systems are already important vaccine targets, while engineered structures are being increasingly utilized in nanobiotechnology for drug delivery. In either case, the exploitation of the self-assembling proteins takes advantage of their repetitive nature and the defined morphological shape. These built-in characteristics can optionally be combined with new functional modules to generate biomaterials of designed properties.

Molecular Assembly in Natural and Engineered Systems is one of the first books to capture this divergent field in its full breadth and depth. The book brings together different disciplines such as structural biology, protein engineering, and materials science to provide a comprehensive account of the multifaceted research activities.

Three main aspects are covered. First, the book summarizes recent progress in the structural analysis of natural protein assemblies with morphologies of spheres, fibers and sheets.

Spherical bacterial microcompartments: The chapter of *Tsai and Yeates* describes the structure and biology of the metabolic compartments that carry out specific reaction cascades in their interiors.

Elongated bacterial pili and flagella: *Van Gerven, Waksman, and Remaut* offer insight into the biology and structure of adhesion pili and secretion systems which are key players in pathogenic processes such as recognition and colonization of target surfaces, biofilm formation, and signaling events.

Sheet-like bacterial S-layers: *Pavkov-Keller, Howorka, and Keller* review the structure of S-layers. These protein lattices cover the cell wall of many biotechnologically relevant or pathogenic bacteria and occur on almost all archaea where they constitute the only cell wall component.

The fibers of the remarkably tough yet light-weight dragline spider silk also belong to ordered protein systems. The spidroin spider proteins assemble into parallel aligned crystallites which are surrounded by elastic nonordered regions. *Humenik, Scheibel, and Smith* examine the structural and biological determinants at this intriguing interface between structural order and disorder.

Representing the second aspect, the book demonstrates how structural understanding has inspired the creation of engineered protein or peptide-based assemblies.

Following the concept of structure-based intervention, *Schulz* explains how otherwise monomeric proteins can be engineered to form homo-oligomers of defined architecture.

In a complementary chapter, *Bromley and Channon* show that alpha-helical peptide assemblies can be rationally designed to add new function to the bottom-up structures useful for applications in cell biology.

The last topic of the book emphasizes how natural and engineered systems can be exploited for a wide range of applications including biocatalysis, biodelivery, or materials science, to name a few.

Sleytr, Schuster, Egelseer, Pum, Horejs, Tscheliessnig, and Ilk demonstrate that S-layer proteins are modular building blocks for the production of new supramolecular materials and nanoscale devices which can be used in molecular nanotechnology, nanobiotechnology, biomimetics, and synthetic biology.

New materials can also be generated using viral capsids, which as shown by *Dedeo, Finley, and Francis*, can be exploited as self-assembling templates to assemble chromophores for use in light harvesting and photocatalytic applications, or as delivery vehicles for drug and imaging cargo.

Several aspects of biotechnological applications are also covered in most of the previous chapters about natural and engineered assembly systems.

The book aims to reach a broad audience of researchers in structural biology, molecular biology, bionanotechnology, biochemistry, biophysics, and related areas, as well as advanced undergraduate and graduate students.

I hope that you enjoy reading this book which demonstrates that protein assemblies are a highly interdisciplinary field which—guided by exciting structural biology—spans the nano- and microscale and can find many bio-related applications while transcending into nonbiological areas.

STEFAN HOWORKA
London, August 2011

Bacterial Microcompartments: Insights into the Structure, Mechanism, and Engineering Applications

SOPHIA J. TSAI AND
TODD O. YEATES

UCLA Department of Chemistry and Biochemistry, UCLA-DOE Institute for Genomics and Proteomics, Los Angeles, California, USA

Bacterial microcompartments are large supramolecular assemblies, resembling viruses in size and shape, found inside many bacterial cells. A protein-based shell encapsulates a series of sequentially acting enzymes in order to sequester certain sensitive metabolic processes within the cell. Crystal structures of the individual shell proteins have revealed details about how they self-assemble and how pores through their centers facilitate molecular transport into and out of the microcompartments. Biochemical and genetic studies have shown that enzymes are directed to the interior in some cases by special targeting sequences in their termini. Together, these findings open up prospects for engineering bacterial microcompartments with novel functionalities for applications ranging from metabolic engineering to targeted drug delivery.

I. Introduction

Bacterial microcompartments (BMCs) are large, protein-based assemblies present inside many bacterial cells.[1] They were first visualized in the late 1950s and 1960s inside cyanobacteria as electron dense, polyhedrally shaped bodies, reminiscent of viruses or phage capsids.[2,3] They are now recognized to be metabolic compartments that carry out specific series of reactions in their interiors.[4–6] Various types of BMCs sequester different metabolic pathways in

Progress in Molecular Biology
and Translational Science, Vol. 103
DOI: 10.1016/B978-0-12-415906-8.00008-X

different bacterial lineages. These different types of BMCs are unified by having an outer shell assembled from protein subunits that are homologous (i.e., evolutionarily related) between the various systems.[7] The three-dimensional structures of numerous shell proteins have recently been elucidated, providing considerable insight into questions of architecture, biochemical mechanisms, and evolutionary relationships.[7–17] Together with biochemical and bioinformatic studies, this structural understanding has opened up exciting prospects for designing BMCs with novel properties, and for using BMCs as a broad framework for systems biology research and applications.

II. Bacterial Microcompartment Form and Function

BMCs vary considerably in size and shape (Fig. 1). The smallest BMCs known are those of the alpha-type carboxysome. In *Halothiobacillus neapolitanus*, diameters of their alpha-carboxysomes have been reported to be in the range 80–120 nm.[18,19] BMCs of this type are also the most geometrically regular among those that have been visualized by electron microscopy. Recent electron cryo-tomography studies have shown that alpha-carboxysome BMCs are nearly icosahedral in shape; 20 roughly flat triangular facets come together at 30 edges and 12 pentagonal vertices.[18,20] According to these dimensions, BMC shells are composed of at least a few thousand shell protein subunits.[13,15,21] It is notable that, even for these most regular BMCs, their sizes vary within a given cell, with an individual cell typically containing between a few and several BMCs. Other types of BMCs (besides carboxysomes) generally appear to be less regular,

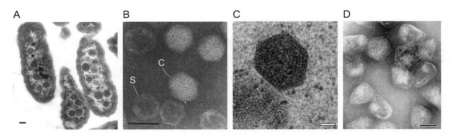

FIG. 1. Visualization of bacterial microcompartments. (A) Transmission electron micrograph of thin-sectioned *H. neapolitanus* cells showing alpha-type carboxysomes (scale bar = 100 nm). (B) Purified carboxysomes from *H. neapolitanus* (scale bar = 100 nm) (S = shell; C = carboxysome). Panels A and B are courtesy of Gordon Cannon and Sabine Heinhorst. (C) beta-type carboxysomes from *Synechocystis* PCC6803 are shown in thin section (scale bar = 50 nm) (courtesy of Robbie Roberson and Allison van de Meene). (D) Isolated *pdu* microcompartments from *Salmonella enterica* spp. Typhimurium LT2, visualized by negative stain (scale bar = 100 nm) (courtesy of Tom Bobik). Figure and legend from Ref. 17.

typically retaining some polyhedral character, such as recognizable edges, but failing to conform to icosahedral shape and symmetry.[22–24] Structural polymorphism appears to be a common feature of most types of BMCs.

The cellular functions of BMCs are to encapsulate a series of sequentially acting enzymes and the metabolic intermediates involved in the pathways they catalyze (Fig. 2). Several distinct types of BMCs are recognized, with the carboxysome being the founding member.[17,25] Carboxysomes are present in all cyanobacteria, and some chemoautotrophs—bacteria that fix inorganic carbon (i.e., CO_2) into organic form.[5,26] Carboxysomes encapsulate two enzymes: carbonic anhydrase (CA) and RuBisCO.[19] Carbonic anhydrase dehydrates bicarbonate to produce CO_2, after which RuBisCO combines CO_2 with the five-carbon molecule 1,6-ribulose bisphosphate (RuBP) to produce two molecules of the three-carbon 3-phosphoglycerate (3-PGA). Colocalization of the two enzymes together is believed to supply the notoriously inefficient RuBisCO enzyme with a high concentration of its CO_2 substrate.[4,27] Based on the overall reaction scheme, bicarbonate and RuBP must cross the outer protein shell into the carboxysome interior, and 3-PGA must exit. While these substrates and products must be able to pass easily across the shell, the CO_2 produced inside the carboxysome must be fixed by RuBisCO before it escapes. Structural data in the past few years has advanced the idea that the protein shell plays a key role in controlling transport into and out of the carboxysome and other BMCs, but numerous questions remain.

Experiments have been conducted on two other types of BMCs that are very different from the carboxysome but are similar to each other in key respects. The propanediol utilizing (Pdu) and ethanolamine utilizing (Eut) BMCs metabolize the three-carbon 1,2-propanediol (1,2-PD) molecule and the two-carbon ethanolamine molecule, respectively[23,28–30] (Fig. 2). These BMCs are widely distributed across the bacterial kingdom, including in enteric bacteria found in human hosts. Both Pdu and Eut microcompartments are present in *Salmonella enterica*, and the Eut operon is present in some strains of *E. coli*. The metabolic pathways for degrading propanediol and ethanolamine both go through aldehyde intermediates, propionaldehyde and acetaldehyde, respectively. Aldehydes are chemically reactive, and experiments have shown that containing them within the BMC so they can be reacted upon in further enzymatic steps avoids the otherwise cytotoxic effects of high concentrations of aldehyde.[31–33] In the case of the Eut BMC, an alternative argument has been put forth based on data showing that the key defect when ethanolamine is metabolized outside of a BMC is evaporative loss of the volatile acetaldehyde from the cell.[34] In either case, the principle is that a small molecule needs to be metabolized in multiple steps without allowing escape of an intermediate species. This general idea applies as well to the carboxysome, where the intermediate that needs to be further metabolized prior to escape is CO_2.

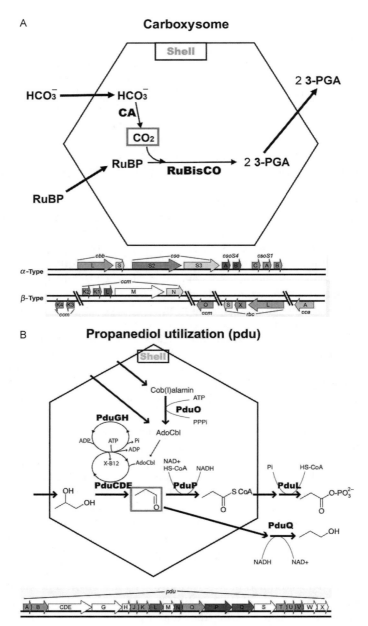

FIG. 2. Gene organization and proposed metabolic pathways for two types of bacterial micro-compartments (BMCs). Genes are colored to indicate their homology. All BMC shell proteins are light blue. For each microcompartment, the key sequestered intermediate is boxed in orange. (A) Function of the carboxysome in enhancing CO_2 fixation. Gene organizations for alpha and beta-type carboxysomes are shown below. (B) A current model for the function of the propanediol utilization (Pdu) microcompartment in metabolizing 1,2-propanediol. The gene organization for the *pdu* operon is shown below. Figure and legend adapted from Ref. 1. (See Color Insert.)

The Pdu and Eut BMCs stand apart from the carboxysome, however, in terms of the complexity of their metabolic enzymes and pathways. In the case of Pdu, as many as eight distinct enzyme subunits are present carrying out five different reaction steps. Furthermore, a number of bulky cofactors are required by the encapsulated enzymes. A firm understanding of which reactions occur inside the Pdu BMC is still incomplete, but it is likely that cobalamin (B_{12}), coenzyme-A, NADH, iron–sulfur clusters, and possibly ATP are required.[6] Some of these turn over or degrade during activity, leading to important questions regarding how such bulky molecules might be able to enter and exit the BMC, while the small molecule intermediate (i.e., propionaldehyde) fails to escape. Structural data have provided some clues to this puzzle.[8,11,14]

The shells of BMCs are formed mainly by proteins from the BMC family, that is, proteins that bear one or more BMC domains approximately 100 amino acids long.[7,21] A few thousand of these small proteins self-assemble to form the outer polyhedral structure (Fig. 3). The sequences of BMC shell proteins were first identified from purified carboxysomes.[19] BMC shell proteins were then shown by sequence homology to be widely distributed across the cyanobacteria, often arranged on the bacterial chromosome in proximity to other proteins involved in carboxysome function, including RuBisCO and carbonic anhydrase.[26,35] In 1994, proteins from the BMC family were found in *Salmonella*, an enteric bacterium, in the large multigene *pdu* operon that encodes enzymes for 1,2-PD metabolism.[28] The presence of BMC proteins in the midst of enzymes for 1,2-PD metabolism suggested that this pathway was compartmentalized in *Salmonella*, which was then established experimentally.[23] The Pdu microcompartment was shown to have a role in preventing aldehyde toxicity in the cytosol, as noted above. A protocol was developed for obtaining homogeneous preparations of Pdu microcompartments, allowing a compositional analysis of this BMC, whose contents are considerably more complex than the carboxysome.[31]

III. Shell Proteins: Structures and Mechanisms

Using modern sequence searching algorithms, some 2000 distinct BMC shell proteins can be found distributed across at least 10 different bacterial phyla; they appear to be absent from eukaryotes and archaea. Their scattered distribution suggests that BMC shell proteins have been spread by horizontal gene transfer, sometimes between highly divergent bacterial species. Based on genomic analysis of enzymes encoded in the vicinity of BMC shell proteins in different bacteria, it has been surmised that there are probably about seven different metabolic categories of BMCs[17,22,25]; only the carboxysome and the Pdu and Eut types have been studied biochemically or genetically, though BMC shell proteins from others have been elucidated structurally.[16]

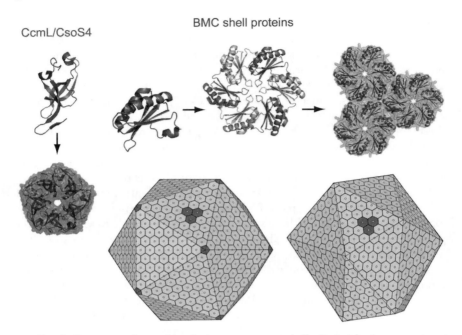

FIG. 3. Structure and assembly of microcompartment shells. Bacterial microcompartment (BMC) shell proteins (top) form hexamers that constitute the main building blocks of the shell. They assemble further to form tightly packed layers with pores in the middle of the hexamers, some of which have gated openings. BMC shell proteins are found widely across the bacteria. Where they occur, they provide a signature for the presence of a microcompartment organelle. In the carboxysome, pentameric proteins (CcmL and CsoS4) likely form the vertices of the shell, which is nearly icosahedral in shape (lower left). Pdu microcompartments and some other types form less geometrically regular shells (lower right) from hexameric BMC proteins; the presence of pentameric shell proteins in those cases has not been established. Figure and legend from Ref. 7. (See Color Insert.)

A common feature seen in essentially all genomes where BMC shell proteins have been found is the presence of multiple homologous (or, more properly, paralogous) copies of the BMC shell protein encoded together.[36] The specialized roles of distinct BMC shell proteins within a given type of BMC have been illuminated in some cases from structural data.

Crystal structures of shell proteins from the carboxysome and other BMCs were determined in our laboratory and others, beginning in 2005,[1,10] leading to fresh insights into architecture and biochemical mechanisms. The BMC domain is an alpha/beta fold, comprising a central beta sheet flanked by short alpha helices. Essentially, all BMC shell proteins self-assemble to form cyclic hexameric oligomers. These hexamers are now understood to represent the basic building blocks of the shell. Most BMC hexamers are disk-like in shape, being relatively flat on one side and bearing a prominent bowl-shaped depression on the other side. In most cases, the N- and C-termini of the protein reside on the latter side. There is considerable variation between BMC proteins at their C-termini. These C-termini often present as flexible extensions that are typically unresolved in crystal structure studies. In numerous crystal structures of BMC shell proteins, the hexamers are found to associate within the crystal in a side-by-side fashion, forming tightly packed molecular layers. In addition, true two-dimensional layers of shell proteins have been assembled *in vitro* and shown by electron microscopy to be arranged in the same way as observed in three-dimensional crystals.[37] This packing within a layer has therefore been postulated to represent the way hexamers are packed within facets of native BMC shell (Figs. 3 and 4).

Most of the BMC hexamers bear a small central pore down the middle. This was postulated from the first structural data to represent the route for molecular transport across the BMC shell. The size and shape of the pore differs between BMC paralogs even within a single type of BMC. Especially in the carboxysome, a tendency for the pore to carry a positive electrostatic potential is notable. This has been argued to confer a transport advantage to negatively charged metabolites, including bicarbonate, which need to enter the carboxysome. Somewhat different pore properties have been observed for the shell proteins of other types of BMCs.[7] The overall porosity of BMC shells has

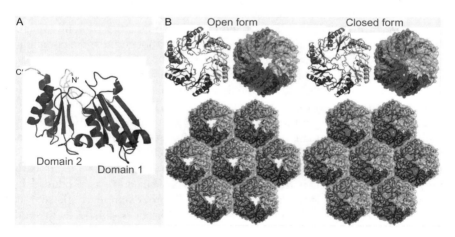

FIG. 4. The structure of the EutL shell protein and its gated pore. (A) A cartoon ribbon diagram of a EutL monomer in its closed form. The first and second BMC domains are colored in blue and purple, respectively. (B) The open (left) and closed (right) configurations of EutL trimers are shown in both ribbon diagram and surface representations. In both configurations, EutL trimers (or pseudohexamers) were observed to pack into tight molecular layers within their respective crystal forms (bottom). Figure and legend from Ref. 14. (See Color Insert.)

been considered from a mathematical perspective and argued to be sufficient to permit adequate movement of small metabolites, particularly when the pore provides an energetically favorable environment for transit of the metabolite.[15]

Minor proteins not belonging to the homologous BMC family are believed to also play a role in shell formation. A suspicion that other proteins might be present in the shell emerged from geometric considerations relating to the flat molecular layers formed by hexameric BMC proteins. A major clue came when crystal structures were determined of functionally uncharacterized proteins encoded within known carboxysome operons. Homologous proteins—named CcmL in the beta-type carboxysome and CsoS4A in the alpha-type carboxysome (previously named OrfA to reflect its unknown function)—were shown to be pentamers.[13] Based on geometric arguments and size compatibility, these pentamers were hypothesized to occupy the 12 vertices of the roughly icosahedral carboxysome, thereby introducing the deformation or "vertex defect" required to close up an otherwise flat layer of hexagons (Fig. 3). The hypothesis is generally consistent with mutational data in which deletion of these genes generally prevents microcompartments from closing up.[38,39] However, the real biological situation may be somewhat more complex than originally hypothesized. In the Eut BMC, the protein homologous to CcmL/CsoS4 is EutN, which forms hexamers instead of pentamers, at least on its own under crystallization conditions.[13] Whether other proteins besides the BMC family and the CcmL/CsoS4/EutN/PduN family might be important in shell formation is not known. It is notable that in several bacterial genomes there are proteins comprising a fusion between a BMC domain and some other protein domain. Little is known about most of these fusion domains, though the extra domain in the EutK shell protein was shown to be a helix–turn–helix domain and was suggested to bind DNA or RNA for an as-yet-unknown function.[14]

Structures of multiple BMC shell proteins have revealed a remarkable number of unexpected topological variations[1,9,11] (Fig. 5). The first reported BMC shell proteins from the carboxysome—and subsequent BMC shell proteins known to be dominant components of their respective shells—adopt what has been referred to as the canonical BMC fold. Other BMC shell proteins, beginning with PduU from *Salmonella*, have been found to have a circularly permuted form in which the same basic tertiary structure is produced by secondary structure elements occurring in a different (circularly permuted) sequence in the protein chain.[9] Among other differences, this leads in some cases to alternate terminal extensions having novel structural motifs. A second prevalent form of structure variation arises from domain duplication. Numerous BMC shell proteins—and apparently at least one in all genomes that encode for BMCs—are composed of two tandem copies of the BMC protein domain.[8,11,12,16] In fact, there are tandem domain versions of both the canonical BMC domain and the circularly permuted form.

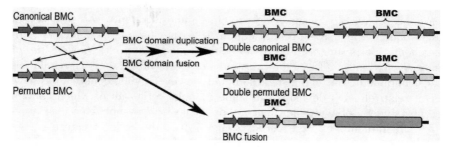

FIG. 5. Variations on the bacterial microcompartment (BMC) protein fold. Secondary structure schematics of BMC proteins in their various arrangements. Individual secondary structure elements are colored. The canonical BMCs include CcmK1/2/3/4, CsoS1A/B/C, PduA/J, and EutM. CcmO likely encodes tandem canonical BMC domains. The permuted, single-domain BMC proteins include PduU and EutS. Tandem permuted BMC proteins include CsoS1D, EutL, and PduT. Proteins in which the BMC domain is fused to another domain include EutK. Figure and legend adapted from Ref. 1. (See Color Insert.)

The numerous variations on the BMC shell protein appear to provide functional specialization. In one case, a circularly permuted BMC protein (EutS) was observed to adopt a bent rather than a flat hexagonal structure.[14] On that basis, it was suggested the EutS shell protein might play a role in allowing the shell to bend at edge positions in the intact polyhedral assembly. The tandem domain shell proteins appear to play special roles in transport. Multiple different tandem BMC domain proteins have been observed to have both open and closed forms[11,14] (Fig. 4). The possibility of gated transport provides a possible clue regarding how large molecules such as bulky cofactors might cross the shell. In another case, a tandem BMC domain protein (PduT) appears to bind an Fe–S cluster in the center of its pore.[8,40] Whether PduT might transport electrons to balance redox reactions inside the Pdu BMC, or whether it might transport intact Fe–S clusters needed by interior enzymes, is unknown. Genetic studies involving deletion of individual Pdu shell genes have been generally consistent with hypotheses based on structural data, but further biochemical studies on shell protein mutants are needed to gain a clearer understanding of the specialized functional roles played by individual shell proteins.

IV. Higher-Level Organization

Isolating BMCs in intact form has presented challenges in a number of cases. Carboxysomes were first isolated from the chemoautotrophic bacterium *Halothiobacillus neapolitanus*.[19] These alpha-type carboxysomes appear largely intact and geometrically regular after cell disruption and purification by sucrose gradient centrifugation. Carboxysomes purified from *H. neapolitanus* have been valuable in analyzing protein composition and were the key to identifying the first shell protein sequences.[41] They have also enabled *in vitro* biochemical experiments, making it possible to begin to probe the activities of encapsulated enzymes and the transport properties of the shell. Genetic tools for manipulating this organism are limited, however. While genetic tools have been relatively well developed in some cyanobacteria—all of which contain carboxysomes—obtaining purified carboxysomes from cyanobacteria has proven more difficult. For the carboxysome BMCs, the disjoint nature of the established systems for preparative biochemical studies compared to genetic studies continues to present a challenge. In contrast with viral capsids, which have evolved under pressure to maintain their integrity under potentially changing conditions, some BMCs may be tailored for a more transient existence. On the positive side, Pdu BMCs have proven to be amenable to isolation; they have been purified from *Salmonella enterica* in the intact form with good yield.[31] This has enabled compositional studies on this BMC, which is arguably the most complex type likely to exist.

Different types of BMCs have different overall shapes. For example, Pdu BMCs tend to appear polyhedral (i.e., to have defined edges and vertices), but they are considerably less regular in shape than carboxysomes; two alpha-type carboxysomes that have been characterized by electron cryo-tomography appear to be nearly regular icosahedra.[18,20] Why some BMCs are more irregular than others is not yet understood. However, the polymorphic oligomerization behavior of the CcmL/CsoS4/PduN/EutN protein family has been implicated.

How BMCs operate biochemically must be governed by how the interior enzymes are arranged with respect to each other, as well as with respect to the pores in the outer shell. The importance of internal organization in the function of the carboxysome was recognized early on, and mathematical models for biochemical function were developed on the basis of different scenarios.[27,42] Recent images from electron cryo-tomography, as well as data on protein–protein interactions, indicate that a few layers of enzymes—the bulk of which must be RuBisCO based on compositional analysis—are organized upon the interior surface of the shell, likely by interactions with the BMC shell proteins.[18,20,43] Unfortunately, cryo-tomography studies on the carboxysome have not been at high enough resolution to provide details about enzyme identities and orientations or their potential relationship to shell proteins and their pores. It might be possible to obtain higher resolution information by electron microscopy, by single particle reconstructions for example, if carboxysomes of greater uniformity and more perfect symmetry could be prepared.

The encapsulation of specific enzymes within BMC shells must rely on molecular recognition between the various protein components. Some of the details of these recognition mechanisms have been partially illuminated. Recent data show that potentially distinct kinds of mechanisms may have evolved in different kinds of BMCs, with major differences evident even between alpha and beta-type carboxysomes. Genetic experiments and protein–protein interaction data show that, in beta carboxysomes, a particular protein, CcmM, plays a key role in organizing the enzymes with respect to the shell.[44,45] The CcmM protein is extremely unusual. Its N-terminal domain has been established to be a redox-sensitive carbonic anhydrase,[46] while its C-terminal region contains multiple sequence repeats of a RuBisCO small sub-unit-like domain,[47] which is likely to interact with RuBisCO large subunits in order to organize that enzyme spatially.

A different mode of recognition has been identified in the Pdu BMC. In this system, it has been shown experimentally that some of the interior enzymes bear special N-terminal amino acid sequences that cause them to be localized inside the BMC.[48,49] In particular, a segment comprised by the first 18 amino acids of the PduP enzyme is both necessary and sufficient for targeting proteins to the Pdu BMC in *Salmonella*. Bioinformatics studies have provided statistical support for terminal sequences as targeting mechanisms in some other BMC-encapsulated enzymes in addition to PduP.[48] This targeting is presumed to be by direct physical interaction between the terminal polypeptides on the enzymes and the inward-facing side of the BMC shell proteins. Attempts to illuminate these interactions in more detail by crystallizing complexes of enzymes with shell proteins have been unsuccessful so far.

V. Future Directions for Research and Design Applications

A flurry of recent research on BMCs has provided new insights into their form and function, but myriad questions remain. Answering these questions will require further studies on many different fronts. A better understanding of the overall organization will require more advanced imaging experiments and additional biophysical studies. For example, based on yeast-two-hybrid inter-action experiments, preliminary data have been provided for which proteins appear to interact with each other.[44,49] Such results will need to be elaborated upon by mutational and structural data in order to paint a clearer three-dimensional picture.

As discussed above, opening and closing conformational transitions in some shell proteins have provided important clues about molecular transport, but most of the details remain obscure. Exactly which metabolites might pass through the pores of which shell proteins is unknown. Across the diverse BMCs, the particu-lar BMC shell protein paralog (or paralogs) believed to constitute the bulk of the shell is a hexamer with a small pore. It has been argued that these small pores that are present throughout the shell surface are likely the transport route for small molecule metabolites, for example, bicarbonate for the carboxysome,

1,2-PD for the PDU BMC, and ethanolamine for the Eut BMC. However, bulkier molecules must cross the shell in all BMC types, and specific BMC shell protein paralogs could be specific for different metabolites and/or cofactors. In addition, the presence of presumptive gating mechanisms prompts the question as to which molecules or signals might trigger pore opening or closing (Fig. 4). This remains unanswered, though the observation of disulfide bonds (and cysteines poised to form disulfide bonds) in some shell proteins hints that cellular redox state could be involved.[8]

Our ability to study BMCs *in vitro* would be advanced considerably if they could be reconstituted from purified enzymes and shell components. The numerous protein components involved makes this a challenging problem even for the carboxysome. For example, while some of the individual shell proteins have been purified with relative ease, other presumptive shell proteins have not been as amenable to large-scale production and purification. It may be possible ultimately to assemble BMCs from partial subsets of the protein components. *In vivo* studies suggest that this might be the case; for both the carboxysome[50] and the Pdu BMC,[49] it appears that the shell can be assembled without the interior enzymes, although the morphology of the Pdu shells produced in this way appear to be somewhat less regular than native shells.

The ability to study native and mutated BMCs *in vitro* would be advanced significantly by methods for stabilizing BMC shells. Possible strategies for stabilizing BMC shells would be improving noncovalent interactions between the various shell proteins, and introducing new covalent bonds. As described above, the individual hexamers of BMC shell proteins appear to be highly stable under diverse conditions. The problem therefore amounts to trying to stabilize the interactions between hexamers where they meet side by side in the molecular layer presumed to comprise the facets of a shell. If mutant shell proteins, designed for enhanced stability, could be engineered into a host organism, the BMCs produced might be stable to purification protocols that otherwise would disrupt native BMCs. Research along this line is under way.

A broad goal in the areas of metabolic engineering and systems biology is to control reaction pathways in a cell. This is essentially the role of BMCs in the bacteria that produce them. BMCs might therefore provide a desirable framework for engineering and controlling nonnative multienzyme pathways. In principle, this would allow a substrate to be metabolized via a nonnative multistep pathway, without the intermediate metabolites being subjected to potential branching reactions. In addition, BMCs could enable the engineering of new pathways with intermediates that would otherwise be cytotoxic. Ultimately, if BMCs can be stabilized outside the cell, they could serve as *ex vivo* reaction chambers for diverse multistep chemical reactions. A number of questions will have to be answered before rational engineering can be applied toward such goals. The question of targeting has already been partly answered, and the targeting of nonnative proteins such as the GFP has already been demonstrated. But how multiple enzymes might be (or should be) organized relative to each other is a more complicated issue. For optimal function, the mechanisms of molecular transport will have to be elucidated in more detail. The numerous crystal structures of BMC shell proteins provide a starting point for understanding and potentially modifying transport properties by directed mutagenesis. However, the evolution of what appear to be complex gating mechanisms in various BMCs implies that this type of sophisticated control could be important for proper function. Ideas for how this could be engineered in a synthetic system are not immediately obvious.

BMCs could be engineered for molecular delivery. In principle, different types of cargo could be encapsulated within BMCs (Fig. 6). BMCs, like viral particles,[51–53] micelles,[54–56] and vault nanoparticles,[57] have the potential to transport and release drugs, spectroscopic probes, or nanoparticles to sites of malignant cellular growth. One potential type of encapsulation involves the

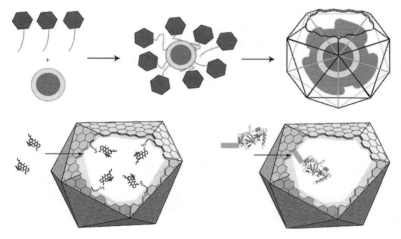

FIG. 6. Proposed methods for using BMCs to encapsulate cargo molecules. (Top) The cargo (a metal nanoparticle) is shown as a red circle. The polyethylene coating, which can be modified to confer negative charge to the particle, is represented by the yellow circle. The dark blue hexagons are the hexamer subunits of the BMC; their light blue tails represent the positive charges engineered into them. The positively charged tails would be attracted to the negatively charged polyethylene glycol coat in order to form a BMC shell around the cargo. (Bottom left) Cysteine thiols are shown engineered on the inside of the BMC shell proteins. A microcompartment is represented in light gray with a cutaway showing the interior of the BMC. Doxorubicin molecules are shown in black, attached to the inside of the microcompartment through reaction with the thiol groups. (Bottom right) The catalytic domain of diptheria toxin (green cartoon) is shown fused with the BMC targeting sequence (orange rectangle), which would direct the cytotoxic protein into the BMC. (For interpretation of the references to color in this figure legend, the reader is referred to the Web version of this chapter.)

creation of metal nanoparticle probes surrounded by a self-assembling protein shell. Such probes can be used in optical and magnetic imaging of tumors. The concept has already been demonstrated using viral capsids[52] by engineering capsid proteins to self-assemble around metal nanoparticles. The nanoparticle to be encapsulated is negatively charged by coating it with short, carboxylated polyethylene glycol (PEG) molecules, while the self-assembling protein is engineered to have a positively charged tail. BMC shell proteins could be adapted to this strategy. Addition of positively charged tails to a BMC shell protein would be expected to drive self-assembly of the hexameric shell proteins around the negatively charged nanoparticle by electrostatic interactions. The use of metal particles[55,56,58] and quantum dots[59] as the core cargo has led to success in the creation of virus-like particles that have useful electromagnetic and fluorescent properties, respectively. Cytotoxic chemicals represent a second type of cargo with medical applications. BMCs could be engineered for targeted delivery of anticancer drugs such as doxorubicin, maytansinoids, and taxoids,[60–62] families of drugs that have been developed specifically for fighting cancer. Reactive analogs of such drugs could be chemically attached to thiols on the interior of the BMC surface by engineering cysteine residues at desired positions. In this way, the cytotoxic drugs would be sequestered from normal cells while being transported to targeted cancer cells, potentially reducing the side effects often associated with the use of cytotoxic drugs. In principle, multiple copies of the drug—for example, one attached to each of the thousands of shell protein subunits—would be encapsulated by one BMC, allowing delivery of a large dose. Cytotoxic proteins constitute a third distinct type of molecules that could be encapsulated. These would be differentiated from smaller cytotoxic chemicals with respect to their mechanism of encapsulation. Cytotoxic proteins could be directed into BMCs by fusing them to an amino acid-targeting sequence. This strategy could convert BMCs from natural metabolic compartments to specialized delivery vehicles for cytotoxic proteins. Molecular cages containing BMC-targeted cytotoxic proteins could be applied for medicinal uses such as cancer therapy by encapsulating the catalytic domain of the diptheria toxin, for example, for delivery to tumor cells. Protein cages of this type could be useful as cellular probes to study the mechanisms of toxin entry into the cytosol of eukaryotic cells, or potentially as novel therapeutics for use in clinical medicine.[63]

BMCs could be engineered to deliver their cargo to specific cells by different strategies (Fig. 7). In conjunction with the above methods for encapsulation, the addition of a targeting moiety to the outer surface of the BMC would open up novel applications involving delivery of molecular probes or drugs. Targeting moieties could be attached to the outside of a BMC shell in two general ways: through chemical attachment and through bioconjugation.

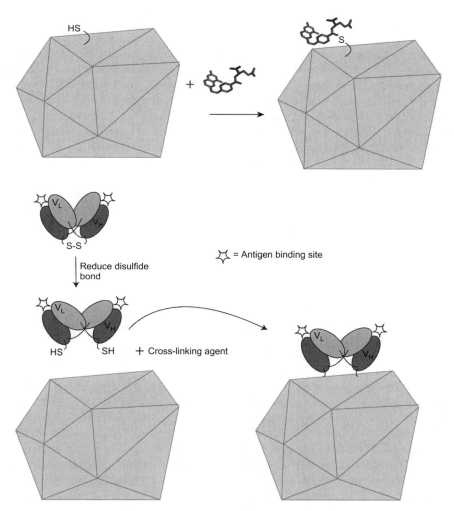

FIG. 7. Proposed methods for using BMCs for targeted cellular delivery. (Top) The BMC shell (gray polygon) is modified with cysteine thiol groups on the outer surface. One of potentially many folate molecules (black) is shown attached to the outside of the BMC shell via a thiol group. (Bottom) The engineered antibody, or "diabody" (dark and light gray ovals) has a disulfide bond on the end distal from the antigen binding site (star), which could be cross-linked to the BMC shell following reduction of the disulfide bond.

In order to attach small molecule nutrients, such as folate, to the BMC shell, a cysteine thiol could be engineered on the outer surface of the shell to serve as the nucleophile in a reaction leading to attachment of the desired small molecule. A thiol is well suited for this purpose since it provides a covalent linkage that is stable under the oxidizing environment of the circulatory system. Nutrients such as folate are of interest because they bind tightly to specific receptors that tend to be upregulated in rapidly dividing tumor cells.[64] Larger molecules could also be attached to the outside of BMCs. In the past several years, the development of monoclonal antibodies (mAbs) and their derivatives has enabled researchers to create a host of new therapeutic and diagnostic tools. To overcome some of the challenges associated with natural antibodies, researchers have developed various types of recombinant antibody fragments for therapeutic applications. Engineered variations on natural antibodies known as diabodies and minibodies have been used in multiple antibody–drug conjugates to date.[65–67] Antibody fragments can be attached to their cargo by covalent or noncovalent conjugation.[68] Minibodies and diabodies have been engineered with sulfhydryl groups in previous studies in order to facilitate covalent attachments. The various strategies possible for attaching targeting molecules to the outside of BMCs, combined with the various possible cargoes they could encapsulate, make them a promising prospect for future efforts in targeted delivery for medical applications.

VI. Closing Remarks

Polyhedral, protein-based microcompartments (BMCs) have been known to exist inside bacterial cells for decades. We are presently in a period of accelerating discovery with respect to their structures and mechanisms. Atomic-resolution structures of various shell proteins have provided key insights and hypotheses for ongoing investigations. However, it could be argued that what we understand about BMCs is still heavily outweighed by what remains to be discovered. The area therefore remains fertile for continuing scientific exploration. In addition, atomic-resolution structures of various shell proteins can now guide engineering studies aimed at exploiting BMCs for various purposes. BMCs are likely to find myriad uses in the broad areas of synthetic and systems biology, including metabolic engineering, targeted molecular delivery, and other areas of nanotechnology.

REFERENCES

1. Yeates TO, Crowley CS, Tanaka S. Bacterial microcompartment organelles: protein shell structure and evolution. *Annu Rev Biophys* 2010;**39**:185–205.
2. Drews G, Niklowitz W. Cytology of Cyanophycea. II. Centroplasm and granular inclusions of *Phormidium uncinatum*. *Arch Mikrobiol* 1956;**24**:147–62.
3. Gantt E, Conti SF. Ultrastructure of blue-green algae. *J Bacteriol* 1969;**97**:1486–93.

4. Cannon GC, Bradburne CE, Aldrich HC, Baker SH, Heinhorst S, Shively JM. Microcompartments in prokaryotes: carboxysomes and related polyhedra. *Appl Environ Microbiol* 2001;**67**:5351–61.
5. Badger MR, Price GD. CO$_2$ concentrating mechanisms in cyanobacteria: molecular components, their diversity and evolution. *J Exp Bot* 2003;**54**:609–22.
6. Cheng S, Liu Y, Crowley CS, Yeates TO, Bobik TA. Bacterial microcompartments: their properties and paradoxes. *Bioessays* 2008;**30**:1084–95.
7. Yeates TO, Thompson MC, Bobik TA. The protein shells of bacterial microcompartment organelles. *Curr Opin Struct Biol* 2011;**21**:223–31.
8. Crowley CS, Cascio D, Sawaya MR, Kopstein JS, Bobik TA, Yeates TO. Structural insights into the mechanisms of transport across the *Salmonella enterica* Pdu microcompartment shell. *J Biol Chem* 2010;**285**:37838–46.
9. Crowley CS, Sawaya MR, Bobik TA, Yeates TO. Structure of the PduU shell protein from the Pdu microcompartment of *Salmonella*. *Structure* 2008;**16**:1324–32.
10. Kerfeld CA, Sawaya MR, Tanaka S, Nguyen CV, Phillips M, Beeby M, et al. Protein structures forming the shell of primitive bacterial organelles. *Science* 2005;**309**:936–8.
11. Klein MG, Zwart P, Bagby SC, Cai F, Chisholm SW, Heinhorst S, et al. Identification and structural analysis of a novel carboxysome shell protein with implications for metabolite transport. *J Mol Biol* 2009;**392**:319–33.
12. Sagermann M, Ohtaki A, Nikolakakis K. Crystal structure of the EutL shell protein of the ethanolamine ammonia lyase microcompartment. *Proc Natl Acad Sci USA* 2009;**106**:8883–7.
13. Tanaka S, Kerfeld CA, Sawaya MR, Cai F, Heinhorst S, Cannon GC, et al. Atomic-level models of the bacterial carboxysome shell. *Science* 2008;**319**:1083–6.
14. Tanaka S, Sawaya MR, Yeates TO. Structure and mechanisms of a protein-based organelle in Escherichia coli. *Science* 2010;**327**:81–4.
15. Tsai Y, Sawaya MR, Cannon GC, Cai F, Williams EB, Heinhorst S, et al. Structural analysis of CsoS1A and the protein shell of the *Halothiobacillus neapolitanus* carboxysome. *PLoS Biol* 2007;**5**:e144.
16. Heldt D, Frank S, Seyedarabi A, Ladikis D, Parsons JB, Warren MJ, et al. Structure of a trimeric bacterial microcompartment shell protein, EtuB, associated with ethanol utilization in *Clostridium kluyveri*. *Biochem J* 2009;**423**:199–207.
17. Yeates TO, Kerfeld CA, Cannon GC, Heinhorst S, Shively JM. Protein-based organelles in bacteria: carboxysomes and related microcompartments. *Nat. Rev. Microbiol.* 2008;**6**:681–91.
18. Schmid MF, Paredes AM, Khant HA, Soyer F, Aldrich HC, Chiu W, et al. Structure of *Halothiobacillus neapolitanus* carboxysomes by cryo-electron tomography. *J Mol Biol* 2006;**364**:526–35.
19. Shively JM, Ball F, Brown DH, Saunders RE. Functional organelles in prokaryotes: polyhedral inclusions (carboxysomes) of *Thiobacillus neapolitanus*. *Science* 1973;**182**:584–6.
20. Iancu CV, Ding HJ, Morris DM, Dias DP, Gonzales AD, Martino A, et al. The structure of isolated *Synechococcus* strain WH8102 carboxysomes as revealed by electron cryotomography. *J Mol Biol* 2007;**372**:764–73.
21. Heinhorst S, Cannon GC, Shively JM. Carboxysomes and carboxysome-like inclusions. In: Shively JM, editor. *Complex intracellular structures in prokaryotes*, vol. 2. Berlin: Springer-Verlag; 2006. p. 141–65.
22. Bobik TA. Bacterial microcompartments. *Microbe* 2007;**2**:25–31.
23. Bobik TA, Havemann GD, Busch RJ, Williams DS, Aldrich HC. The propanediol utilization (*pdu*) operon of *Salmonella enterica* serovar Typhimurium LT2 includes genes necessary for formation of polyhedral organelles involved in coenzyme B$_{12}$-dependent 1, 2-propanediol degradation. *J Bacteriol* 1999;**181**:5967–75.

24. Brinsmade SR, Paldon T, Escalante-Semerena JC. Minimal functions and physiological conditions required for growth of *Salmonella enterica* on ethanolamine in the absence of the metabolosome. *J Bacteriol* 2005;**187**:8039–46.
25. Kerfeld CA, Heinhorst S, Cannon GC. Bacterial microcompartments. *Annu Rev Microbiol* 2010;**64**:391–408.
26. Cannon GC, Heinhorst S, Bradburne CE, Shively JM. Carboxysome genomics: a status report. *Funct Plant Biol* 2002;**29**:175–82.
27. Reinhold L, Zviman X, Kaplan A. A quantitative model for carbon fluxes and photosynthesis in cyanobacteria. *Plant Physiol Biochem* 1989;**27**:945–54.
28. Chen P, Andersson DI, Roth JR. The control region of the *pdu/cob* regulon in *Salmonella typhimurium*. *J Bacteriol* 1994;**176**:5474–82.
29. Kofoid E, Rappleye C, Stojiljkovic I, Roth J. The 17-gene ethanolamine (eut) operon of *Salmonella typhimurium* encodes five homologues of carboxysome shell proteins. *J Bacteriol* 1999;**181**:5317–29.
30. Stojiljkovic I, Baumler AJ, Heffron F. Ethanolamine utilization in *Salmonella typhimurium*: nucleotide sequence, protein expression, and mutational analysis of the cchA cchB eutE eutJ eutG eutH gene cluster. *J Bacteriol* 1995;**177**:1357–66.
31. Havemann GD, Bobik TA. Protein content of polyhedral organelles involved in coenzyme B12-dependent degradation of 1,2-propanediol in *Salmonella enterica* serovar Typhimurium LT2. *J Bacteriol* 2003;**185**:5086–95.
32. Rondon MR, Kazmierczak R, Escalante-Semerena JC. Glutathione is required for maximal transcription of the cobalamin biosynthetic and 1,2-propanediol utilization (*cob/pdu*) regulon and for the catabolism of ethanolamine, 1,2-propanediol, and propionate in *Salmonella typhimurium* LT2. *J Bacteriol* 1995;**177**:5434–9.
33. Sampson EM, Bobik TA. Microcompartments for B12-dependent 1,2-propanediol degradation provide protection from DNA and cellular damage by a reactive metabolic intermediate. *J Bacteriol* 2008;**190**:2966–71.
34. Penrod JT, Roth JR. Conserving a volatile metabolite: a role for carboxysome-like organelles in *Salmonella enterica*. *J Bacteriol* 2006;**188**:2865–74.
35. Price GD, Sültemeyer D, Klughammer B, Ludwig M, Badger MR. The functioning of the CO_2 concentrating mechanism in several cyanobacterial strains: a review of general physiological characteristics, genes, proteins and recent advances. *Can J Bot* 1998;**76**:973–1002.
36. Beeby M, Bobik TA, Yeates TO. Exploiting genomic patterns to discover new supramolecular protein assemblies. *Protein Sci* 2009;**18**:69–79.
37. Dryden KA, Crowley CS, Tanaka S, Yeates TO, Yeager M. Two-dimensional crystals of carboxysome shell proteins recapitulate the hexagonal packing of three-dimensional crystals. *Protein Sci* 2009;**18**:2629–35.
38. Price GD, Howitt SM, Harrison K, Badger MR. Analysis of a genomic DNA region from the cyanobacterium *Synechococcus* sp. strain PCC7942 involved in carboxysome assembly and function. *J Bacteriol* 1993;**175**:2871–9.
39. Parsons JB, Dinesh SD, Deery E, Leech HK, Brindley AA, Heldt D, et al. Biochemical and structural insights into bacterial organelle form and biogenesis. *J Biol Chem* 2008;**283**:14366–75.
40. Pang A, Warren MJ, Pickersgill RW. Structure of PduT, a trimeric bacterial microcompartment protein with a 4Fe-4S cluster-binding site. *Acta Crystallogr D Biol Crystallogr* 2011;**67**:91–6.
41. English RS, Lorbach SC, Qin X, Shively JM. Isolation and characterization of a carboxysome shell gene from *Thiobacillus neapolitanus*. *Mol Microbiol* 1994;**12**:647–54.
42. Reinhold L, Kosloff R, Kaplan A. A model for inorganic carbon fluxes and photosynthesis in cyanobacterial carboxysomes. *Can J Bot* 1991;**69**:984–8.

43. Iancu CV, Morris DM, Dou Z, Heinhorst S, Cannon GC, Jensen GJ. Organization, structure, and assembly of alpha-carboxysomes determined by electron cryotomography of intact cells. *J Mol Biol* 2010;**396**:105–17.

44. Cot SS, So AK, Espie GS. A multiprotein bicarbonate dehydration complex essential to carboxysome function in cyanobacteria. *J Bacteriol* 2008;**190**:936–45.

45. Long BM, Badger MR, Whitney SM, Price GD. Analysis of carboxysomes from *Synechococcus* PCC7942 reveals multiple Rubisco complexes with carboxysomal proteins CcmM and CcaA. *J Biol Chem* 2007;**282**:29323–35.

46. Pena KL, Castel SE, de Araujo C, Espie GS, Kimber MS. Structural basis of the oxidative activation of the carboxysomal gamma-carbonic anhydrase, CcmM. *Proc Natl Acad Sci USA* 2010;**107**:2455–60.

47. Ludwig M, Sultemeyer D, Price GD. Isolation of ccmKLMN genes from the marine cyano-bacterium, *Synechococcus* sp. PCC7002 (Cyanophyceae), and evidence that CcmM is essential for carboxysome assembly. *J Phycol* 2000;**36**:1109–19.

48. Fan C, Cheng S, Liu Y, Escobar CM, Crowley CS, Jefferson RE, et al. Short N-terminal sequences package proteins into bacterial microcompartments. *Proc Natl Acad Sci USA* 2010;**107**:7509–14.

49. Parsons JB, Frank S, Bhella D, Liang M, Prentice MB, Mulvihill DP, et al. Synthesis of empty bacterial microcompartments, directed organelle protein incorporation, and evidence of fila-ment-associated organelle movement. *Mol Cell* 2010;**38**:305–15.

50. Menon BB, Dou Z, Heinhorst S, Shively JM, Cannon GC. *Halothiobacillus neapolitanus* carboxysomes sequester heterologous and chimeric RubisCO species. *PLoS One* 2008;**3**:e3570.

51. Goicochea NL, De M, Rotello VM, Mukhopadhyay S, Dragnea B. Core-like particles of an enveloped animal virus can self-assemble efficiently on artificial templates. *Nano Lett* 2007;**7**:2281–90.

52. Sun J, DuFort C, Daniel MC, Murali A, Chen C, Gopinath K, et al. Core-controlled polymor-phism in virus-like particles. *Proc Natl Acad Sci USA* 2007;**104**:1354–9.

53. Aniagyei SE, Dufort C, Kao CC, Dragnea B. Self-assembly approaches to nanomaterial encapsulation in viral protein cages. *J Mater Chem* 2008;**18**:3763–74.

54. McPherson A. Micelle formation and crystallization as paradigms for virus assembly. *Bioessays* 2005;**27**:447–58.

55. Chen C, Daniel MC, Quinkert ZT, De M, Stein B, Bowman VD, et al. Nanoparticle-templated assembly of viral protein cages. *Nano Lett* 2006;**6**:611–5.

56. Chen C, Kwak ES, Stein B, Kao CC, Dragnea B. Packaging of gold particles in viral capsids. *J Nanosci Nanotechnol* 2005;**5**:2029–33.

57. Buehler DC, Toso DB, Kickhoefer VA, Zhou ZH, Rome LH. Vaults engineered for hydropho-bic drug delivery. *Small* 2011;**7**:1432–9.

58. Dragnea B, Chen C, Kwak ES, Stein B, Kao CC. Gold nanoparticles as spectroscopic enhan-cers for in vitro studies on single viruses. *J Am Chem Soc* 2003;**125**:6374–5.

59. Dixit SK, Goicochea NL, Daniel MC, Murali A, Bronstein L, De M, et al. Quantum dot encapsulation in viral capsids. *Nano Lett* 2006;**6**:1993–9.

60. Bar H, Yacoby I, Benhar I. Killing cancer cells by targeted drug-carrying phage nanomedicines. *BMC Biotechnol* 2008;**8**:37.

61. Chari RV. Targeted cancer therapy: conferring specificity to cytotoxic drugs. *Acc Chem Res* 2008;**41**:98–107.

62. Chen J, Chen S, Zhao X, Kuznetsova LV, Wong SS, Ojima I. Functionalized single-walled carbon nanotubes as rationally designed vehicles for tumor-targeted drug delivery. *J Am Chem Soc* 2008;**130**:16778–85.

63. Murphy JR. Mechanism of diphtheria toxin catalytic domain delivery to the eukaryotic cell cytosol and the cellular factors that directly participate in the process. *Toxins* 2011;**3**:294–308.

64. Leamon CP, Reddy JA, Vetzel M, Dorton R, Westrick E, Parker N, et al. Folate targeting enables durable and specific antitumor responses from a therapeutically null tubulysin B analogue. *Cancer Res* 2008;**68**:9839–44.
65. Olafsen T, Cheung CW, Yazaki PJ, Li L, Sundaresan G, Gambhir SS, et al. Covalent disulfide-linked anti-CEA diabody allows site-specific conjugation and radiolabeling for tumor targeting applications. *Protein Eng Des Sel* 2004;**17**:21–7.
66. Sirk SJ, Olafsen T, Barat B, Bauer KB, Wu AM. Site-specific, thiol-mediated conjugation of fluorescent probes to cysteine-modified diabodies targeting CD20 or HER2. *Bioconjug Chem* 2008;**19**:2527–34.
67. Barat B, Sirk SJ, McCabe KE, Li J, Lepin EJ, Remenyi R, et al. Cys-diabody quantum dot conjugates (immunoQdots) for cancer marker detection. *Bioconjug Chem* 2009;**20**:1474–81.
68. Xing Y, Chaudry Q, Shen C, Kong KY, Zhau HE, Chung LW, et al. Bioconjugated quantum dots for multiplexed and quantitative immunohistochemistry. *Nat Protoc* 2007;**2**:1152–65.

Pili and Flagella: Biology, Structure, and Biotechnological Applications

Nani Van Gerven,* Gabriel Waksman,† and Han Remaut*

*Structural & Molecular Microbiology, VIB/Vrije Universiteit Brussel, Brussels, Belgium

†Institute of Structural and Molecular Biology, University College London and Birkbeck College, London, United Kingdom

Bacteria and Archaea expose on their outer surfaces a variety of thread-like proteinaceous organelles with which they interact with their environments. These structures are repetitive assemblies of covalently or non-covalently linked protein subunits, organized into filamentous polymers known as pili ("hair"), flagella ("whips") or injectisomes ("needles"). They

Progress in Molecular Biology
and Translational Science, Vol. 103
DOI: 10.1016/B978-0-12-415906-8.00005-4

21

serve different roles in cell motility, adhesion and host invasion, protein and DNA secretion and uptake, conductance, or cellular encapsulation. Here we describe the functional, morphological and genetic diversity of these bacterial filamentous protein structures. The organized, multi-copy build-up and/or the natural function of pili and flagella have lead to their biotechnological application as display and secretion tools, as therapeutic targets or as molecular motors. We review the documented and potential technological exploitation of bacterial surface filaments in light of their structural and functional traits.

I. Introduction

Bacteria and Archaea are able to assemble a wide range of multisubunit fibers on their cell surfaces, serving versatile biological roles. The first such appendages to be spotted were hair-like structure on Gram-negative bacteria, dubbed pili. Since their discovery in 1949, different terms including "bristles," "cilia," "filaments," "fimbriae," "fibrillae," "pili," or "needles" have been given to them.[1] It is now clear that the diverse appendages observed on bacterial surfaces serve various functions and stem from different biosynthetic pathways.[2] The terms "pili" (hair-like structures) and "fimbriae" (threads) are now generally being used to indicate exterior appendages involved in adhesion, while "filaments" and "needles" designate surface structures associated with secretion apparatuses. Both pili and secretion system appendages are key players in numerous essential biological processes such as, for example, recognition and colonization of target surfaces, biofilm formation, shielding and host subversion, motility, and signaling events.[3] Flagella represent the second main type of linear multisubunit surface organelles on bacteria and Archaea. They are considered unique motility organelles not only used for swimming but also essential for swarming. Visualized by electron microscopy (EM), flagella are thicker, longer, and less numerous than pili. Invariably, these two types of surface appendages are built up of one or a few repeating (glyco)protein subunits that are covalently or noncovalently attached to linear or branched structures. The structural and physical characteristics of these repeating units and the self-assembling properties for some of them have raised interest for their use in bio(nano)technological applications. Here we review the various classes of bacterial surface appendages along with their biosynthetic pathways and structural properties, in light of their potential or proven use in synthetic applications.

II. Pili in Gram-Negative Bacteria

A. Introduction

Characteristic for Gram-negative bacteria is that they have both an inner and an outer cell membrane, enclosing in between a peptidoglycan layer that maintains cell wall structure. This double membrane system constitutes a formidable barrier for proteins destined for the cell exterior. Consequently, it is not surprising that structure–function analysis of pili in Gram-negative bacteria reveal multiple, dedicated pathways for pilus biosynthesis. Five classes of pili—chaperone/usher (CU) pili, curli, type IV pili, type III secretion needle, and type IV secretion pili—each having their own intricate mechanisms for folding, secretion, and ordered assembly, will be discussed for Gram-negative bacteria.

B. Chaperone/Usher Pili

The most widely distributed group of bacterial cell surface appendages are the pili synthesized by the CU pathway. This pathway is a terminal branch of the general secretory pathway involving just two proprietary assembly proteins: a specialized periplasmic chaperone and an outer membrane (OM) pilus assembly platform, called the usher.[4,5] Most of the CU pili mediate microbial attachment by binding to specific receptors via a fimbrial adhesin.[6] In pathogenic strains, pili constitute important virulence factors that allow host recognition and immune evasion via specific attachment, cell invasion, and/or biofilm formation.[7,8]

1. Morphology and Structure

CU-assembled organelles are formed by linear, unbranched polymers of several hundreds to thousands of pilus subunits, also called pilins, which range in size from ~ 10 to ~ 20 kDa. Subunits polymerize through noncovalent head-to-tail interactions mediated by a fold complementation mechanism (see below).[4,5] The resulting fibers form single-start filaments that range in morphology from relatively thick and rigid, helically wound pilus rods (~ 7–8 nm) which radiate outward from the bacteria to extended, very thin and flexible fibrillae (~ 2–3 nm) that collapse into a tangled mass on the outer surface.[6] CU organelles also differ widely in complexity. The canonical ancestral system is composed of a long polymer of a single subunit type, capped at its distal end by a single copy of an adhesive subunit.[9] In some clusters, this is reduced to just a single polymerizing subunit, which can itself have adhesive characteristics.[7] However, in most systems the basic structure is extended with a small flexible tip composed of one to four different minor subunit types that link the tip adhesin to the major pilus rod (Fig. 1). Finally, some clusters found in animal enterotoxigenic *Escherichia coli* (ETEC) strains form fimbriae that are heteropolymeric along the length of the fiber.[9]

2. PILUS ASSEMBLY

The ordered secretion, folding, and assembly of thousands of pilin subunits into surface-anchored fibers is orchestrated by two dedicated proteins, a periplasmic chaperone and an OM usher. These proteins are specific to the different fimbriae and are encoded inside the operons comprising the pilus subunit genes. Based on the wealth of structural and molecular information available, P and type 1 pili have become the prototypes and will also be used in this review as models to explain assembly of pili by the CU pathway.

Pilus subunits cross the inner membrane (IM) via the Sec translocon, followed by cleavage of their amino terminal signal sequence. All pilus subunits consist of an incomplete immunoglobulin-like (Ig-like) β-sandwich fold that characteristically lacks its C-terminal β-strand, strand G. This compromised Ig fold makes folding of the subunits dependent on a periplasmic chaperone (FimC in the case of type 1 and PapD in the case of P pili). The chaperone consists of two Ig-like domains stabilized by a conserved salt bridge. Chaperones complement for the missing strand by providing a donor strand, the G1 strand (strand G in the chaperone N-terminal domain), to complete the Ig-like fold of the subunit, a process termed "donor-strand complementation" (DSC).[10,11] The mechanism of DSC provides the steric information necessary for proper folding and prevents premature subunit–subunit interactions. Binary chaperone–subunit complexes are then targeted to the usher, which forms the pilus assembly platform and translocation channel in the OM (FimD for type 1 pili or PapC for P pili) (Fig. 1). The order of subunit incorporation is based on their differential affinity for the usher and for each other.[12,13] In case of type 1 and P pili, fiber polymerization is initiated by binding of the chaperone–adhesin complex (respectively, FimC:FimH and PapD:PapG). Ushers form 24-stranded β-barrels that constitute the translocation channel and are flanked by periplasmic subunit-binding domains at the N- and C-terminus.[14,15a] In the absence of subunits, the pore is blocked by a plug domain. Once bound by an initiating chaperone–subunit complex, the usher undergoes a conformational change to a form in which the plug is displaced, such that the polymerized subunits can transit the channel. This assembly competent state is maintained throughout pilus formation. At the usher, the mature pilus fiber is assembled by a donor-strand exchange reaction, in which the G1 β-strand of the chaperone is replaced by a 10–20-residue-long peptide, called the N-terminal extension or Nte, that is present on the N-terminus of the neighboring subunit and is exposed on incoming chaperone–subunit complexes.[15] The chaperone is released from the chaperone–subunit complex

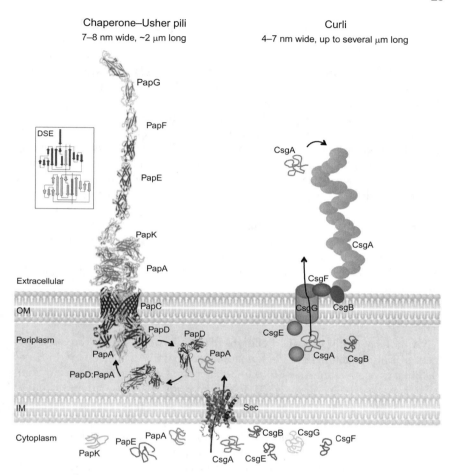

FIG. 1. Chaperone–Usher pili (left). Schematic representation of the biosynthetic pathway of chaperone–usher (CU) pili, here exemplified by P pili. P pili are composed of two major subassemblies: a thick and rigid rod made up of subunits of PapA (light blue) arranged in a right-handed helical cylinder; and a thin, flexible tip fiber in the distal end of the rod. In P pili, the tip fiber is composed of 5–10 PapE subunits (dark green) flanked by two minor subunits, PapF (orange) and PapK (yellow). The adhesin subunit PapG is responsible for interaction with the host tissues and present at the tip of the fiber. Pilus assembly is orchestrated by a periplasmic chaperone PapD (red), and an outer membrane (OM) assembly platform, or usher, PapC (dark blue). Pilus subunits cross the inner membrane (IM) to the periplasm via SecYEG (pink). The chaperone then captures subunits via a mechanism called "donor-strand complementation" whereby the incomplete immunoglobulin fold of subunits is completed by donation of a chaperone's strand. Subunits are then targeted to the OM dimeric usher where subunits polymerize via a mechanism called "donor-strand exchange" (DSE) whereby the missing strand of subunits is provided by the N-terminal extension of the subunit coming next in assembly (see DSE diagram).

coincidently to the Nte of the pilus subunit zippering into the hydrophobic groove of the previously assembled subunit, to complete the Ig-like fold.[16] Upon exchange, the pilus subunit undergoes a topological transition that triggers the closure of its groove, cementing its neighbor's Nte into part of its own Ig fold.[17,18] The mature pilus would thus consist of an array of Ig domains, each of which contributes a strand to the fold of the preceding subunit to produce a highly stable organelle. The process of polymerization and extrusion of the subunits through the OM does not require external energy input. Purified chaperone–subunit complexes will also spontaneously turn over into donor-strand-exchanged fibers *in vitro*, though at rates at least two orders of magnitude lower than observed in the usher-catalyzed reaction.[19] The energy that drives the process is provided by the conformational transition from a high-energy folding intermediate stabilized by the interaction with the chaperone to a collapsed subunit-bound form during donor-strand exchange.[17,18] Once in donor-strand exchange, the subunits form formidably strong noncovalent interactions, with extrapolated half-lives of 3×10^9 years and which are stable under conditions usually considered harsh for biomolecules.[20]

C. Curli

A more recently identified class of filaments expressed on the outer surfaces of Enterobacteriaceae are functional amyloid fibers, called curli in *E. coli* or thin aggregative fimbriae (Tafi) in *Salmonella*. Outer surface amyloid fibers have now also been observed in Pseudomonads, the biosynthetic route of which is unclear at present.[21] The discovery of functional amyloid fibers demonstrates that amyloids can be a productive part of cellular biology, and research in controlled amyloid formation could provide valuable insight in off-pathway, disease-associated amyloidogenesis and cytotoxicity seen in human diseases.[22] First observed in 1989 in fibronectin-binding *E. coli* isolates from bovine fecal samples,[23] curli fibers have been shown to be important for

---

In P pili, incorporation of a termination subunit PapH prevents further subunit additions, thus arresting pilus growth (not shown). PDB entry codes used in this figure are 3RFZ (PapC model based on FimD:FimC:FimH structure), 2J2Z (PapD–PapH complex), 2UY7 (PapA), 2UY6 (PapD: PapA$_2$ complex), 1N12 (PapE bound to Nte of PapK), 2W07 (PapF), and 1J8R (PapG). Curli (right). Curli biosynthesis pathway in *E. coli*. The major curli subunit CsgA (light green) is secreted from the cell as a soluble monomeric protein. The minor curli subunit CsgB (dark green) is associated with the OM and acts as nucleator for the conversion of CsgA from a soluble protein to amyloid deposit. CsgG (light blue) assembles into an oligomeric curli-specific translocation channel in the OM. CsgE (dark blue) and CsgF (purple) form soluble accessory proteins required for productive CsgA and CsgB transport and deposition. All curli proteins have putative Sec signal sequences for transport across the cytoplasmic (inner) membrane. (See Color Insert.)

interactions between individual bacteria and between bacteria and host tissues, for biofilm formation, and even for colonization of inert surfaces that often resist bacterial colonization, such as Teflon and stainless steel.[24] They can increase amyloidosis in the host by acting as cross-seeding nuclei, are potent promoters of a proinflammatory response, and can also enhance hypotension and bleeding disorders during sepsis by their absorption of contact-phase proteins and fibrinogen.[25] Curli further provide bacteria significant protection against environmental stresses, such as desiccation and antimicrobial agents.[24]

1. MORPHOLOGY AND STRUCTURE

In *Escherichia* and *Salmonella* species, curli form highly aggregative flexible fibers of ~ 4–7 nm diameter, which are so abundant that up to 40% of the biomass in activated sludge may consist of amyloid-like filaments.[26] Much of our knowledge on amyloid deposits comes from studies of eukaryotic protein misfolding diseases. Whether disease-associated or the product of a dedicated assembly pathway, amyloid fibers are characterized by a "cross-β-strand" structure, where the β-sheets are oriented antiparallel to the fiber axis.[27] This gives rise to several typical physical properties, such as binding to amyloid-specific dyes Congo red and thioflavin T as well as extreme resistance to chemical and temperature denaturation and to digestion by proteinases.

In *E. coli*, curli are composed of two proteins: CsgA and CsgB, both approximately 100 residues in size. The mature CsgA protein has two distinct functional domains, a 22-amino acid N-terminal domain required for secretion[28] and an amyloid protease-resistant core domain[29] (Fig. 1). The core domains of both CsgA and CsgB contain five imperfect repeating units distinguished by the consensus sequence Ser-X5-Gln-X4-Asn-X5-Gln, which is commonly observed in amyloidogenic peptides.[30,31] No structures are currently available for curli and tafi subunits. Molecular modeling studies suggest a parallel β-helix comprised of five repeat units.[29,32] Curli fibers are believed to form an extended β-helix by the head-to-tail stacking of the curlin subunits.

2. CURLI ASSEMBLY

Curli are assembled by the extracellular nucleation/precipitation pathway. The genes involved in curli production are organized into two adjacent divergently transcribed operons, in *E. coli* called *csgBAC* and *csgDEFG* (curli-specific genes) and in *Salmonella agfBAC* and *agfDEFG* (*aggregative fimbriae*). Both operons are required for biosynthesis and assembly. The *csgBAC* operon encodes two homologous proteins (CsgA and CsgB) that constitute, respectively, the major and minor curli subunits.[33] The third gene in the operon, *csgC*, has no described function, nor has a transcript or protein been detected in *E. coli*. However, the conservation of *agfC* with *agfBA* throughout *Salmonella*,

E. coli, and even in *Shigella*[34] suggests that *csgC/agfC* is functionally important, and at least in *Salmonella* spp. AgfC is shown to both direct AgfA assembly traffic and influence the structural characteristics of curli.[35] The *csgDEFG* operon encodes CsgD, a transcriptional activator of curli synthesis, and three nonstructural assembly factors, CsgE, CsgF, and CsgG.

CsgB, CsgA, CsgE, CsgF, and CsgG each have putative Sec translocation sequences for translocation across the cytoplasmic (inner) membrane.[24] CsgG proteins assemble into barrel-shaped oligomeric complexes that constitute a curli-specific translocation channel in the OM.[28] Secretion and stability of CsgA is dependent on 22 specific amino acids at the N-terminus of CsgA, which is believed to target the subunit to the curlin translocation complex formed by CgsG.[28] CsgE and CsgF form soluble accessory proteins required for productive CsgA and CsgB transport and deposition (Fig. 1). Both CsgE and CsgF interact with CsgG; and at least CsgE has been called a chaperone.[33] An inherent problem in functional amyloid systems is maintaining control of a self-propagating reaction such that it does not occur at the wrong time or in the wrong place. The curli system has achieved this control by separating the nucleation and elongation properties into two distinct proteins (the minor curli subunit CsgB and the major subunit CsgA, respectively) and then dictating when and where those proteins interact. CsgA is secreted from the cell as a soluble monomeric protein and, although CsgA polymerization can occur *in vitro*, this requires the presence of curli seeds or the CsgB subunit. CsgB does not itself polymerize, but rather acts as nucleator for the conversion of CsgA from a soluble protein to amyloid deposit.[33,36] In CsgB, the first four repeating units contain an amyloidogenic domain and are necessary for nucleation of CsgA. The fifth domain is vital for its OM association. CsgB-based nucleation marks the first step of curli biogenesis and, once fiber formation is initiated, CsgA can then use the growing fiber tip as a template for amyloid conversion[33,36] (Fig. 1).

D. Type IV Pili

The type IV pili form a third distinct group of pili presented by a wide variety of Gram-negative bacteria, including human, animal, and plant pathogenic species such as *Neisseria gonorrhoeae*, *N. meningitidis*, *Pseudomonas aeruginosa*, *Salmonella enterica*, and *Legionella pneumophila*. Some bacteria such as *Vibrio cholerae* and EPEC express bundled type IV pili.[37] Type IV pili are critical for an array of functions including adherence to both biotic and abiotic surfaces, biofilm formation, cellular invasion, competence for DNA uptake during natural transformation, motility, bacteriophage infection, and electron transfer.[2,38,39] Unique to type IV pilus biology is their ability to retract through the bacterial cell wall via disassembly of pilin subunits at the base of the fiber. When in contact with a solid surface, repeated rounds of pilus extension, adhesion to the substrate, and pilus retraction cause a flagella-independent form of movement known as "twitching motility."[40,41] The forces exerted during pilus retraction are of the order of 100 pN, making it one of the strongest molecular machines characterized to date.[42]

1. MORPHOLOGY AND STRUCTURE

Type IV pili are typically 5–7 nm in diameter and can extend to several micrometers in length. The filaments are polymers composed of thousands of subunits of the major 15–20-kDa pilin subunit, usually termed PilA or pilin, which are arranged in a helical conformation with five subunits per turn and which may be glycosylated and/or phosphorylated. Type IV pilins have a highly distinct primary structure, notably a positively charged leader sequence and a conserved and highly hydrophobic ~ 25-residue N-terminal α-helical domain that will form the core of the pilus fiber.[43,44] The leader sequence is cleaved, and the resulting N-terminal residue is methylated by a dedicated leader peptidase (prepilin peptidase or PilD). Downstream of the conserved 25 residue N-terminal region, the sequences of the major pilin diverge.[40] Type IV pilins are classified into two subgroups, type IVa and type IVb, based on the lengths of their signal peptide (type IVa pilins have ~ 6 residue leaders, type IVb pilins ~ 15–30) and the mature sequence (type IVb are larger). Type IVa pili are present on a variety of bacteria with broad host ranges, whereas type IVb pili are found almost exclusively on enteric pathogens such as *Salmonella*, *Vibrio*, and *E. coli* species.[43] The major pilins of the two subtypes have limited sequence identity, and polymerized pili from type IVa and type IVb subunits differ significantly in diameter as well as in helical structure. The variations in the major subunits are further mirrored in the architecture of their respective assembly systems, where type IVa pilins are assembled by complex systems encoded across the genome of the host organism and type IVb assembly systems are composed of fewer components that are typically encoded in single gene clusters, often located on plasmids.[45]

Structures of pilins from both the IVa and IVb subclasses have been solved and reveal a similar overall architecture[46] (Fig. 2). Both have the conserved structural core mixed α/β domain and a long, hydrophobic N-terminal α-helix. The N-terminal half of this helix, α1-N, protrudes from the protein core and retains individual subunits in the IM until assembly. The C-terminal half, α1-C, packs against the antiparallel four-to-five-stranded β-sheet of the structural domain. Two conserved cysteines form a disulfide bond that links this C-terminal segment to the β-sheet. This N-terminal α-helix is flanked by two regions that vary substantially amongst different pilins: the $\alpha\beta$-loop, which is situated between α1 and the β-sheet; and the D-region, encompassing the conserved cysteines (Fig. 2). Despite their general similarities, close examination of type IVa and IVb pilin structures illustrates that the topology of their β-sheets differs. The type

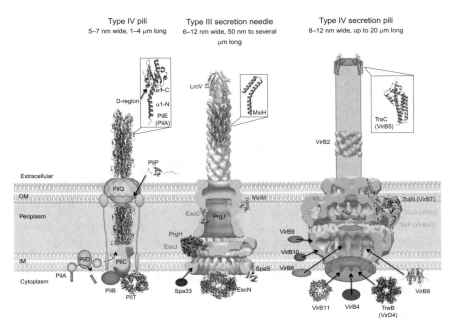

FIG. 2. Type IV pili (left). Schematic representation of the Type IV pili (nomenclature as for
the *P. aeruginosa pil* cluster). Type IV filaments are polymers composed of thousands of helically
arranged subunits of the major pilin subunit, called PilA (green). After transport via the Sec
translocon and maturation by a dedicated bifunctional endopeptidase PilD (orange), PilA remains
anchored in the inner membrane (IM) prior to incorporation at the base of the pilus (IM). Pilus
assembly and disassembly is powered by two cytoplasmic ATPases, PilB (purple) and PilT (red),
respectively, and further requires an IM protein, PilC (light blue). The pilus (here shown using the
model from *N. gonorrhoeae* PilE) is composed of the helical arrangement of the pilins through
coiled-coil interactions via the N-terminal α-helical tail (α1-N). The pilus crosses the OM through
an oligomeric channel formed by the secretin PilQ (light blue) associated with the lipoprotein PilP
(dark blue). The identity of the factor(s) joining the IM and outer membrane (OM) components is
presently unclear. Protein Data Bank (PDB) entry codes used in this figure are 2GSZ (PilT), 2HIL
(PilE, *N. gonorrhoeae* homolog of PilA), 2IVW (PilP), and 3JVU (pilT). Type III Secretion pili
(middle). Schematic model for the Type III secretion (T3S) machinery and associated appendages.
The extracellular structure is a hollow needle composed of a single polymerizing subunit, called
MxiH in *Shigella* (green). At its distal end, the needle carries a capping structure that regulates
needle assembly and provides a scaffold (LrcV in *Yersinia pestis*, yellow) for pore-forming proteins
that contact and insert into the host cytoplasmic membrane (YopB and YopD in *Yersinia*, not
shown). The basal body forms the assembly machinery of T3S filaments and is shown here by the
cryo-electron microscopy 3D structure of the needle complex of *S. typhimurium* (light blue). It
consists of two double rings (EscC in light blue and EscJ/PrgH in orange and dark blue) that span
the IM and OM and are linked by a hollow structure (PrgJ in light green) that crosses the
periplasmic space. At its cytoplasmic side, the basal body is contacted by ATPases (EscN in light
red) and chaperone-like accessory proteins (SpaS in purple and Spa33 in red) responsible for
driving and ordering protein secretion and filament assembly. PDB entry codes used in this figure

IVa pilins display four contiguous β-strands, as shown for gonococcal (GC) pilin from *N. gonorrhoeae*.[46] In contrast, structures of *V. cholerae* TcpA and EPEC BfpA show that the β-sheet connectivity for the type IVb pilins is more complex, containing five to seven β-strands with variable β-sheet topology.[43,44]

The first model for type IV pilus architecture was proposed on the basis of the structure of *N. gonorrhoeae* GC pili,[46] which revealed the hydrophobic N-terminal α-helices pack together in a coiled arrangement inside the filament core (Fig. 2). The globular domains pack with their β-sheets against this central coiled α-helix bundle and thus decorate the filament surface. A similar packing principle has also been observed for the TCP, PAK, and the bundle-forming pili (BFP) from *V. cholera*, *P. aeruginosa*, and EPEC, respectively,[37,44] where crystal structures, fiber diffraction analysis, and EM reconstructions all show filaments with their pilins arranged in three-start helices, packed along the pilus core via coiled α-helix interactions. The extent of the interactions between the α/β domains varies between pili, giving shape to fibers decorated with smooth or carved surfaces. In all models, however, the αβ-loops and the D-regions are optimally exposed and shape the filaments' molecular surface, thus defining pilus function. The exposed αβ-loop often undergoes posttranslational modifications and the D-region displays hypervariable regions, both of which could help pathogens to evade the host immune system.[43]

2. ASSEMBLY–DISASSEMBLY

Unlike CU pili and curli, type IV pilus assembly requires an elaborate an assembly–disassembly machinery that is ATP-driven and spans both IM and OM. The exact mechanism by which type IV pili are put together is still poorly understood.[43] Based on the current state of knowledge, it is proposed that, after transport via the Sec translocon and maturation by a dedicated bifunctional endopeptidase (PilD; nomenclature as in the *P. aeruginosa* system), the pilin subunits (PilA) remain anchored in the IM. The process by which the pilin

---

are 3GR5 (EscC), 1YJ7 (EscJ), 2OBM (EscN), 2CA5 (MxiH), 3GR0 (PrgH), 2VT1 (Spa40), and 1R6F (LcrV). Type IV Secretion pili (right). The archetypical type IV secretion system (T4SS) as found in the *Agrobacterium tumefaciens* T plasmid transfer system (*vir* genes) and the *E. coli* conjugative plasmid *pKM101* (*Tra* genes) is composed of 12 proteins named VirB1-B11 and VirD4 (*A. tumefaciens* nomenclature). The extracellular pilus consists of a processed form of VirB2, the major structural component, and of an additional minor component, VirB5. The biogenesis of the T4SS assembly has been proposed to start with the formation of a core complex (VirB6–VirB10), followed by recruitment of the ATPases, VirB4, VirB11, and VirD4 (red). The cryo-electron microscopy 3D reconstruction of the core complex and the extracellular filament are shown in light blue and green, respectively. PDB entry codes used in this figure are 1GL7 (TrwB, VirD4 homologue), 1NLZ (VirB11), 1R8I (TraC, VirB5 homologue), 2CC3 (VirB8), and 3JQO (TraF/TraO/TraN complex, homologous to VirB10/VirB9/VirB7). (See Color Insert.)

subunits translocate from this IM reservoir into the growing filament is unclear, but polymerization requires an assembly ATPase (PilB)[47] and an IM protein (possibly PilC).[48] For retractile pili, a retraction ATPase (PilT) is required to rapidly depolymerize the pili, presumably repopulating the membrane-bound reservoir. Once a subunit is inserted, the filament is extruded a short distance into the periplasm. Secretion of the filament across the OM is dependent on a secretin (PilQ), which together with a lipoprotein "pilot" protein (PilP) forms an oligomeric channel in the OM.[49–51] To facilitate correct docking of the filament to its OM portal, proteins involved in pilus assembly likely form a large dynamic complex that spans the periplasm, connecting the IM and OM components. Recent data demonstrate that the minor pilins PilV/PilW/PilE are incorporated into the fiber and support a role for them in the initiation, but not termination, of pilus assembly.[52]

E. Type III Secretion-Related Organelles

A fourth type of surface appendages in Gram-negative bacteria are tubular Type III secretion-related organelles, which were first observed in *Salmonella typhimurium* as needle-like surface structures and as a pilus-like structure in *Pseudomonas syringae* and other plant pathogens.[53,54] Since their discovery, filaments belonging to the type III secretion system (T3SS) have been identified in a range of animal and plant pathogens, where they enable the bacteria to translocate substrates (effectors) directly into the cytoplasm of the host cell to exert a broad range of virulence functions.[55] To do so, T3SSs form complex, supramolecular structures which span the IM, the periplasmic space, the OM, the extracellular space, and a host cellular membrane (Fig. 2). Because of their shape and their ability to translocate proteins in a cell-contact-dependent manner, type III secretion needles are also referred to as "injectisomes" or "molecular needles." So far, no adhesive role has been observed. T3SS are genetically, structurally, and functionally related to the bacterial flagellum assembly system.[56]

1. MORPHOLOGY AND STRUCTURE

The extracellular structure in T3SSs is found in three different morphologies depending on the specific system. The archetypical "needle" seen in the *Salmonella* SPI-1 T3SS and *Shigella* injectisomes[54,57,58] is a short, rigid structure of about 60 nm in length and an inner diameter of 2–3 nm. The needle is composed of a single polymerizing subunit: PrgJ, MxiH, YscF, and Escf in *Salmonella, Shigella, Yersinia*, and EPEC, respectively. Although T3SS needle subunits are significantly smaller (9 kDa compared to 45 kDa) and share no sequence homology with the flagellar hook or filament, they display a common helical architecture (5.6 units per turn, 24 Å helical pitch[59]; Fig. 2). Available structures of the needle subunits (*Shigella* MxiH,

Burkholderia BsaL, and *Salmonella* PrgI) show that these consist of a two-helix bundle motif that is linked by a conserved PxxP turn.[60] In the needle, 100–150 subunits are believed to polymerize through superhelical coiled-coil interactions[61] (Fig. 2). In the EPEC, the needle is extended by a longer flexible "filament" up to 600 nm in length that is composed of an additional polymerizing subunit EspA.[62,63] In plant pathogens, the needles are replaced altogether by a long, flexible "pilus," the Hrp pilus, in agreement with the requirement to traverse the thick cell wall in order to come into contact with the cytoplasmic membrane.[53,64] Needle subunits comprise a helix–loop–helix structure similar to the D0 portion of flagellin (Figs. 2 and 3)[65]. Like flagella, T3SS filaments invariably form hollow tubes and grow from the distal end by the addition of subunits that are transported from the cytoplasm through the lumen of the fibers.[66] After assembly, T3SS filaments function as transport conducts for the delivery of proteinaceous toxins into the host cytosol.[55] At their distal end, T3SS needles carry a distinct tip complex involved in pore formation in the target cytoplasmic membrane.[67] The tip complex is composed of three proteins (LcrV, YopB, and YopD in *Yersinia*) that assemble into a protein-conducting channel upon contact with the target membrane.[65,68,69]

2. ASSEMBLY

Stepwise assembly of the T3SS involves the tightly coordinated interplay of more than 25 proteins, including structural elements, secreted proteins (pore-forming translocators and effectors), intracellular chaperones, and regulatory components.[61] T3SS filaments are assembled by a periplasm-spanning protein complex reminiscent of the flagellar basal body. The T3SS basal body is formed by two oligomeric ring complexes in the OM and IM, joined together by a vestibule spanning the periplasmic space[54] (Fig. 2). The OM ring complex consists of secretins, which are believed to exert a stabilizing and anchoring function for the needle complex.[70] At its cytoplasmic side, the basal body is contacted by ATPases and accessory chaperone-like proteins, responsible for driving and ordering protein secretion and filament assembly.[71,72] Assembly of T3SS appendages is thought to follow a mechanism similar to that seen in flagella. During morphogenesis, the formation of the basal body precedes the insertion of the needle. The components forming the transmembrane rings are assembled in a Sec-dependent manner. Subsequent filament subunits and secretion substrates are sequentially exported through the nascent secretion apparatus itself.[73,74] How the system controls the length of the filaments and the substrate switching during needle assembly and protein secretion is a matter of intensive investigation.[55,75]

F. Type IV Secretion Pili

A last type of pili found in Gram-negative bacteria are formed by the type IV secretion systems (T4SS), which are specialized machinery used for one-step transport of protein, nucleic acid, or nucleoprotein substrates across cell membranes. T4SS are evolutionarily related to bacterial conjugation systems and can be found as plasmid transfer systems (conjugative systems) or adapted conjugative systems. In plant and human pathogens such as *Agrobacterium tumefaciens*, *L. pneumophila*, and *Helicobacter pylori*, T4SS are involved in diverse processes relevant to virulence, ranging from the delivery of virulence factors into eukaryotic cells to conjugative transfer of genetic material and the uptake or release of DNA (natural competence).[76,77] Conjugative transfer of antibiotic-resistance genes via T4SSs is a leading cause for the emergence of multidrug resistance.[78]

1. MORPHOLOGY AND STRUCTURE

Based on homology, T4SSs can be grouped as type IVa, IVb, or "other." Though the secretion apparatus and function are homologous, little sequence similarity is found between Type Iva and IVb systems. The archetypical type IVa system is the *A. tumefaciens* VirB/D4 cluster, comprising 11 VirB proteins (VirB1–11) and one VirD protein (VirD4).[79] The pili formed by IVa T4SSs are short (< 1 μm), rigid rods with a diameter of 8–12 nm, called "IncP-like pili," and are exemplified by TrbC from the IncP plasmid RP4 and by VirB2 from *A. tumefaciens*. The type IVb systems are assembled from subunits related to the prototypical *L. pneumophila* Dot/Icm system.[80] Compared to IncP-type T4SSs, type IVb systems carry about twice as many genes. Type IVb secretion pili are also called IncF-like pili and form long (2–20 μm), flexible appendages with a diameter of 8–9 nm. They contain pili produced by the IncF, IncH, IncT, and IncJ systems. The "other" systems are often not involved in bacterial conjugation and bear little or no discernible ancestral relatedness to the types IVa or IVb systems. Examples are the *A. tumefaciens* T-pilus, which has a diameter of 10 nm but is flexible and variable in length, and the 100–200 nm needle-like type IV secretion-related appendages produced by *H. pylori*.[79]

T4SS pilins, the protein subunits that build up the extracellular appendages associated with T4SSs, are homologous to one another and display a number of common physical properties. Typically, these proteins are synthesized as pre-proteins with an unusually long signal peptide of 25–50 residues and need extensive processing by dedicated cellular factors to mature (see below). The structure of the F-pilus was examined using X-ray diffraction[81] and cryo-EM.[82] F-pili are shown to be cylindrical filaments with a central lumen of 2 nm and an external diameter of 8 nm. Two different subunit packing arrangements that seem to coexist in the pilus structure were observed: a stack of pilin rings of C4 symmetry and a one-start helical symmetry with an axial rise of 3.5 Å per subunit and a pitch of 12.2 Å. The 2-nm lumen diameter seems large enough to afford passage of single-stranded DNA but not folded accessory proteins, such as the F-pilus conjugative transfer relaxase TraI.[79]

The T-pilus of A. *tumefaciens* consists of a processed form of VirB2, the major structural component, and of an additional minor component, VirB5 (Fig. 2). No structural information for VirB2-like proteins alone is available. However, a crystal structure is available for the VirB5 homolog in pKM101, TraC, which shows that the structure is composed of a three-helix bundle capped and a smaller globular appendage.[83] Structure-based mutagenesis studies suggested a role of VirB5 in adhesion and host recognition.[83] VirB5 has homologs in other pathogenic bacteria such as *Brucella suis* or *Bartonella henselae* (called VirB5 in both species) and in conjugative plasmids such as pKM101 (TraC/), RP4 (TrbF), R388 (TrwJ/), or F factor (TraE).

2. ASSEMBLY

Little is known about the assembly of type IV secretion pili. All T4SS pilins that have been characterized undergo several processing reactions during maturation. The pro-pilins are first targeted to the IM by a signal peptide that is cleaved by a dedicated protease. Additional maturation steps depend on the type of pilin. Most IncF-like pilins are acetylated at the N-terminus and inserted into the IM. By contrast, the IncP-like conjugative pilins are proteolytically processed at one or both of their termini and then undergo a cyclization reaction resulting in the formation of an intramolecular covalent head-to-tail peptide bond.[79]

Most studies aimed at deciphering the mechanism of action of the T4S machines have focused on the A. *tumefaciens* VirB/D4 system.[84] T4SSs are formed by at least 12 proteins, termed VirB1–VirB11 and VirD4, although some T4SSs are composed of just a subset of these proteins. The VirB/VirD4 proteins can be grouped on the basis of their function and location into energizing proteins (the ATPases, VirB4, VirB11, and VirD4), translocation pore core complex (VirB6 to VirB10), pilus-associated proteins (VirB2 and VirB5), and other T4SS proteins (the putative lytic transglycosylase VirB1 and membrane protein VirB3). The biogenesis of the T4SS assembly has been proposed to start with the formation of a core complex (VirB6, B7, B8, B9, and B10), followed by recruitment of the ATPases and/or pilus biogenesis.[85] Structural studies of the pKM101 conjugative system reveal that the subunits TraN (VirB7), TraO (VirB9), and TraF (VirB10) form a 14-fold symmetrical megadalton core complex of 185 Å diameter that spans the IM and OM and forms a hollow vestibule constricted to a 55 and 10 Å channel at the cytoplasmic and OM, respectively[86] (Fig. 2). In flagella and T3SS, the extracellular filaments are built up from a base ring inside a similar periplasm-spanning vestibular complex and are extended from the distal end through deposition of subunits that traverse the growing filament (Figs. 2 and 3).[66] How pili are assembled in T4SSs is presently unknown. Secretion and pilus biogenesis might require two different complexes and, thus, after assembly of the core complex, there might be a divergence in the assembly process leading to formation of one complex used for secretion and the other used for pilus biogenesis.[87]

III. Pili in Gram-Positive Bacteria

Gram-positive bacteria lack an OM. Surface proteins are embedded in the cytoplasmic membrane or can be covalently or noncovalently attached to the cell wall.[88] Pilus-like structures in Gram-positive bacteria were first detected in 1968 in *Corynebacterium renale*.[89] Surface appendages have now been reported to be present in multiple Gram-positive bacteria, including *Actinomyces* spp. and important human pathogens such as *Corynebacterium* spp., *Streptococcus* spp., *Pneumococcus* spp., and *Enterococcus* spp.[90] The pili of Gram-positive pathogens have a similar role in adhesion and attachment to host cells as those in Gram-negative bacteria, and are therefore also important virulence factors.[91,92] In addition to mediating binding, Gram-positive pili have been shown to modulate host immune responses and promote the development of bacterial biofilms.[93] Two types of pili have been identified in Gram-positive bacteria by EM. The first are sortase-assembled pili, generally called "Gram-positive pili." In contrast to Gram-negative pili, the major pilus subunits are covalently connected into linear or branched fibers anchored to the cell wall by a sortase enzyme. Second, some Gram-positive bacteria express type IV pili similar to those of Gram-negative organisms. These pili can bind to cellulose and confer gliding motility.[94]

A. Sortase-Assembled Pili

1. MORPHOLOGY AND STRUCTURE

Sortase-assembled pili are thin rods that range from 1 to 5 nm wide and can be as long as 5 μm. Gram-positive pili are composed of multiple copies of a single major pilin subunit and usually one or two accessory subunits (Fig. 3). These latter are not required for the integrity or synthesis of the pilus.[95] In part, pilus length is determined by availability of the major pilus subunit, and overexpression leads to longer pili. The pili genes are clustered in genetic loci, always containing a sortase gene, encoding a specific membrane-associated transpeptidase enzyme.[96] Several Gram-positive organisms harbor more than one pilin locus, each encoding components specific for the assembly of a unique pilus type.[93,96] Pilin proteins are synthesized with an N-terminal Sec signal, which is cleaved during Sec-mediated transit through the cytoplasmic membrane, and a C-terminal cell wall sorting signal (CWSS) that contains an LPXTG (or similar) amino acid motif, followed by a hydrophobic region and a positively charged C-terminus.[97] An additional motif (the "E box") has been identified in the proteins that constitute the shafts of the known pili. In some pili, but not all, this E box is needed for the incorporation of the minor pilus subunits.[96]

2. ASSEMBLY

Insight into the mechanisms of pilus assembly in Gram-positive bacteria initially came from studies of *Corynebacterium diphtheria*,[95] but the proposed model is thought to be applicable for most sortase-assembled pili.[90] After Sec-dependent secretion, the pilus components are retained in the cell membrane by their hydrophobic domain and charged tail. Next, a pilus-specific sortase cleaves the cell-anchored proteins at the LPXTG motif, between the threonine and glycine residue.[98] This reaction leads to the formation of acyl-enzyme intermediates.[97] Oligomerization of the pilus subunits involves the release of the sortase enzyme from the pilin by nucleophilic attack of the ε amino group of a lysine side chain in a conserved "pilin motif" (WxxxVxVYPKN) of the neighboring subunit[97] (Fig. 3). According to the prevailing model, pilus growth occurs by subunit addition at the base of the pilus. Sequential transpeptidation reactions result in a pilus fiber consisting of a linear array of pilin molecules, the bulk of which is composed of the major pilin subunit. Finally, anchoring of the membrane-proximal pilus subunit to the cell wall occurs through nucleophilic attack by the amino group of the peptidoglycan precursor, this time catalyzed by the "housekeeping" sortase, resulting in the covalent linkage of the pilus to the cell wall[97,98] (Fig. 3). Several aspects of the pilus assembly process, such as the incorporation of the ancillary proteins, remain to be elucidated.[93] It is also noted that an alternative model can be envisaged, in which pilus growth occurs by the addition of subunits to the top of a bended pilus fiber.[90]

B. Type IV-Like Pili

Pili presenting characteristics typical of the type IV pili of Gram-negative bacteria have been discovered in the Gram-positive genera *Clostridia*[99] and *Ruminococcus*.[100] In *Clostridium perfringens*, they are involved in biofilm formation[101] and to mediate a gliding motility similar to twitching in Gram-negative bacteria.[102] Type IV-like pili are thin (5–8 nm in width), long (several micrometer in length), and flexible filaments consisting of polymers of a single peptide subunit, PilA[45] (Fig. 3). These pilin subunits contain a hydrophilic leader sequence ending with a glycine that is cleaved by a unique leader peptidase, an unusually modified amino acid as the first residue of the mature pilin (N-methylated phenylalanine) and an extremely hydrophobic, highly conserved N-terminal α-helical domain.[45] The pilins are polymerized via a

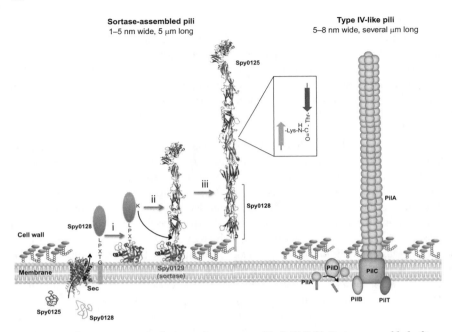

FIG. 3. Pili in Gram-positive bacteria. Sortase-assembled pili (left). Sortase-assembled pili are composed of multiple copies of a single major pilin subunit (Spy0128 in green; nomenclature as in *Streptoccocus pyogenes* Spy cluster) and usually one or two accessory subunits (Spy0125 in blue). After Sec-dependent secretion, the pilus components are retained in the cell membrane by their hydrophobic domain and charged tail. A pilus-specific sortase (Spy0129 in orange) then cleaves the cell-anchored proteins at a conserved motif: LPxTG to form an acyl-enzyme intermediate. (i) Oligomerization of the pilus subunits involves nucleophylic attack by a Lys amine of the acyl-sortase complex at the base of the growing. (ii) Stepwise covalent additions form linear or branched fibers, finally covalently anchored to the cell wall via a housekeeping sortase (not shown) and nucleophilic attack by the peptidoglycan precursor lipid II. (iii) PDB entry codes used in this figure are 2XIC (Spy0125), 3B2M (Spy0128), and 3PSQ (Spy0129). Type IV-like pili (right). Type IV-like pili in Gram-positive bacteria present characteristics typical of the type IV pili of Gram-negative bacteria. They are flexible filaments consisting of polymers of a single peptide subunit, PilA (green). The pilins are polymerized via a complex of proteins, including a signal peptidase PilD (orange), an extension motor PilB (light blue), a membrane protein PilC (purple), and a retraction motor PilT (dark blue; Nomenclature as in *Clostridium perfringens*). (See Color Insert.)

complex of proteins, including PilD (a signal peptidase that recognizes PilA), PilB (an extension motor), PilC (a membrane protein), and PilT (a retraction motor; Fig. 3).[103] These proteins are highly conserved, and the proposed 3D fold of PilA shows remarkable similarity to pilins from Gram-negative bacteria, suggesting a common origin and mechanism of pilus assembly.[45] In Gram-negative bacteria, the OM secretin PilQ forms ring-like oligomers to regulate the extension of the pilus through the OM. As Gram-positive bacteria lack an OM, the absence of a PilQ homolog is expected. However, *C. perfringens* does have a thick peptidoglycan layer in which an opening must be maintained to allow the pilus through. Some of the predicted proteins seen associated with the type IV genes may play a role in this function.[45]

Ruminococcus albus plays a role in fiber breakdown in the rumen of herbivores and produces pili involved in the bacterium's adhesion to cellulose.[104] Pili subunit proteins have been identified in strain 8 (CbpC) and strain 20 (GP25) and both are type IV pili proteins.[99] Downstream of the *cbpC/gp25* gene (now referred to as *pilA1*) is a second pilin gene (*pilA2*), which plays a role in the synthesis and assembly of type IV pili, but its role is restricted to cell-associated functions, rather than being part of the externalized pili structure. An assembly pathway for pili in ruminococci remains to be established, but most likely it will resemble the other type IV pathways.[105]

IV. Surface Filaments in Archaea

Archaea are best known for their ability to thrive in extreme environments, but recent studies identified them in nearly every habitat accommodating life, including garden soil, coral reefs, oral cavities, and the human gut.[106] Very early, it was evident that the structure of the archaeal cell envelope differs substantially from that of bacteria.[107] With the exception of *Ignicoccus* species, which exhibits an OM enclosing a huge periplasmic space,[108] all other known Archaea possess only a single membrane.[109] This cytoplasmic membrane is enclosed by a surface layer (S-layer), consisting of glycoprotein subunits that form a two-dimensional paracrystalline structure and containing pores that are permeable to solutes and small molecules.[109] Weiss reported in as early as 1973 that *Sulfolobus* cells taken freshly from a hot spring are attached to sulfur particles by numerous 5-nm-wide pili. Pilus-like structures have since been observed on the surface of many other Archaea, but the functional, genetic, molecular, and structural investigations of archaeal pili have only recently begun (Fig. 4). Consequently, it is presently unclear if Archaea possess the vast diversity of pilus types, with assorted functions and assembly mechanisms, presently known to occur in bacteria.[110]

A. Archaeal Type IV-Like Pili

Genome analysis identified diverse genes as being similar to bacterial type IV pilins, including a subset of nonflagellar proteins that are specifically cleaved by a novel prepilin peptidase.[111] Their structural resemblance with homologs of bacterial pilin-like substrates and the observed colocalization with genes

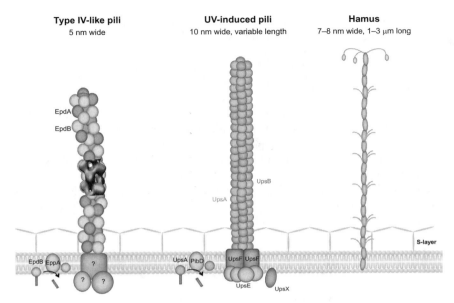

Fig. 4. Pili in Archaea. Archaeal Type IV-like pili (left). Archaeal Type IV-like pili are formed by two subunits, EpdA (light green) and EpdB (dark green). Precursor proteins are processed by EppA (orange). The assembly system further consists of an integral membrane protein and a cytoplasmic ATPase. UV-induced pili (middle), UV-induced pili of the *ups* operon (*UV*-inducible *pili* operon of *Sulfolobus*) are straight surface appendages. They are composed of two prepilins containing a class III signal peptide UpsA and UpsB (light and dark green, respectively), which are processed by PipD (orange). Pilus formation involves a potential secretion ATPase UpsE (light blue), a putative transmembrane protein UpsF (dark blue), and a protein of unknown function UpsX (purple). Hamus (right). Hami are mainly composed of 120-kDa subunits and the end of filament is formed by a tripartite, barbed grappling hook. (For interpretation of the references to color in this figure legend, the reader is referred to the Web version of this chapter.)

homologous to bacterial type IV pilin assembly genes suggest that these might assemble into archaeal cell surface structures, such as pili.[111] However, recent structural studies on the *Methanococcus maripaludis* pili indicated that, even though the pilins had similarities to bacterial type IV pilins, the quaternary structure formed by these archaeal pilins was entirely different from both the bacterial type IV pilus and the archaeal flagella[112] (Fig. 4). Two subunit packing arrangements were found (a one-start helical filament and a ring structure of four subunits), both coexisting within the same pili filaments. In addition, the pili contain a central channel only slightly smaller than that observed in bacterial flagella and T3SS needle structures. Further research will be necessary to determine whether the lumen of this pilus is involved in the secretion of pilus subunits or other substances.[112,113]

B. UV-Induced Pili

Sulfolobales ssp. also express UV-induced pili encoded by the *ups* operon (*UV*-inducible *pili* operon of *Sulfolobus*).[114] The Ups pili are straight surface appendages with 10 nm in diameter, variable in length, not bundled or polarized, and composed of three evenly spaced helices, thereby clearly being distinguishable from archaeal flagella (Fig. 4). Interestingly, the Ups pili are essential for cellular aggregation and are hypothesized to enhance DNA transfer among *Sulfolobus* cells to provide increased repair of damaged DNA via homologous recombination. The *ups* operon encodes a potential secretion ATPase (UpsE), two prepilins containing a class III signal peptide (UpsA and UpsB), a putative transmembrane protein (UpsF), and a protein of unknown function (UpsX)[114] (Fig. 4).

C. Hamus

Another unusual archaeal cell surface appendage is the hamus. This structure represents a novel filamentous cell appendage of unexpectedly high complexity, found on Archaea growing in cold sulfidic springs.[115] These coccoidal cells are surrounded by a halo of about 100 hami that each measures 1–3 μm in length and 7–8 nm in diameter and has a helical structure with three prickles (each 4 nm in diameter) emanating from the filament at periodic distances (Fig. 4). The end of filament is formed by a tripartite, barbed grappling hook. The hami are mainly composed of 120-kDa subunits and remain stable in a broad temperature (0–70 °C) and pH (pH 0.5–11.5) range. It is proposed that the hami function in surface attachment and biofilm initiation, much like flagella and pili in bacterial biofilm formation, but in addition provide a strong means of anchoring.[115]

1. OTHER PILI

Methanothermobacter thermoautotrophicus possesses surface filaments that are distinct from type IV pili, as they are composed of a 16-kDa glycoprotein Mth60 lacking a class III signal peptide.[116] These pili can be up to 10 μm long, have a diameter of 4–5 nm, and proved adhering to solid (organic and inorganic) surfaces thereby acting as adhesins. However, studies need to be undertaken to further characterize this system.

V. Flagella

The perception that bacteria are motile dates back to the very beginning of microscopy, when Antonie van Leeuwenhoek envisioned the "little living animalcules" to possess tiny "paws" that enable a "strong and swift motion." Now, over 330 years later, these "paws" are called "flagella" and it is clear that they are used by many bacteria and Archaea for motility.[117] Besides being a chemo-taxis-navigated swimming system used in liquid environments, flagella also perform a role in swarming across solid surfaces. Additionally, it was shown that in many species a functional flagellar system is required for adhesion[118-120] and biofilm formation,[121,122] and that the flagellum is also implicated in mea-suring environmental conditions such as viscosity or wetness.[123] Flagella from Gram-positive and Gram-negative bacteria are essentially conserved, except that in Gram-negative bacteria they extend through a second, OM that is absent in Gram-positive bacteria.[124] While superficially similar to the bacterial flagellum, archaeal flagella have a different mechanism of flagellin secretion and a high number of posttranslational modifications. Instead, the archaeal flagellar filament resemble the bacterial type IV pilus, as the flagellins have similar signal peptides processed by a type IV prepilin peptidase-like enzyme, and assembly in both systems involves hydrophobic N-terminal α-helices that form three-stranded coils in the center of these filaments.[109]

A. Bacterial Flagella

The bacterial flagellum is a large macromolecular complex composed of about 30 different proteins with copy numbers ranging from a few to tens of thousands. Flagella are genetically, structurally, and functionally related to T3SSs and consist of three major substructures: the filament, the hook, and the basal body.[56] The basal body is found at the base of each flagellum and works as a reversible rotary motor, which is powered by a proton (or sodium) motive force (PMF). The motor consists of two parts: the rotating part, or rotor, which is connected to the hook and the filament, and the nonrotating part, or stator, which conducts coupling ions and is responsible for energy conversion. The basal body is imbedded within the cell envelope, while the hook and filament, which function as a universal joint and a propeller, respectively, extend outward in the cell exterior.[125]

1. MORPHOLOGY AND STRUCTURE

As in other T3SS, the flagellar basal body provides an anchor for the (flagellar) filament and is formed by oligomeric ring complexes joined together by a rod. Gram-negative organisms have five such rings: a cytoplasmic ring (C ring), two for anchoring the flagellum to the cytoplasmic membrane (M–S rings), one for the peptidoglycan (P ring), and one for the OM (L ring; Fig. 5).[117] In Gram-positive bacteria, the flagellar structure is less complex, lacking the P and L rings.[126] The M ring anchors the flagellum to the plasma membrane, while a second ring structure (S) is embedded in the thick pepti-doglycan layer.[127] MotA and MotB proteins structure the nonrotating stator segment of the basal body and form a channel that is crossed by a flow of

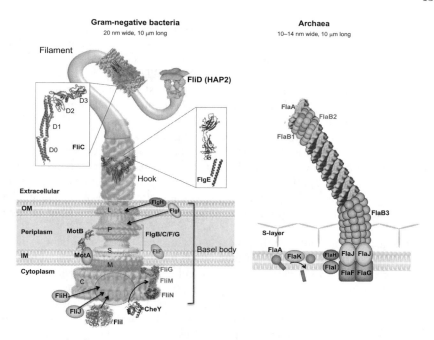

Fig. 5. Flagella. Bacterial flagella (left). The bacterial flagellum is a large macromolecular complex composed of about 30 different proteins and consisting of three major substructures: the filament, the hook, and the basal body. The flagellar basal body is formed by oligomeric ring complexes joined together by a rod and represented here by the cryo-electron microscopy 3D structure (light blue). MotA and MotB proteins structure the nonrotating stator segment of the basal body. A cytoplasmic ATPase complex consists of FliI, FliH, and FliJ (red). The cytoplasmic (C) ring consists of several proteins including three switch proteins (FliG, FliM, and FliN), which allow the flagellum to change the direction of rotation. The hook region connects the filament to the basal body and is composed of approximately 20 copies of a single protein FlgE. The filament itself consists of a single flagellin protein. The flagellin subunit of *S. typhimurium* FliC has an overall shape of the Greek letter "Γ" and consists of four linearly connected domains, D0 to D3. At the tip, the flagellum is capped by a pentamer of FliD proteins. PDB entry codes used in this figure are 3A5X (FliC), 3A69 (FlgE), 1LKV (FliG), 2HP7 (FliM), 1O6A (FliN), 2ZVY (MotB), 2DPY (FliI), and 2B1J (CheY). Archaeal flagella (right). Upon removal of the signal peptide by FlaK (orange), the processed flagellins (FlaA, FlaB1, and FlaB2 in different shades of green) are incorporated into the base of the structure, likely with the help of the ATPase FlaI (red) and the integral membrane protein FlaJ (purple). FlaB3 (brown) is localized proximal to the cell surface forming a curved structure with similarity to the bacterial hook structure. (See Color Insert.)

protons (or sodium ions in some alkalophilic and marine *Vibrio* species; Fig. 5). The strength of the PMF or sodium motive force generated by this flow determines the rotation speed of the flagellum. The flagellar type III protein export apparatus is an ATPase complex thought to be localized at the cytoplasmic side of the basal body. It consists of the ATPase that drives export (FliI), a negative regulator of the ATPase (FliH), and a general flagellar-secretion chaperone (FliJ) that recognizes and unfolds flagellar components prior to secretion. The primary role of the FliHIJ ATPase complex is to shuttle substrates to the integral-membrane secretion system and present them as unfolded substrates (flagellar subunits) for efficient PMF-dependent secretion.[124] The rotor is connected to the hook and the filament and extends into the cytoplasm forming the cytoplasmic (C) ring (Fig. 5). This C ring consists of several proteins including three switch proteins (FliG, FliM, and FliN), which allow the flagellum to change the direction of rotation almost instantaneously, thereby adjusting the path of swimming in response to attractants or repellents in the environment[125] (Fig. 5). A complex chemotaxis system results in the direct contact of a phosphorylated CheY protein with the FliM switch protein.[128]

The filament is a hollow tube composed of up to 20,000 subunits of a single flagellin protein, having a 2-nm-diameter lumen through which unfolded flagellins can be transported. Less commonly, the filament is composed of several different flagellins, as in *Campylobacter jejuni*, where it is composed of a minor flagellin, FlaB, and a major flagellin, FlaA. Each flagellar filament is ~ 10 μm long, and its length is probably limited by hindered diffusion of the filament subunits. This theory is supported by the observation that the rate of *in vivo* polymerization exponentially slows down with increasing length, whereas there is no effect of increasing length on the polymerization rate *in vitro*.[124] The flagellin subunit of *S. typhimurium* FliC has an overall shape of the Greek letter "Γ" and consists of four linearly connected domains, D0 to D3 (Fig. 5). The D3 domain is surface-exposed and corresponds to the highly variable region to which antibody responses are directed. D2 is also surface-exposed, whereas the conserved domains D0 and D1 compose the core of the flagellum, interacting with neighboring flagellins, and are required for filament assembly[129] (Fig. 5). Every flagellar filament is composed of different protofilaments, each composed of thousands of flagellin molecules, one on top of the other, forming coiled-coil interactions through the D0 region of the subunits (Fig. 5). Each of the protofilaments can be in a left-handed (L) or a right-handed (R) state, with all the subunits within that protofilament being in the same state. The L state has a slightly longer (by 0.8 Å) intersubunit distance. A filament will contain a mix of both L and R protofilaments. This nonequivalence of all protofilaments breaks the strict helical symmetry of the flagellin subunits and leads to a curvature that generates thrust when the flagellar filament is rotated by the flagellar motor. At the tip, the flagellum is capped by a pentamer of HAP2 (hook-associated protein 2, also called FliD) proteins. Cryo-EM revealed the cap to be a plate-like structure forming a lid on top of the filament, with five legs protruding downwards interacting with the filament.[130] The highly conserved N-terminal 40 amino acids and C-terminal 50 residues of HAP2 form a "leg domain" that mediates contact with the flagellin subunits at the tip of the filament (Fig. 5). The central part of HAP2 that builds up the cap plate is important for the polymerization of FliD.[127]

The hook region connects the filament to the basal body and is composed of ~20 copies of a single protein FlgE. The junction of the hook and filament requires the presence of a small number of two hook-associated proteins called HAP1 and HAP3. The hook region has a defined length (55 nm) and different models exist to explain how this length is regulated. One of the models postulates that hook length is controlled by the C ring which acts as a measuring cup. In this model, FlgE subunits fill the C ring and are then exported and assembled *en masse*. The length of the hook is determined by the capacity of the C ring and its binding sites for the hook subunits.[127] Other models are identical to those proposed for the needle length control in the T3SS, and include the molecular ruler model and a kinetic regulation model.[55]

2. Assembly

The assembly of the flagellum is a well-known sequential process. The components of the basal body and the export apparatus are transported by the Sec system. The more distal components are secreted through the hollow core of the basal body and the nascent filament, and assemble at their exit point from the channel. A cap at the tip of the growing flagellum ensures the efficient polymerization of the different subunits as they reach the tip. In the absence of the cap, the subunits are exported normally but are shed into the surrounding environment rather than undergoing polymerization.[124,131]

One of the unusual variations on bacterial flagellation is that found in Spirochaetes, where the flagella are located in the periplasmic space. These flagella are believed to have vital skeletal and motility functions, as their rotation causes the entire bacterium to move forward in a corkscrew-like motion.[127]

B. Archaeal Flagella

While the archaeal flagellum resembles the bacterial flagellum in terms of being a reversible, rotating organelle responsible for swimming motility and biofilm formation, its composition, structure, and mode of assembly are all very different.[132] Furthermore, Archaea do not possess any homologs of bacterial flagellar genes. In contrast to bacterial flagellar subunits, the secretion and assembly of archaeal flagellins resembles those of the bacterial type IV pilus system.[110] These similarities include the lack of a central channel in the flagellar filament, the fact that the flagellin subunits possess class III signal peptides that are cleaved before the incorporation of the protein into the flagellar filament, and the presence of several homologous proteins in the assembly machinery, including a type IV prepilin peptidase-like enzyme (FlaK) and a VirB11-like ATPase (FlaI).[111,133]

1. Morphology and Structure

Most archaeal flagella are composed of several types of flagellin subunits, although in *Sulfolobus* spp. the flagellum consists of a single type of protein. The *fla* operon of *Methanococcus voltae* contains four structural flagellin genes: *flaA*, *flaB1*, *flaB2*, and *flaB3*. FlaB1 and FlaB2 are the major components of the flagellum, while FlaA is distributed throughout the flagellum as a minor component. FlaB3 is localized proximal to the cell surface, forming a curved structure with similarity to the bacterial hook structure, which may indicate that a similar torque-driven motion is generated by the *M. voltae* flagellum[132] (Fig. 5). The flagellin subunits contain a type IV pilin-like signal peptide that is necessary for their membrane insertion and assembly into the flagellum. Their H-domains likely fold into an extended hydrophobic α-helix, which participates in interactions between subunits. Additionally, the flagellins are frequently posttranslationally modified, usually by glycosylation, which may be necessary for proper assembly and/or function.[127]

The archaeal flagellum has a right-handed helical subunit packaging with a diameter of approximate 10–14 nm, which is much thinner than the 20 nm observed for the bacterial flagellum. While a hook structure is evident, convincing evidence for a basal body with rings has been lacking. The impossibility to isolate a basal body may, however, be caused by a different anchoring mechanism in Archaea or because the structure is more fragile than its bacterial counterpart. The archaeal flagellum is not hollow and the inner space is most probably formed by coiled-coil interaction of the N-terminal hydrophobic domains of the flagellins similar to the assembled type IV pilus.[132]

2. Assembly

Archaeal preflagellin subunits contain a class III signal sequence, which directs them to the Sec system for translocation and stably anchors them in the cytoplasmic membrane. Type IV pilin-like signal peptidases proteolytically remove the signal sequence, allowing dislocation from the membrane and subsequent assembly into the growing flagellum. Unlike bacterial type IV pili, there is no evidence for methylation or any other type of modification of the newly formed N-terminus. The flagellar accessory proteins FlaI, a cytoplasmic ATPase, and FlaJ, an integral-membrane protein, seem to form a minimal system that is required for assembly of the flagellin subunits into the flagellum.[109] Recent studies suggest that the energy required for the rotation of the *Halobacterium salinarum* flagellum is directly gained from ATP hydrolysis and not from the PMF, indicating a fundamentally different mechanism of rotation than that of the bacterial system.[134] In the archaeal flagellum, the N-terminal hydrophobic domains of the flagellin subunits are oriented toward the central core of the filament, resulting in a compact structure with no lumen. Therefore, assembly of the archaeal flagellum takes place from the base, similar to bacterial pilus assembly, and cannot be done by a mechanism analogous to bacterial flagella, where the subunits travel through the hollow flagellum to the tip.[109] The exact mechanism of assembly of the archaeal flagellum is, however, still a matter of debate, as it is not known whether the accessory proteins FlaI and FlaJ only assemble the flagellum or whether they also form components of the motor. In *H. salinarum*, a link to the sensory apparatus was provided by the flagellar accessory proteins FlaCE and FlaD, which interact indirectly with proteins from the Che signaling system.[135]

VI. Applications

As outlined in the previous section, both Eubacteria and Archaea have developed elegant mechanisms to efficiently assemble multisubunit fibrillar structures on their outer surfaces. Their natural function and/or highly organized multisubunit feature make them attractive biotechnological tools. The various types of surface appendages differ fundamentally in their structure, assembly, and function, and accordingly the choice of the best suited organelle will vary depending on the type of application desired. Several applications exploit the intrinsic properties of bacterial appendages, but they can also be engineered to obtain new characteristics. Here we discuss some general considerations regarding the use of pili and flagella as biotechnological instruments and provide selected examples to illustrate their use in current and future applications.

A. DNA Transfer and Uptake

Arguably, some of the bacterial filaments with the longest history of biotechnological application are type IV secretion pili. Conjugative pili account for horizontal DNA transfer among bacteria. This process, called "conjugation," was thought to require the formation of secure intercellular contacts before DNA transfer can occur,[40] but that notion has recently been contested.[136] The genes responsible for conjugation are generally organized in clusters located on plasmids and code for proteins forming three macromolecular assemblies: (i) a relaxosome responsible for recognition of the plasmid's origin of transfer (OriT) region and covalent attachment of DNA at the OriT; (ii) a mating-pair formation complex or T4SS responsible for DNA transport; and (iii) a pilus, the function of which is not entirely clear but likely responsible for either mate recognition (or adhesion) or forming a conduit through which the DNA is threaded.[137] Occasionally, plasmids such as the F plasmid integrate into a random position in the bacterial chromosome and, during conjugation, some of the donor's chromosomal DNA may also be transferred with the plasmid DNA and integrate into the recipient genome via homologous recombination.[138] From a human perspective, because plasmids are the main carrier of antibiotics resistance genes, one of the dramatic consequences of a bacterium's ability to pass genetic information via conjugation is the rapid spread of these genes among prokaryotes.[139]

In some pathogenic bacteria, adapted conjugative systems transfer DNA into eukaryotic host cells. One of the best studied examples, *A. tumefaciens,* is capable of transferring a specific DNA segment (T-DNA) of the tumor-inducing (Ti) plasmid into the nucleus of infected plant cells where it is subsequently stably integrated into the host genome and transcribed, causing the crown gall disease. The biotechnological potential soon became clear as it turned out that any foreign DNA placed between the T-DNA borders can be transferred to plant cells, no matter its origin.[140] This finding has had, and continues to have, a profound impact on plant biology, agriculture, and biotechnology. Plant transformation mediated by bacterial conjugative systems is being used extensively for the transfer of various traits to an ever-increasing list of plant species including crops such as tobacco, potato, cotton, rice, wheat, tomato, rapeseed, and asparagus.[141] Another major contribution of *Agrobacterium* to plant research has been the use of T-DNA as a mutagen to determine gene functions. Libraries containing a large number of independently transformed seeds, with T-DNA insertions in almost every open reading frame, can be created, as was shown first for *Arabidopsis*. The resultant mutant plants can be screened by phenotype and using reverse genetics to identify specific genes of interest.[142]

It has now been observed that other bacterial conjugation systems can transport DNA into various types of eukaryotic cells, albeit at low frequencies.[143] With growing knowledge of its molecular mechanism, this "natural" potential could one day be used as an *in vivo* DNA delivery tool in any eukaryotic cell. A key characteristic of bacterial conjugation is that the DNA enters bound to a functional relaxase, which is used to catalyze site-specific integration of the transferred DNA into the eukaryotic genome.[138] It is tempting to imagine that, if this strand-transferase activity could be engineered to integrate selected genes into specific chromosomal locations, this would provide an attractive alternative to existing methods for eukaryotic genomic manipulation or even gene therapy.[143]

B. Protein Transport and Delivery

Expression technologies facilitating export of foreign proteins to the extracellular medium or into eukaryotic cells are gaining increasing interest. These methods rely on genetic fusions of the protein or peptide of interest to genes encoding efficiently secreted carrier proteins or to their secretion signals (Table I). Potential applications include high-level expression of proteins, injection of therapeutically important proteins into target cells, and the use of live bacteria for antigen delivery in vaccinations.[151]

Type III as well as some T4SSs (e.g., Cag in *H. pylori*) secrete proteins directly into the cytosol of their eukaryotic hosts. Upon bacterial contact with a eukaryotic cell, the T3SS creates a hypodermic-needle-like apparatus that spans the bacterial envelope and injects proteins into the host cell. The T3SSs of *Salmonella*, *Yersinia*, and *Shigella* spp. have successfully been exploited for the delivery of biologically active heterologous molecules such as cytokines[144] and foreign antigens, including the p60 protein of *Listeria monocytogenes*,[145] the NY-ESO-1 tumor antigen,[146] and epitopes from different viruses, such as the influenza virus,[147] the measles virus,[148] and the lymphocytic choriomeningitis virus (LCMV).[149] Effector proteins of the *Salmonella* T3SS, such as SopE, directly access the MHC class I-restricted antigen-processing pathways, making them excellent antigen carriers for vaccinations.[152] For the p60 protein, it was further shown that the time point and duration of hybrid protein translocation during the *Salmonella* infection cycle can be modulated by employing various type III carrier molecules.[145]

Biogenesis of bacterial flagella is also based on a T3SS and, in addition to their functions as a motility organelle, adhesin, sensor for wetness, or promoter of biofilm formation, flagella have been shown to function as an export apparatus for virulence-associated nonflagellar proteins.[153–155] Extracellular secretion of foreign polypeptides using the *E. coli* flagellar-secretion apparatus was achieved by making an N-terminal fusion to the flagellin FliC or by fusing both the transcriptional terminator and an untranslated DNA fragment upstream of the *fliC* gene into the gene of interest.[150] Secretion of (fusion) proteins up to 434 residues in length (including the green fluorescent protein, the Peb1 adhesin from *C. jejuni*, the alpha-enolase (eno) from *Streptococcus pneumonia*, and the fibronectin-binding D repeats of FnBPA from *Staphylococcus aureus*) was achieved. The foreign polypeptides were found to represent > 50% of the total secreted protein, with concentrations in the growth medium ranging from 1 to 15 mg/l.[150] The flagellar T3SS, therefore, makes a noteworthy candidate for high-level expression of recombinant proteins.

C. Charge Transport

1. Conductive Pili

Geobacter species are known to ferment acetate using extracellular, insoluble Fe(III) and Mn(IV) oxides as terminal electron acceptors.[156] Conducting-probe atomic force microscopy revealed that type IV pili of *Geobacter sulfurreducens* are highly conductive and serve as biological nanowires between the

TABLE I

Protein Transport and Delivery

Organism	Fusion partner	Exported polypeptide/epitope	References
Type III secretion			
Shigella flexneri	IpaH9.8	Cytokines	144
Salmonella enterica serovar *Typhimurium*	SspH2	p60 protein of *Listeria monocytogenes*	145
S. typhimurium	SopE	NY-ESO-1 tumor antigen	146
S. typhimurium	SptP	Influenza virus epitope	147
Yersinia enterocolitica	YopE	Measles virus epitope	148
S. typhimurium	SptP	Lymphocytic choriomeningitis virus (LCMV) epitope	149

	Exported polypeptide		References
Flagella			
Escherichia coli	Green fluorescent protein		150
E. coli	Fibronectin-binding D repeats of FnBPA from *Staphylococcus aureus*		150
E. coli	Peb1 the major cell-binding factor of *Campylobacter jejuni*		150
E. coli	Alpha-enolase of *Streptococcus pneumoniae* (Eno)		150
S. typhimurium	p60 Protein of *L. monocytogenes*		145

cell surface and the metal oxides.[157] Electrically conductive appendages are, however, not exclusive to metal-reducing bacteria such as *Geobacter* and *Shewanella* species. They have also been observed in the oxygenic phototrophic cyanobacterium *Synechocystis* and the thermophilic, fermentative bacterium *Pelotomaculum thermopropionicum*, revealing that such "nanowires" may represent a common bacterial strategy for efficient electron transfer and energy distribution outside the cell.[158] Although recent observations indicate that conductive type IV pili in *G. sulfurreducens* are decorated with an extracellular c-Type cytochrome, the mechanism and extent of the conductivity are not well understood.[159] Nonetheless, electron transfer through pili suggests possibilities for unique cell–surface and cell–cell (communication) interactions, and for bioengineering of novel conductive materials in nanoelectronic applications.[157]

Microbes that can transfer electrons to extracellular electron acceptors are essential in organic matter degradation and nutrient cycling in soils and sediments. This distinct property makes them important anode catalysts in new-generation microbial fuel cells (MFCs), which provide the sustainable production of energy from biomass-derived fuels by microbial consortia. A variety of readily degradable compounds such as glucose and acetate, and diverse types of wastewater such as domestic, starching, and paper recycling plant wastewater, have operated successfully as substrate to harvest electricity in MFC. Apart from electricity generation, MFCs offer efficient treatment of wastewater by simultaneously removing multiple pollutants.[160] MFCs have also been studied for applications as biosensors such as sensors for biological oxygen demand monitoring,[160] and the first demonstration of an MFC as a viable power supply was achieved by powering a meteorological buoy that measures air temperature, pressure, relative humidity, and water temperature.[161]

It should be mentioned, however, that the role of type IV pili as nanowires has been questioned, as, contrary to previous observations, the *Shewanella oneidensis* MR-1 type IV pili mutants generated more current in MFCs than the wild type. These results indicate that, under the conditions studied, pili are not essential for extracellular electron transfer to metals or anode surfaces, and the majority of extracellular electron transfer to insoluble electron acceptors occurs through direct contact with the surface-exposed OM cytochromes.[162]

D. Protein and Peptide Display

As discussed above, the assembly machineries of bacterial surface filaments can be used for the secretion of heterologous proteins, often in the form of fusion proteins with the filamentous subunits. When made in permissive positions, such fusions can also be built in into the surface organelles themselves. The first papers to describe surface display fall within the field of vaccine development using a filamentous phage-coat protein[163] and the *E. coli* OM protein, LamB.[164] Soon after, it was realized that this surface display provided a promising technique that offered many new and exciting applications in bio-technology. As a result, there was a corresponding boost in the published literature in the 1990s concerning diverse host cells, carrier proteins, and applications (Table II). Now, microbial cell surface display has a myriad of potential applications, such as the scaffolding of industrial catalysts, sorbents, sensors, vaccine-delivery vehicles, screening platforms, remediators of pollu-tants, and biofuels production machinery.[197]

1. EPITOPE DISPLAY IN VACCINOLOGY

Thus far, most of the work dealing with flagellar and pili-assisted peptide display has been focused on the development of recombinant vaccines. Pili and flagella are particularly attractive candidates for epitope display, as they are present in high numbers, are strong immunogens because of their repetitive nature, can often be easily purified, and possess adhesive properties.[198] Fla-gella and several different pili have been used to display immune-relevant sectors of various foreign proteins. Pioneering work considering flagellum display as a vaccination tool dates back to 1989, when antigenic determinants of both the cholera toxin and hepatitis B virus were shown to be immunogenic in mice immunized with attenuated *Salmonella* expressing these chimeric flagellins.[183,184] Since then, many other peptides or proteins have been fused, such as 15 aa of the M protein of *Streptoccocus pyogenes*[185] or an HIV1-gp41 epitope.[188] On the major flagellar subunit FliC, foreign polypeptides have to be inserted in the surface-exposed variable D3 region. In addition, variable sur-face-exposed regions of the minor cap protein FliD can be used for insertion of heterologous polypeptides, even in combination with fusions to the major flagellar subunit FlaC.[196] In pili, both the major structural subunit and the minor tip protein allow grafting of heterologous peptides. As in flagella, the availability of two structurally tolerant proteins makes simultaneous display of multiple foreign peptides possible. The potential of the major structural pro-tein of type 1 pili to incorporate heterologous peptides was assessed using a sequence mimicking a neutralizing epitope of the cholera toxin B chain as a reporter epitope.[165] The cholera toxin epitope was authentically displayed, and immunization of rabbits with purified chimeric fimbriae resulted in a serum that specifically recognized cholera toxin B chain. The same strategy was followed using the major subunit of Dr pili and a herpes simplex virus epitope.[172]

As in the case of these last two examples, chimeric pili and flagella can be delivered as purified filaments.[165,172] However, the potential of exploiting this technology for the development of recombinant live attenuated vaccines was also recognized in early experiments.[183,184] In this context, the cell surface display of heterologous antigenic determinants has been considered

TABLE II
PROTEIN AND PEPTIDE DISPLAY

Pilus	Subunit	Displayed polypeptide/epitope	References
Pili			
Chaperone/usher			
Type 1	FimA	Surface antigen epitopes of Hepatitis B, Poliovirus, Foot-and-mouth disease (FMD) virus; cholera toxin B epitope	165
	FimH	PreS2 segment of Hepatitis B, cholera toxin B epitopes	166
		Heavy-metal-binding sequences	167
		Random peptide libraries	167,168
F11	FelA	HIV, FMD virus, *Mycobacterium leprae* 65 kDa protein, *Plasmodium falsiparum* surface protein epitopes	169
		Gonadotropin releasing hormone	170
F4 (K88)	FaeG	HIV, Human influenza virus, Hepatitis B, FMD virus, *N. gonorrhoeae* pilin, Human somatostatin epitopes	171
Dr	DraE	*Herpes simplex* virus epitope	172
987P	FasA	*H. simplex* virus and transmissible gastroenteritis virus (TGEV) epitopes	173,174
SEF14	SefA	*Leishmania major* gp63 epitope	175
CS31A	ClpG	TGEV epitopes	176–178
Curli			
Tafi	AgfA	*L. major* gp63 epitope	175
Type IV pili			
Dichelobacter nodosus		Mimick of antigenic determinant of the VP1 coat protein of the FMD Virus	179
Type IV secretion pili			
F	TraA	Filamentous bacteriophage adhesion epitopes	180,181
Sortase-assembled pili			
Streptococcus pyogenes T3	Cpa	*E. coli* maltose-binding protein	182

Subunit	Displayed polypeptide/epitope	References
Flagella		
FliC	Cholera toxin epitope	183
	Hepatitis B virus epitope	184
	M protein epitope of *Streptococcus pyrogenes*	185,186
	Influenza virus epitopes	186,187
	HIV1-gp41 epitope	188
	HIV1-gp120 epitope	189
	T-cell epitopes of listeriolysin of *Listeria monocytogenes*	190
	Enhanced green fluorescent protein (EGFP)	191
	P10 epitpope of cell wall protein (gp43) of fungus *Paracoccidioides brasiliensis*	192
	Eptipe from *Plasmodium yoelii* circumsporozoite (CS) protein	193
	YadA adhesin	194
	Peptide library for nickel biosorption	195

(Continues)

TABLE II (*Continued*)

Subunit	Displayed polypeptide/epitope	References
FliD	Fibronectin-binding D repeats of FnBPA from *Staphylococcus aureus*	196
	Collagen-binding YadA adhesin from *Yersinia enterocolitica*	196
	SlpA S-layer protein of *Lactobacillus brevis*	196

advantageous over cytoplasmic expression[199]: for example, expressing a heterologous *E. coli* 987P fimbrial antigen on the surface of a live *Salmonella* vaccine vector was more beneficial than intracellular expression as demonstrated by a stronger immune response to the antigen.[200] Furthermore, expressing epitopes from the transmissible gastroenteritis virus (TGEV) fused to 987P fimbriae was shown to elicit a higher anti-TGEV antibody response than surface expression of the same epitopes by an autotransporter, indicative of the advantage provided by a polymeric display system.[173] The flagellum has long been used in immunization experiments as an immunoadjuvant, as it activates both the adaptive and the innate arms of the immune system.[201–204]

Although most of the research concerning surface display on filamentous appendages was performed in Gram-negative bacteria, there is an increased interest in heterologous protein expression on pili of Gram-positive bacteria. As commensal or food-grade bacteria, such as *Streptococcus gordonii* and several staphylococcal and lactic acid bacteria, do not invade the host, they are generally inefficient in generating strong antibody responses. However, vaccine delivery can be significantly improved by the codisplay of adhesins that will assist in the targeting to the mucosal epithelium.[205] As an example, a recent report showed that it was possible to fuse the *E. coli* maltose-binding protein to the tip protein of the *S. pyogenes* T3 pilus. Immunization of mice with *Lactococcus lactis* expressing this fusion protein resulted in production of both a systemic and a mucosal response against the foreign antigen.[182]

2. RANDOM PEPTIDE LIBRARIES

Work has been performed demonstrating the amenability of pili toward the powerful technology of random polypeptide display[168,198] (Table II). By introducing random peptide libraries in the FimH adhesin of type 1 pili, it was, for example, possible to create and select designer adhesins with specific sequences conferring binding of the recombinant cells to various metal oxides (PbO_2, CoO, MnO_2, Cr_2O_3).[167] Also, flagellar proteins have been used for the rapid high-throughput screening of libraries, as, for example, a peptide library for nickel biosorption.[195] One application involves fusion into a disulfide loop of *E. coli* thioredoxin that has been inserted into flagellin, facilitating the expression of random peptides in a conformationally constrained manner readily accessible on the flagellar surface.[206] Random peptide libraries have been applied in antibody epitope mapping and are suitable for biopanning procedures in the study of ligand–receptor interactions, potentially with altered substrate specificity.[207] Bacterial display can constitute an alternative to the common phage display technology with some potentially beneficial features: there is only one host for propagation of the library; there is no need for reinfection in order to amplify the selected variants; and direct screening using fluorescence-activated cell sorting (FACS) is possible.[205]

3. LIMITATIONS OF FLAGELLAR/PILI DISPLAY

While shown successful for several proteins, not all proteins can be efficiently exposed on the bacterial surface using multisubunit fibers. One of the problems usually encountered with flagellar or fimbrial display systems is the limited size of heterologous grafts that can be displayed without causing detrimental effects on the structure and/or function of the carrier protein. Most reports claim that in flagella the permissible size is less than 60 aa,[199] but there have been reports where a 302-aa YadA adhesin was expressed in the H7 flagellin.[194] For pili the upper size limit seems to be lower, being 34 aa and 52 aa for, respectively, the major and minor tip subunits.[205] Depending on the system of choice, one should consider whether fusions should be either carboxy or amino terminal, or in permissive loops, to remain polymerization competent. In the flagellar system, fusions to the major subunit must be done in the variable-surface-exposed D3 part of the flagellin.[196] In CU pili, the canonical C-terminus is required for interaction with the assembly machinery. Short fusions to the N-terminus of the mature protein or in loop regions of the Ig-like fold can be tolerated. In case of the type 1 pilus tip adhesin FimH, insertions of Ig-like domains N-terminal of the lectin domain were surface-exposed in pili.[208] However, in the case of insertion into the minor subunits, biotechnological applications are often hampered by the low copy number of the displayed fusion protein. These problems are overcome by using the major subunit of pili or flagella as fusion partner, which makes it possible to display thousands of copies.[207] Also, expression of many surface proteins is restricted to confined host backgrounds, which can limit the versatility of the corresponding display systems. Clearly, an ideal display system should combine the ability to accommodate large inserts with a high copy number and a broad host range.

Additional and possibly more serious limitations derive from the characteristics of the displayed proteins themselves. Any peptide or protein whose properties prevent it from crossing the assembly machinery will stop the correct transfer of the hybrid protein, resulting in incorrectly assembled fibers. For example, type III secretion substrates traverse the needle in an unfolded manner, and rapidly and tightly folding fusions were found unable to be secreted.[71] In systems with periplasmic intermediates, subunits cross the IM through the Sec translocon. Thus, in either case, fusion proteins undergo an unfolding step during translocation, preventing the functional display of cargo proteins that require cytoplasmically derived cofactors. Other restraints derive from the chemical characteristics of the periplasmic and the exterior environment which may affect the folding and stability of the protein to be displayed.[197] As the flagellar export system bypasses the periplasmic space, catalyzed formation and isomerization of disulfide bonds will be lacking in flagellar fusion proteins.[207] Surface expression is undoubtedly an inexpensive way to produce immobilized enzymes, but bacteria might suffer from practical drawbacks such as inhibition of growth or cell lysis. The fact that in some systems polymerization can occur without the need of accessory proteins or pathways allows one to work outside the biological system (e.g., in the case of curli) and might overcome these last drawbacks. Current challenges in surface display research include expressing large multimeric proteins greater than 60 kDa, displaying multiple proteins, and alleviating the spread of genetically modified organisms (GMOs) in live-vaccine and field applications.[197]

Finally, the capacity of some Archaea to assemble surface structures in extreme conditions that bacterial structures often fail to withstand is of great interest to numerous biotechnical applications. However, the research on archaeal pili and flagella is still in its infancy, and the number of experiments performed cannot be matched to the experience accumulated in the last years by several hundred laboratories working on bacterial systems.

E. Bacterial Filaments as Therapeutic Targets

Adhesion of pathogenic organisms to host tissues is the prerequisite for the initiation of most, if not all, infectious diseases. Targeting the bacteria's ability to adhere to host cells is therefore a very attractive way to prevent bacterial infections[209,210] (Table III). In many cases, the adhesive subunits are incorporated in surface filaments. For this purpose, pili have for long been regarded as important vaccine candidates, and different purified pili have been tested as immunogens as early as the 1970s when Charles Brinton and colleagues tested purified GC pili as vaccine candidates in humans. Unfortunately, this vaccine failed in large-scale efficacy trials, probably owing to high antigenic variation of gonococcus pili and poor immunogenicity of the actual adhesin.[236] A more positive outcome was observed in case of K88 (F4) and K99 (F5) pili of pig enterotoxigenic *E. coli*, where vaccination of the sows with isolated pili resulted in passive immunity in suckling piglets.[215] However, it should be noted that in these pili the major subunit provides the adhesive phenotype, as opposed to most pili, where only a minor tip protein fulfills this function.[215] Most efforts to examine the use of entire pili in vaccines have been thwarted by findings that the major immunodominant component of pilus fibers is often

TABLE III

BACTERIAL FILAMENTS AS THERAPEUTIC TARGET

Organelle	Vaccination with	Organism	References
Vaccination			
CU pili			
987P	Whole pili	Enterotoxigenic *Escherichia coli*	211–213
F4 (K88)	Whole pili	Enterotoxigenic *E. coli*	213,214
F5 (K99)	Whole pili	Enterotoxigenic *E. coli*	211,213,215
Type 1	Adhesin FimH in a complex with its chaperone FimC	Uropathogenic *E. coli*	216
Saf	Adhesin SafD in a complex with its chaperone SafB	*Salmonella enteritidis*	217
Type IV pili			
	Whole pili	*Dichelobacter nodosus*	218,219
	Whole pili	*Moraxella bovis*	220,221
	Peptide from the receptor-binding domain	*Pseudomonas aeruginosa*	222
Flagella			
	Flagellin	*S. enteritidis*	223,224
	Flagellin	*P. aeruginosa*	225
	Hook-associated proteins	*S. enteritidis*	223
	Crude flagella preparation	*Vibrio cholerae*	226
	Crude flagella preparation	*P. aeruginosa*	225,227–230
Flagella as immunoadjuvant			201–204

Organelle	Compound and target	Organism	References
Chemical targeting			
Type 1	Alkyl and aryl mannosides as FimH receptor analogs	Uropathogenic *E. coli*	231,232
P pili	Galabiose derivatives as PapG receptor analogs	Uropathogenic *E. coli*	233
Type 1, P pili	2-Pyridone pilicides as inhibitors of chaperone:subunit recruitment at the usher	Uropathogenic *E. coli*	234,235

antigenically variable and offers limited protection, as seen in vaccination studies using type IV pili of *N. meningitidis*.[209] The adhesins and other minor pilus proteins are typically well conserved, but, as a result of their low abundance, intact whole pili do not elicit a strong antibody response to them.[210,237] The tremendous advancements in the structural molecular characterization of pilus components now make it possible to produce stable adhesin formulations in large quantities. As an example, a subunit vaccine composed of the type 1 pili adhesin FimH in a complex with its chaperone FimC has been shown to prevent urinary tract infections by pathogenic *E. coli* in mice and monkeys.[216] In another approach, monoclonal antibodies directed against the CFA/I and F4 (K88) fimbrial adhesins from human and pig ETEC strains have also been found to prevent host colonization and have been suggested as food additives for passive immunization.[214,238]

Besides vaccination, bacterial pili have also formed the subject for chemical targeting. Receptor analogs directed against fimbrial adhesins have been evaluated as anti-adhesives for the prevention of bacterial attachment and biofilm formation.[231–233] In the best characterized example, alkyl and aryl mannosides form high-affinity competitive inhibitors for the FimH tip adhesin and have been shown to prevent type 1 pili-mediated attachment of uropathogenic *E. coli* (UPEC) to bladder epithelial cells.[231,232] The P pilus adhesin PapG, responsible for UPEC binding to globoside-series glycosphyngolipids in kidney epithelium, has also been successfully targeted using galabiose derivatives.[233] Because these chemical anti-adhesives are based on receptor analogs, their use is therefore often restricted to specific fimbriae, restricting their broad-range use. Indeed, pathogens contain a plethora of fimbrial adhesins, even amongst different *E. coli* strains: for example, over 17 different fimbrial clusters have been found.[9] The biosynthetic pathways on the other hand are more conserved, such that targeting pilus assembly could provide a route toward more broad-acting compounds. Following such approach, Svensson *et al.* found that a class of ring-fused 2-pyridones were able to disrupt CU pilus biogenesis.[234] These compounds were found to interfere with chaperone–subunit recruitment to the usher, the pilus assembly machinery in the OM.[235] Targeting bacterial virulence systems has gained interest as a nonantibiotic alternative to the static or lytic antimicrobials in common use.[239,240] *In vitro* screens of chemical libraries have now also identified a number of acylated hydrazones of salicylaldehydes as specific inihibitors of Type III secretion in *Yersinia pseudotuberculosis*, *S. enterica*, and *Chlamydia trachomatis*.[241–243]

Finally, surface filaments and their assembly machineries are often targeted by bacteriophages for host-specific attachment and adsorption.[244] Conjugative F-pili, for example, serve as adsorption organelles for several classes of bacteriophage. Notably, the F1/M13 class of filamentous bacteriophages, which adhere to the tip of the F-pili, are worth mentioning.[245] Flagellum-dependent phages were recently shown to mediate generalized transduction with high efficiency between *Serratia* and *Pantoea*, both primarily environmental organisms that can, however, cause disease in plants, insects, or animals.[246] Defining phage host ranges and understanding bacterial mechanisms of phage resistance are essential for applications in medicine, such as phage therapy.[245] Further, as phage infection results in substantial economic losses in food fermentation industries, researchers are trying to find strategies to implement resistance against bacteriophages that infect food fermentation starters.[247–249]

F. Tools for Targeting

The primary function of many CU pili is binding to specific receptors. For this aim, pili have specialized adhesins that recognize only a small number of carbohydrate or proteinaceous moieties. Consequently, these adhesins can be employed to direct substances to cells or surfaces expressing the correct receptors. Potential applications similar to those of antibody targeting can be envisaged, ranging across basic research, disease diagnostics, and the development of therapeutics. In medicine, antibodies have been explored as delivery agents of various conjugated molecules, including chemotherapeutic drugs and protein-based toxins for cancer treatment; radioisotopes for diagnostic and therapeutic use; enzymes that convert nontoxic systemic prodrugs to active cytotoxic forms; cytokines such as tumor necrosis factor-α, interferon γ, or interleukin-2; and colloids, such as nanoparticles and liposomes, which are used for drug delivery.[250]

G. Motility, Artificial Bacterial Flagella

Finally, a remarkable, accomplishment that draws attention is the development of artificial bacterial flagella (ABFs). Inspired by the natural design of bacterial flagella, researchers recently reported the first demonstration of artificial swimmers that use helical nanobelt propulsion.[251] ABFs consist of helical tails fabricated by the self-scrolling of helical nanobelts and soft-magnetic heads composed of Cr/Ni/Au stacked thin films. They have comparable geometries and dimensions to their organic counterparts and can swim at speeds of 1–2 μm/s in a controllable fashion in three dimensions using weak applied magnetic fields.[252] Preliminary results have further shown that ABFs are capable of performing micro-object manipulation either directly by mechanical contact or indirectly by generating a localized fluid flow.[253] Although nanobots (tiny robots that can be injected into the body to perform medical procedures, such as targeted drug delivery and implantation or removal of tissues and other objects) still are subjects of science fiction, swimming microrobots propelled by artificial flagella bring that fantasy closer to reality.[254]

REFERENCES

1. Duguid JP, Anderson ES. Terminology of bacterial fimbriae, or pili, and their types. *Nature* 1967;**215**:89–90.
2. Fronzes R, Remaut H, Waksman G. Architectures and biogenesis of non-flagellar protein appendages in Gram-negative bacteria. *EMBO J* 2008;**27**:2271–80.
3. Klemm P, Schembri MA. Bacterial adhesins: function and structure. *Int J Med Microbiol* 2000;**290**:27–35.
4. Waksman G, Hultgren SJ. Structural biology of the chaperone-usher pathway of pilus biogenesis. *Nat Rev Microbiol* 2009;**7**:765–74.
5. Sauer FG, Remaut H, Hultgren SJ, Waksman G. Fiber assembly by the chaperone-usher pathway. *Biochim Biophys Acta* 2004;**1694**:259–67.
6. Zav'yalov V, Zavialov A, Zav'yalova G, Korpela T. Adhesive organelles of Gram-negative pathogens assembled with the classical chaperone/usher machinery: structure and function from a clinical standpoint. *FEMS Microbiol Rev* 2010;**34**:317–78.
7. Zavialov A, Zav'yalova G, Korpela T, Zav'yalov V. FGL chaperone-assembled fimbrial poly-adhesins: anti-immune armament of Gram-negative bacterial pathogens. *FEMS Microbiol Rev* 2007;**31**:478–514.
8. Mulvey MA, Lopez-Boado YS, Wilson CL, Roth R, Parks WC, Heuser J, et al. Induction and evasion of host defenses by type 1-piliated uropathogenic *Escherichia coli*. *Science* 1998;**282**:1494–7.
9. Nuccio SP, Baumler AJ. Evolution of the chaperone/usher assembly pathway: fimbrial classification goes Greek. *Microbiol Mol Biol Rev* 2007;**71**:551–75.
10. Choudhury D, Thompson A, Stojanoff V, Langermann S, Pinkner J, Hultgren SJ, et al. X-ray structure of the FimC-FimH chaperone-adhesin complex from uropathogenic *Escherichia coli*. *Science* 1999;**285**:1061–6.
11. Sauer FG, Futterer K, Pinkner JS, Dodson KW, Hultgren SJ, Waksman G. Structural basis of chaperone function and pilus biogenesis. *Science* 1999;**285**:1058–61.
12. Thanassi DG, Saulino ET, Hultgren SJ. The chaperone/usher pathway: a major terminal branch of the general secretory pathway. *Curr Opin Microbiol* 1998;**1**:223–31.
13. Rose RJ, Verger D, Daviter T, Remaut H, Paci E, Waksman G, et al. Unraveling the molecular basis of subunit specificity in P pilus assembly by mass spectrometry. *Proc Natl Acad Sci USA* 2008;**105**:12873–8.
14. Remaut H, Tang C, Henderson NS, Pinkner JS, Wang T, Hultgren SJ, et al. Fiber formation across the bacterial outer membrane by the chaperone/usher pathway. *Cell* 2008;**133**:640–52.
15. Sauer FG, Knight SD, Waksman GJ, Hultgren SJ. PapD-like chaperones and pilus biogenesis. *Semin Cell Dev Biol* 2000;**11**:27–34.
15a. Phan G, Remaut H, Wang T, Allen WJ, Pirker KF, Lebedev A, et al. Crystal structure of the FimD usher bound to its cognate FimC-FimH substrate. *Nature* 2011;**474**:49–53.
16. Remaut H, Rose RJ, Hannan TJ, Hultgren SJ, Radford SE, Ashcroft AE, et al. Donor-strand exchange in chaperone-assisted pilus assembly proceeds through a concerted beta strand displacement mechanism. *Mol Cell* 2006;**22**:831–42.
17. Sauer FG, Pinkner JS, Waksman G, Hultgren SJ. Chaperone priming of pilus subunits facilitates a topological transition that drives fiber formation. *Cell* 2002;**111**:543–51.
18. Zavialov AV, Tischenko VM, Fooks LJ, Brandsdal BO, Aqvist J, Zav'yalov VP, et al. Resolving the energy paradox of chaperone/usher-mediated fibre assembly. *Biochem J* 2005;**389**:685–94.
19. Nishiyama M, Ishikawa T, Rechsteiner H, Glockshuber R. Reconstitution of pilus assembly reveals a bacterial outer membrane catalyst. *Science* 2008;**320**:376–9.

20. Puorger C, Eidam O, Capitani G, Erilov D, Grutter MG, Glockshuber R. Infinite kinetic stability against dissociation of supramolecular protein complexes through donor strand complementation. *Structure* 2008;**16**:631–42.

21. Dueholm MS, Petersen SV, Sonderkaer M, Larsen P, Christiansen G, Hein KL, et al. Functional amyloid in Pseudomonas. *Mol Microbiol* 2010;**77**:1009–20.

22. Wang X, Chapman MR. Curli provide the template for understanding controlled amyloid propagation. *Prion* 2008;**2**:57–60.

23. Olsen A, Jonsson A, Normark S. Fibronectin binding mediated by a novel class of surface organelles on Escherichia coli. *Nature* 1989;**338**:652–5.

24. Epstein EA, Chapman MR. Polymerizing the fibre between bacteria and host cells: the biogenesis of functional amyloid fibres. *Cell Microbiol* 2008;**10**:1413–20.

25. Herwald H, Morgelin M, Olsen A, Rhen M, Dahlback B, Muller-Esterl W, et al. Activation of the contact-phase system on bacterial surfaces—a clue to serious complications in infectious diseases. *Nat Med* 1998;**4**:298–302.

26. Larsen P, Nielsen JL, Dueholm MS, Wetzel R, Otzen D, Nielsen PH. Amyloid adhesins are abundant in natural biofilms. *Environ Microbiol* 2007;**9**:3077–90.

27. Nelson R, Sawaya MR, Balbirnie M, Madsen AO, Riekel C, Grothe R, et al. Structure of the cross-beta spine of amyloid-like fibrils. *Nature* 2005;**435**:773–8.

28. Robinson LS, Ashman EM, Hultgren SJ, Chapman MR. Secretion of curli fibre subunits is mediated by the outer membrane-localized CsgG protein. *Mol Microbiol* 2006;**59**:870–81.

29. Collinson SK, Parker JM, Hodges RS, Kay WW. Structural predictions of AgfA, the insoluble fimbrial subunit of Salmonella thin aggregative fimbriae. *J Mol Biol* 1999;**290**:741–56.

30. Wang X, Smith DR, Jones JW, Chapman MR. In vitro polymerization of a functional *Escherichia coli* amyloid protein. *J Biol Chem* 2007;**282**:3713–9.

31. Hammer ND, Schmidt JC, Chapman MR. The curli nucleator protein, CsgB, contains an amyloidogenic domain that directs CsgA polymerization. *Proc Natl Acad Sci USA* 2007;**104**: 12494–9.

32. White AP, Collinson SK, Banser PA, Gibson DL, Paetzel M, Strynadka NC, et al. Structure and characterization of AgfB from *Salmonella enteritidis* thin aggregative fimbriae. *J Mol Biol* 2001;**311**:735–49.

33. Chapman MR, Robinson LS, Pinkner JS, Roth R, Heuser J, Hammar M, et al. Role of *Escherichia coli* curli operons in directing amyloid fiber formation. *Science* 2002;**295**:851–5.

34. Sakellaris H, Hannink NK, Rajakumar K, Bulach D, Hunt M, Sasakawa C, et al. Curli loci of Shigella spp. *Infect Immun* 2000;**68**:3780–3.

35. Gibson DL, White AP, Rajotte CM, Kay WW. AgfC and AgfE facilitate extracellular thin aggregative fimbriae synthesis in *Salmonella enteritidis*. *Microbiology* 2007;**153**:1131–40.

36. Hammar M, Bian Z, Normark S. Nucleator-dependent intercellular assembly of adhesive curli organelles in *Escherichia coli*. *Proc Natl Acad Sci USA* 1996;**93**:6562–6.

37. Ramboarina S, Fernandes PJ, Daniell S, Islam S, Simpson P, Frankel G, et al. Structure of the bundle-forming pilus from enteropathogenic *Escherichia coli*. *J Biol Chem* 2005;**280**:40252–60.

38. Gerlach RG, Hensel M. Protein secretion systems and adhesins: the molecular armory of Gram-negative pathogens. *Int J Med Microbiol* 2007;**297**:401–15.

39. Coureuil M, Mikaty G, Miller F, Lecuyer H, Bernard C, Bourdoulous S, et al. Meningococcal type IV pili recruit the polarity complex to cross the brain endothelium. *Science* 2009;**325**:83–7.

40. Mattick JS. Type IV pili and twitching motility. *Annu Rev Microbiol* 2002;**56**:289–314.

41. Merz AJ, So M, Sheetz MP. Pilus retraction powers bacterial twitching motility. *Nature* 2000;**407**:98–102.

42. Maier B, Koomey M, Sheetz MP. A force-dependent switch reverses type IV pilus retraction. *Proc Natl Acad Sci USA* 2004;**101**:10961–6.
43. Craig L, Li J. Type IV pili: paradoxes in form and function. *Curr Opin Struct Biol* 2008;**18**:267–77.
44. Craig L, Taylor RK, Pique ME, Adair BD, Arvai AS, Singh M, et al. Type IV pilin structure and assembly: X-ray and EM analyses of Vibrio cholerae toxin-coregulated pilus and *Pseudomonas aeruginosa* PAK pilin. *Mol Cell* 2003;**11**:1139–50.
45. Pelicic V. Type IV pili: e pluribus unum? *Mol Microbiol* 2008;**68**:827–37.
46. Parge HE, Forest KT, Hickey MJ, Christensen DA, Getzoff ED, Tainer JA. Structure of the fibre-forming protein pilin at 2.6 A resolution. *Nature* 1995;**378**:32–8.
47. Turner LR, Lara JC, Nunn DN, Lory S. Mutations in the consensus ATP-binding sites of XcpR and PilB eliminate extracellular protein secretion and pilus biogenesis in *Pseudomonas aeruginosa*. *J Bacteriol* 1993;**175**:4962–9.
48. Crowther LJ, Anantha RP, Donnenberg MS. The inner membrane subassembly of the enteropathogenic *Escherichia coli* bundle-forming pilus machine. *Mol Microbiol* 2004;**52**:67–79.
49. Balasingham SV, Collins RF, Assalkhou R, Homberset H, Frye SA, Derrick JP, et al. Interactions between the lipoprotein PilP and the secretin PilQ in *Neisseria meningitidis*. *J Bacteriol* 2007;**189**:5716–27.
50. Drake SL, Sandstedt SA, Koomey M. PilP, a pilus biogenesis lipoprotein in *Neisseria gonorrhoeae*, affects expression of PilQ as a high-molecular-mass multimer. *Mol Microbiol* 1997;**23**:657–68.
51. Hardie KR, Lory S, Pugsley AP. Insertion of an outer membrane protein in *Escherichia coli* requires a chaperone-like protein. *EMBO J* 1996;**15**:978–88.
52. Giltner CL, Habash M, Burrows LL. *Pseudomonas aeruginosa* minor pilins are incorporated into type IV pili. *J Mol Biol* 2010;**398**:444–61.
53. Roine E, Wei W, Yuan J, Nurmiaho-Lassila EL, Kalkkinen N, Romantschuk M, et al. Hrp pilus: an hrp-dependent bacterial surface appendage produced by *Pseudomonas syringae* pv. tomato DC3000. *Proc Natl Acad Sci USA* 1997;**94**:3459–64.
54. Kubori T, Matsushima Y, Nakamura D, Uralil J, Lara-Tejero M, Sukhan A, et al. Supramolecular structure of the *Salmonella typhimurium* type III protein secretion system. *Science* 1998;**280**:602–5.
55. Cornelis GR. The type III secretion injectisome. *Nat Rev Microbiol* 2006;**4**:811–25.
56. Blocker A, Komoriya K, Aizawa S. Type III secretion systems and bacterial flagella: insights into their function from structural similarities. *Proc Natl Acad Sci USA* 2003;**100**:3027–30.
57. Tamano K, Aizawa S, Katayama E, Nonaka T, Imajoh-Ohmi S, Kuwae A, et al. Supramolecular structure of the Shigella type III secretion machinery: the needle part is changeable in length and essential for delivery of effectors. *EMBO J* 2000;**19**:3876–87.
58. Blocker A, Jouihri N, Larquet E, Gounon P, Ebel F, Parsot C, et al. Structure and composition of the Shigella flexneri "needle complex", a part of its type III secreton. *Mol Microbiol* 2001;**39**:652–63.
59. Cordes FS, Komoriya K, Larquet E, Yang S, Egelman EH, Blocker A, et al. Helical structure of the needle of the type III secretion system of *Shigella flexneri*. *J Biol Chem* 2003;**278**:17103–7.
60. Moraes TF, Spreter T, Strynadka NC. Piecing together the type III injectisome of bacterial pathogens. *Curr Opin Struct Biol* 2008;**18**:258–66.
61. Blocker AJ, Deane JE, Veenendaal AK, Roversi P, Hodgkinson JL, Johnson S, et al. What's the point of the type III secretion system needle? *Proc Natl Acad Sci USA* 2008;**105**:6507–13.

62. Knutton S, Rosenshine I, Pallen MJ, Nisan I, Neves BC, Bain C, et al. A novel EspA-associated surface organelle of enteropathogenic *Escherichia coli* involved in protein translocation into epithelial cells. *EMBO J* 1998;**17**:2166–76.

63. Sekiya K, Ohishi M, Ogino T, Tamano K, Sasakawa C, Abe A. Supermolecular structure of the enteropathogenic *Escherichia coli* type III secretion system and its direct interaction with the EspA-sheath-like structure. *Proc Natl Acad Sci USA* 2001;**98**:11638–43.

64. Li CM, Brown I, Mansfield J, Stevens C, Boureau T, Romantschuk M, et al. The Hrp pilus of *Pseudomonas syringae* elongates from its tip and acts as a conduit for translocation of the effector protein HrpZ. *EMBO J* 2002;**21**:1909–15.

65. Deane JE, Roversi P, Cordes FS, Johnson S, Kenjale R, Daniell S, et al. Molecular model of a type III secretion system needle: implications for host-cell sensing. *Proc Natl Acad Sci USA* 2006;**103**:12529–33.

66. Erhardt M, Namba K, Hughes KT. Bacterial nanomachines: the flagellum and type III injectisome. *Cold Spring Harb Perspect Biol* 2010;**2**:a000299.

67. Mueller CA, Broz P, Muller SA, Ringler P, Erne-Brand F, Sorg I, et al. The V-antigen of Yersinia forms a distinct structure at the tip of injectisome needles. *Science* 2005;**310**:674–6.

68. Deane JE, Abrusci P, Johnson S, Lea SM. Timing is everything: the regulation of type III secretion. *Cell Mol Life Sci* 2010;**67**:1065–75.

69. Derewenda U, Mateja A, Devedjiev Y, Routzahn KM, Evdokimov AG, Derewenda ZS, et al. The structure of *Yersinia pestis* V-antigen, an essential virulence factor and mediator of immunity against plague. *Structure* 2004;**12**:301–6.

70. Yip CK, Strynadka NC. New structural insights into the bacterial type III secretion system. *Trends Biochem Sci* 2006;**31**:223–30.

71. Akeda Y, Galan JE. Chaperone release and unfolding of substrates in type III secretion. *Nature* 2005;**437**:911–5.

72. Muller SA, Pozidis C, Stone R, Meesters C, Chami M, Engel A, et al. Double hexameric ring assembly of the type III protein translocase ATPase HrcN. *Mol Microbiol* 2006;**61**:119–25.

73. Sukhan A, Kubori T, Wilson J, Galan JE. Genetic analysis of assembly of the *Salmonella enterica serovar Typhimurium* type III secretion-associated needle complex. *J Bacteriol* 2001;**183**:1159–67.

74. He SY, Nomura K, Whittam TS. Type III protein secretion mechanism in mammalian and plant pathogens. *Biochim Biophys Acta* 2004;**1694**:181–206.

75. Galan JE, Wolf-Watz H. Protein delivery into eukaryotic cells by type III secretion machines. *Nature* 2006;**444**:567–73.

76. Christie PJ, Vogel JP. Bacterial type IV secretion: conjugation systems adapted to deliver effector molecules to host cells. *Trends Microbiol* 2000;**8**:354–60.

77. Cascales E, Christie PJ. The versatile bacterial type IV secretion systems. *Nat Rev Microbiol* 2003;**1**:137–49.

78. Llosa M, Roy C, Dehio C. Bacterial type IV secretion systems in human disease. *Mol Microbiol* 2009;**73**:141–51.

79. Fronzes R, Christie PJ, Waksman G. The structural biology of type IV secretion systems. *Nat Rev Microbiol* 2009;**7**:703–14.

80. Segal G, Feldman M, Zusman T. The Icm/Dot type-IV secretion systems of *Legionella pneumophila* and *Coxiella burnetii*. *FEMS Microbiol Rev* 2005;**29**:65–81.

81. Marvin DA, Folkhard W. Structure of F-pili: reassessment of the symmetry. *J Mol Biol* 1986;**191**:299–300.

82. Wang YA, Yu X, Silverman PM, Harris RL, Egelman EH. The structure of F-pili. *J Mol Biol* 2009;**385**:22–9.

83. Yeo HJ, Yuan Q, Beck MR, Baron C, Waksman G. Structural and functional characterization of the VirB5 protein from the type IV secretion system encoded by the conjugative plasmid pKM101. *Proc Natl Acad Sci USA* 2003;**100**:15947–52.

84. Christie PJ, Atmakuri K, Krishnamoorthy V, Jakubowski S, Cascales E. Biogenesis, architecture, and function of bacterial type IV secretion systems. *Annu Rev Microbiol* 2005;**59**:451–85.

85. Cascales E, Christie PJ. Definition of a bacterial type IV secretion pathway for a DNA substrate. *Science* 2004;**304**:1170–3.

86. Fronzes R, Schafer E, Wang L, Saibil HR, Orlova EV, Waksman G. Structure of a type IV secretion system core complex. *Science* 2009;**323**:266–8.

87. Rego AT, Chandran V, Waksman G. Two-step and one-step secretion mechanisms in Gram-negative bacteria: contrasting the type IV secretion system and the chaperone-usher pathway of pilus biogenesis. *Biochem J* 2010;**425**:475–88.

88. Navarre WW, Schneewind O. Surface proteins of gram-positive bacteria and mechanisms of their targeting to the cell wall envelope. *Microbiol Mol Biol Rev* 1999;**63**:174–229.

89. Yanagawa R, Otsuki K, Tokui T. Electron microscopy of fine structure of *Corynebacterium renale* with special reference to pili. *Jpn J Vet Res* 1968;**16**:31–7.

90. Telford JL, Barocchi MA, Margarit I, Rappuoli R, Grandi G. Pili in gram-positive pathogens. *Nat Rev Microbiol* 2006;**4**:509–19.

91. Barocchi MA, Ries J, Zogaj X, Hemsley C, Albiger B, Kanth A, et al. A pneumococcal pilus influences virulence and host inflammatory responses. *Proc Natl Acad Sci USA* 2006;**103**:2857–62.

92. Gibbons RJ, Hay DI. Human salivary acidic proline-rich proteins and statherin promote the attachment of *Actinomyces viscosus* LY7 to apatitic surfaces. *Infect Immun* 1988;**56**:439–45.

93. Mandlik A, Swierczynski A, Das A, Ton-That H. Pili in Gram-positive bacteria: assembly, involvement in colonization and biofilm development. *Trends Microbiol* 2008;**16**:33–40.

94. Proft T, Baker EN. Pili in Gram-negative and Gram-positive bacteria—structure, assembly and their role in disease. *Cell Mol Life Sci* 2009;**66**:613–35.

95. Ton-That H, Schneewind O. Assembly of pili on the surface of *Corynebacterium diphtheriae*. *Mol Microbiol* 2003;**50**:1429–38.

96. Scott JR, Zahner D. Pili with strong attachments: gram-positive bacteria do it differently. *Mol Microbiol* 2006;**62**:320–30.

97. Ton-That H, Marraffini LA, Schneewind O. Sortases and pilin elements involved in pilus assembly of *Corynebacterium diphtheriae*. *Mol Microbiol* 2004;**53**:251–61.

98. Navarre WW, Schneewind O. Proteolytic cleavage and cell wall anchoring at the LPXTG motif of surface proteins in gram-positive bacteria. *Mol Microbiol* 1994;**14**:115–21.

99. Rakotoarivonina H, Jubelin G, Hebraud M, Gaillard-Martinie B, Forano E, Mosoni P. Adhesion to cellulose of the Gram-positive bacterium *Ruminococcus albus* involves type IV pili. *Microbiology* 2002;**148**:1871–80.

100. Morrison M, Miron J. Adhesion to cellulose by *Ruminococcus albus*: a combination of cellulosomes and Pil-proteins? *FEMS Microbiol Lett* 2000;**185**:109–15.

101. Varga JJ, Therit B, Melville SB. Type IV pili and the CcpA protein are needed for maximal biofilm formation by the gram-positive anaerobic pathogen Clostridium perfringens. *Infect Immun* 2008;**76**:4944–51.

102. Varga JJ, Nguyen V, O'Brien DK, Rodgers K, Walker RA, Melville SB. Type IV pili-dependent gliding motility in the Gram-positive pathogen *Clostridium perfringens* and other Clostridia. *Mol Microbiol* 2006;**62**:680–94.

103. Nudleman E, Kaiser D. Pulling together with type IV pili. *J Mol Microbiol Biotechnol* 2004;**7**:52–62.

104. Pegden RS, Larson MA, Grant RJ, Morrison M. Adherence of the gram-positive bacterium *Ruminococcus albus* to cellulose and identification of a novel form of cellulose-binding protein which belongs to the Pil family of proteins. *J Bacteriol* 1998;**180**:5921–7.

105. Rakotoarivonina H, Larson MA, Morrison M, Girardeau JP, Gaillard-Martinie B, Forano E, et al. The Ruminococcus albus pilA1-pilA2 locus: expression and putative role of two adjacent pil genes in pilus formation and bacterial adhesion to cellulose. *Microbiology* 2005;**151**:1291–9.

106. Chaban B, Ng SY, Jarrell KF. Archaeal habitats—from the extreme to the ordinary. *Can J Microbiol* 2006;**52**:73–116.

107. Kandler O, Konig H. Cell wall polymers in Archaea (Archaebacteria). *Cell Mol Life Sci* 1998;**54**:305–8.

108. Rachel R, Wyschkony I, Riehl S, Huber H. The ultrastructure of Ignicoccus: evidence for a novel outer membrane and for intracellular vesicle budding in an archaeon. *Archaea* 2002;**1**:9–18.

109. Albers SV, Szabo Z, Driessen AJ. Protein secretion in the Archaea: multiple paths towards a unique cell surface. *Nat Rev Microbiol* 2006;**4**:537–47.

110. Ng SY, Zolghadr B, Driessen AJ, Albers SV, Jarrell KF. Cell surface structures of archaea. *J Bacteriol* 2008;**190**:6039–47.

111. Szabo Z, Stahl AO, Albers SV, Kissinger JC, Driessen AJ, Pohlschroder M. Identification of diverse archaeal proteins with class III signal peptides cleaved by distinct archaeal prepilin peptidases. *J Bacteriol* 2007;**189**:772–8.

112. Wang YA, Yu X, Ng SY, Jarrell KF, Egelman EH. The structure of an archaeal pilus. *J Mol Biol* 2008;**381**:456–66.

113. Albers SV, Pohlschroder M. Diversity of archaeal type IV pilin-like structures. *Extremophiles* 2009;**13**:403–10.

114. Frols S, Ajon M, Wagner M, Teichmann D, Zolghadr B, Folea M, et al. UV-inducible cellular aggregation of the hyperthermophilic archaeon *Sulfolobus solfataricus* is mediated by pili formation. *Mol Microbiol* 2008;**70**:938–52.

115. Moissl C, Rachel R, Briegel A, Engelhardt H, Huber R. The unique structure of archaeal 'hami', highly complex cell appendages with nano-grappling hooks. *Mol Microbiol* 2005;**56**:361–70.

116. Thoma C, Frank M, Rachel R, Schmid S, Nather D, Wanner G, et al. The Mth60 fimbriae of *Methanothermobacter thermoautotrophicus* are functional adhesins. *Environ Microbiol* 2008;**10**:2785–95.

117. Thormann KM, Paulick A. Tuning the flagellar motor. *Microbiology* 2010;**156**:1275–83.

118. Erdem AL, Avelino F, Xicohtencatl-Cortes J, Giron JA. Host protein binding and adhesive properties of H6 and H7 flagella of attaching and effacing *Escherichia coli*. *J Bacteriol* 2007;**189**:7426–35.

119. Tasteyre A, Barc MC, Collignon A, Boureau H, Karjalainen T. Role of FliC and FliD flagellar proteins of *Clostridium difficile* in adherence and gut colonization. *Infect Immun* 2001;**69**:7937–40.

120. Arora SK, Ritchings BW, Almira EC, Lory S, Ramphal R. The *Pseudomonas aeruginosa* flagellar cap protein, FliD, is responsible for mucin adhesion. *Infect Immun* 1998;**66**:1000–7.

121. Harshey RM. Bacterial motility on a surface: many ways to a common goal. *Annu Rev Microbiol* 2003;**57**:249–73.

122. Pruss BM, Besemann C, Denton A, Wolfe AJ. A complex transcription network controls the early stages of biofilm development by *Escherichia coli*. *J Bacteriol* 2006;**188**:3731–9.

123. Wang Q, Suzuki A, Mariconda S, Porwollik S, Harshey RM. Sensing wetness: a new role for the bacterial flagellum. *EMBO J* 2005;**24**:2034–42.

124. Chevance FF, Hughes KT. Coordinating assembly of a bacterial macromolecular machine. *Nat Rev Microbiol* 2008;**6**:455–65.

125. Minamino T, Imada K, Namba K. Molecular motors of the bacterial flagella. *Curr Opin Struct Biol* 2008;**18**:693–701.
126. Kobayashi K, Saitoh T, Shah DS, Ohnishi K, Goodfellow IG, Sockett RE, et al. Purification and characterization of the flagellar basal body of Rhodobacter sphaeroides. *J Bacteriol* 2003;**185**:5295–300.
127. Bardy SL, Ng SY, Jarrell KF. Prokaryotic motility structures. *Microbiology* 2003;**149**:295–304.
128. Bourret RB, Stock AM. Molecular information processing: lessons from bacterial chemotaxis. *J Biol Chem* 2002;**277**:9625–8.
129. Yonekura K, Maki-Yonekura S, Namba K. Complete atomic model of the bacterial flagellar filament by electron cryomicroscopy. *Nature* 2003;**424**:643–50.
130. Yonekura K, Maki S, Morgan DG, DeRosier DJ, Vonderviszt F, Imada K, et al. The bacterial flagellar cap as the rotary promoter of flagellin self-assembly. *Science* 2000;**290**:2148–52.
131. Minamino T, Imada K, Namba K. Mechanisms of type III protein export for bacterial flagellar assembly. *Mol Biosyst* 2008;**4**:1105–15.
132. Ellen AF, Zolghadr B, Driessen AM, Albers SV. Shaping the archaeal cell envelope. *Archaea* 2010;**2010**:608243.
133. Trachtenberg S, Cohen-Krausz S. The archaeabacterial flagellar filament: a bacterial propeller with a pilus-like structure. *J Mol Microbiol Biotechnol* 2006;**11**:208–20.
134. Streif S, Staudinger WF, Marwan W, Oesterhelt D. Flagellar rotation in the archaeon *Halobacterium salinarum* depends on ATP. *J Mol Biol* 2008;**384**:1–8.
135. Schlesner M, Miller A, Streif S, Staudinger WF, Muller J, Scheffer B, et al. Identification of Archaea-specific chemotaxis proteins which interact with the flagellar apparatus. *BMC Microbiol* 2009;**9**:56.
136. Babic A, Lindner AB, Vulic M, Stewart EJ, Radman M. Direct visualization of horizontal gene transfer. *Science* 2008;**319**:1533–6.
137. Alvarez-Martinez CE, Christie PJ. Biological diversity of prokaryotic type IV secretion systems. *Microbiol Mol Biol Rev* 2009;**73**:775–808.
138. Llosa M, Gomis-Ruth FX, Coll M, de la Cruz FF. Bacterial conjugation: a two-step mechanism for DNA transport. *Mol Microbiol* 2002;**45**:1–8.
139. Barlow M. What antimicrobial resistance has taught us about horizontal gene transfer. *Methods Mol Biol* 2009;**532**:397–411.
140. Hooykaas PJ, Schilperoort RA. Agrobacterium and plant genetic engineering. *Plant Mol Biol* 1992;**19**:15–38.
141. Broothaerts W, Mitchell HJ, Weir B, Kaines S, Smith LM, Yang W, et al. Gene transfer to plants by diverse species of bacteria. *Nature* 2005;**433**:629–33.
142. Valentine L. *Agrobacterium tumefaciens* and the plant: the David and Goliath of modern genetics. *Plant Physiol* 2003;**133**:948–55.
143. Llosa M, de la Cruz F. Bacterial conjugation: a potential tool for genomic engineering. *Res Microbiol* 2005;**156**:1–6.
144. Chamekh M, Phalipon A, Quertainmont R, Salmon I, Sansonetti P, Allaoui A. Delivery of biologically active anti-inflammatory cytokines IL-10 and IL-1ra in vivo by the Shigella type III secretion apparatus. *J Immunol* 2008;**180**:4292–8.
145. Panthel K, Meinel KM, Sevil D, Trulzsch VK, Russmann H. Salmonella type III-mediated heterologous antigen delivery: a versatile oral vaccination strategy to induce cellular immunity against infectious agents and tumors. *Int J Med Microbiol* 2008;**298**:99–103.
146. Nishikawa H, Sato E, Briones G, Chen LM, Matsuo M, Nagata Y, et al. In vivo antigen delivery by a *Salmonella typhimurium* type III secretion system for therapeutic cancer vaccines. *J Clin Invest* 2006;**116**:1946–54.
147. Russmann H, Shams H, Poblete F, Fu Y, Galan JE, Donis RO. Delivery of epitopes by the Salmonella type III secretion system for vaccine development. *Science* 1998;**281**:565–8.

148. Gundel I, Weidinger G, ter Meulen V, Heesemann J, Russmann H, Niewiesk S. Oral immunization with recombinant *Yersinia enterocolitica* expressing a measles virus CD4 T cell epitope protects against measles virus-induced encephalitis. *J Gen Virol* 2003;**84**:775–9.

149. Shams H, Poblete F, Russmann H, Galan JE, Donis RO. Induction of specific CD8+ memory T cells and long lasting protection following immunization with *Salmonella typhimurium* expressing a lymphocytic choriomeningitis MHC class I-restricted epitope. *Vaccine* 2001;**20**:577–85.

150. Majander K, Anton L, Antikainen J, Lang H, Brummer M, Korhonen TK, et al. Extracellular secretion of polypeptides using a modified *Escherichia coli* flagellar secretion apparatus. *Nat Biotechnol* 2005;**23**:475–81.

151. Mergulhao FJ, Summers DK, Monteiro GA. Recombinant protein secretion in *Escherichia coli*. *Biotechnol Adv* 2005;**23**:177–202.

152. Kotton CN, Hohmann EL. Enteric pathogens as vaccine vectors for foreign antigen delivery. *Infect Immun* 2004;**72**:5535–47.

153. Konkel ME, Klena JD, Rivera-Amill V, Monteville MR, Biswas D, Raphael B, et al. Secretion of virulence proteins from *Campylobacter jejuni* is dependent on a functional flagellar export apparatus. *J Bacteriol* 2004;**186**:3296–303.

154. Lee SH, Galan JE. Salmonella type III secretion-associated chaperones confer secretion-pathway specificity. *Mol Microbiol* 2004;**51**:483–95.

155. Young GM, Schmiel DH, Miller VL. A new pathway for the secretion of virulence factors by bacteria: the flagellar export apparatus functions as a protein-secretion system. *Proc Natl Acad Sci USA* 1999;**96**:6456–61.

156. Lovley DR, Phillips EJ. Novel mode of microbial energy metabolism: organic carbon oxidation coupled to dissimilatory reduction of iron or manganese. *Appl Environ Microbiol* 1988;**54**:1472–80.

157. Reguera G, McCarthy KD, Mehta T, Nicoll JS, Tuominen MT, Lovley DR. Extracellular electron transfer via microbial nanowires. *Nature* 2005;**435**:1098–101.

158. Gorby YA, Yanina S, McLean JS, Rosso KM, Moyles D, Dohnalkova A, et al. Electrically conductive bacterial nanowires produced by *Shewanella oneidensis* strain MR-1 and other microorganisms. *Proc Natl Acad Sci USA* 2006;**103**:11358–63.

159. Leang C, Qian X, Mester T, Lovley DR. Alignment of the c-type cytochrome OmcS along pili of *Geobacter sulfurreducens*. *Appl Environ Microbiol* 2010;**76**:4080–4.

160. Du Z, Li H, Gu T. A state of the art review on microbial fuel cells: a promising technology for wastewater treatment and bioenergy. *Biotechnol Adv* 2007;**25**:464–82.

161. Tender LM, Gray SA, Groveman E, Lowy DA, Kauffman P, Melhado J, et al. The first demonstration of a microbial fuel cell as a viable power supply: powering a meteorological buoy. *J Power Sources* 2008;**179**:571–5.

162. Bouhenni R, Vora G, Biffinger J, Shirodkar S, Brockman K, Ray R, et al. The role of *Shewanella oneidensis* MR-1 outer surface structures in extracellular electron transfer. *Electroanalysis* 2010;**22**:856–64.

163. Smith GP. Filamentous fusion phage: novel expression vectors that display cloned antigens on the virion surface. *Science* 1985;**228**:1315–7.

164. Charbit A, Boulain JC, Ryter A, Hofnung M. Probing the topology of a bacterial membrane protein by genetic insertion of a foreign epitope; expression at the cell surface. *EMBO J* 1986;**5**:3029–37.

165. Stentebjerg-Olesen B, Pallesen L, Jensen LB, Christiansen G, Klemm P. Authentic display of a cholera toxin epitope by chimeric type 1 fimbriae: effects of insert position and host background. *Microbiology* 1997;**143**(Pt 6):2027–38.

166. Pallesen L, Poulsen LK, Christiansen G, Klemm P. Chimeric FimH adhesin of type 1 fimbriae: a bacterial surface display system for heterologous sequences. *Microbiology* 1995;**141**(Pt 11):2839–48.
167. Schembri MA, Kjaergaard K, Klemm P. Bioaccumulation of heavy metals by fimbrial designer adhesins. *FEMS Microbiol Lett* 1999;**170**:363–71.
168. Kjaergaard K, Schembri MA, Klemm P. Novel Zn(2+)-chelating peptides selected from a fimbria-displayed random peptide library. *Appl Environ Microbiol* 2001;**67**:5467–73.
169. van Die I, van Oosterhout J, van Megen I, Bergmans H, Hoekstra W, Enger-Valk B, et al. Expression of foreign epitopes in P-fimbriae of *Escherichia coli*. *Mol Gen Genet* 1990;**222**:297–303.
170. van der Zee A, Noordegraaf CV, van den Bosch H, Gielen J, Bergmans H, Hoekstra W, et al. P-fimbriae of *Escherichia coli* as carriers for gonadotropin releasing hormone: development of a recombinant contraceptive vaccine. *Vaccine* 1995;**13**:753–8.
171. Thiry G, Clippe A, Scarcez T, Petre J. Cloning of DNA sequences encoding foreign peptides and their expression in the K88 pili. *Appl Environ Microbiol* 1989;**55**:984–93.
172. Zalewska B, Piatek R, Konopa G, Nowicki B, Nowicki S, Kur J. Chimeric Dr fimbriae with a herpes simplex virus type 1 epitope as a model for a recombinant vaccine. *Infect Immun* 2003;**71**:5505–13.
173. Chen H, Schifferli DM. Comparison of a fimbrial versus an autotransporter display system for viral epitopes on an attenuated Salmonella vaccine vector. *Vaccine* 2007;**25**:1626–33.
174. Rani DB, Bayer ME, Schifferli DM. Polymeric display of immunogenic epitopes from herpes simplex virus and transmissible gastroenteritis virus surface proteins on an enteroadherent fimbria. *Clin Diagn Lab Immunol* 1999;**6**:30–40.
175. White AP, Collinson SK, Banser PA, Dolhaine DJ, Kay WW. *Salmonella enteritidis* fimbriae displaying a heterologous epitope reveal a uniquely flexible structure and assembly mechanism. *J Mol Biol* 2000;**296**:361–72.
176. Der Vartanian M, Girardeau JP, Martin C, Rousset E, Chavarot M, Laude H, et al. An *Escherichia coli* CS31A fibrillum chimera capable of inducing memory antibodies in outbred mice following booster immunization with the entero-pathogenic coronavirus transmissible gastroenteritis virus. *Vaccine* 1997;**15**:111–20.
177. Mechin MC, Der VM, Martin C. The major subunit ClpG of Escherichia coli CS31A fibrillae as an expression vector for different combinations of two TGEV coronavirus epitopes. *Gene* 1996;**179**:211–8.
178. Bousquet F, Martin C, Girardeau JP, Mechin MC, Der VM, Laude H, et al. CS31A capsule-like antigen as an exposure vector for heterologous antigenic determinants. *Infect Immun* 1994;**62**:2553–61.
179. Jennings PA, Bills MM, Irving DO, Mattick JS. Fimbriae of Bacteroides nodosus: protein engineering of the structural subunit for the production of an exogenous peptide. *Protein Eng* 1989;**2**:365–9.
180. Malmborg AC, Soderlind E, Frost L, Borrebaeck CA. Selective phage infection mediated by epitope expression on F pilus. *J Mol Biol* 1997;**273**:544–51.
181. Rondot S, Anthony KG, Dubel S, Ida N, Wiemann S, Beyreuther K, et al. Epitopes fused to F-pilin are incorporated into functional recombinant pili. *J Mol Biol* 1998;**279**:589–603.
182. Quigley BR, Hatkoff M, Thanassi DG, Ouattara M, Eichenbaum Z, Scott JR. A foreign protein incorporated on the Tip of T3 pili in *Lactococcus lactis* elicits systemic and mucosal immunity. *Infect Immun* 2010;**78**:1294–303.
183. Newton SM, Jacob CO, Stocker BA. Immune response to cholera toxin epitope inserted in *Salmonella flagellin*. *Science* 1989;**244**:70–2.

184. Wu JY, Newton S, Judd A, Stocker B, Robinson WS. Expression of immunogenic epitopes of hepatitis B surface antigen with hybrid flagellin proteins by a vaccine strain of Salmonella. *Proc Natl Acad Sci USA* 1989;**86**:4726–30.

185. Newton SM, Kotb M, Poirier TP, Stocker BA, Beachey EH. Expression and immunogenicity of a streptococcal M protein epitope inserted in *Salmonella flagellin. Infect Immun* 1991;**59**:2158–65.

186. Stocker BA, Newton SM. Immune responses to epitopes inserted in Salmonella flagellin. *Int Rev Immunol* 1994;**11**:167–78.

187. Levi R, Arnon R. Synthetic recombinant influenza vaccine induces efficient long-term immunity and cross-strain protection. *Vaccine* 1996;**14**:85–92.

188. Newton SM, Joys TM, Anderson SA, Kennedy RC, Hovi ME, Stocker BA. Expression and immunogenicity of an 18-residue epitope of HIV1 gp41 inserted in the flagellar protein of a Salmonella live vaccine. *Res Microbiol* 1995;**146**:203–16.

189. Cattozzo EM, Stocker BA, Radaelli A, De Giuli MC, Tognon M. Expression and immunogenicity of V3 loop epitopes of HIV-1, isolates SC and WMJ2, inserted in Salmonella flagellin. *J Biotechnol* 1997;**56**:191–203.

190. Verma NK, Ziegler HK, Wilson M, Khan M, Safley S, Stocker BA, et al. Delivery of class I and class II MHC-restricted T-cell epitopes of listeriolysin of *Listeria monocytogenes* by attenuated Salmonella. *Vaccine* 1995;**13**:142–50.

191. Cuadros C, Lopez-Hernandez FJ, Dominguez AL, McClelland M, Lustgarten J. Flagellin fusion proteins as adjuvants or vaccines induce specific immune responses. *Infect Immun* 2004;**72**:2810–6.

192. Braga CJ, Rittner GM, Munoz Henao JE, Teixeira AF, Massis LM, Sbrogio-Almeida ME, et al. Paracoccidioides brasiliensis vaccine formulations based on the gp43-derived P10 sequence and the *Salmonella enterica* FliC flagellin. *Infect Immun* 2009;**77**:1700–7.

193. Braga CJ, Massis LM, Sbrogio-Almeida ME, Alencar BC, Bargieri DY, Boscardin SB, et al. CD8+ T cell adjuvant effects of Salmonella FliCd flagellin in live vaccine vectors or as purified protein. *Vaccine* 2010;**28**:1373–82.

194. Westerlund-Wikstrom B, Tanskanen J, Virkola R, Hacker J, Lindberg M, Skurnik M, et al. Functional expression of adhesive peptides as fusions to *Escherichia coli* flagellin. *Protein Eng* 1997;**10**:1319–26.

195. Dong J, Liu C, Zhang J, Xin ZT, Yang G, Gao B, et al. Selection of novel nickel-binding peptides from flagella displayed secondary peptide library. *Chem Biol Drug Des* 2006;**68**:107–12.

196. Majander K, Korhonen TK, Westerlund-Wikstrom B. Simultaneous display of multiple foreign peptides in the FliD capping and FliC filament proteins of the *Escherichia coli* flagellum. *Appl Environ Microbiol* 2005;**71**:4263–8.

197. Wu CH, Mulchandani A, Chen W. Versatile microbial surface-display for environmental remediation and biofuels production. *Trends Microbiol* 2008;**16**:181–8.

198. Klemm P, Schembri MA. Fimbriae-assisted bacterial surface display of heterologous peptides. *Int J Med Microbiol* 2000;**290**:215–21.

199. Georgiou G, Stathopoulos C, Daugherty PS, Nayak AR, Iverson BL, Curtiss III R. Display of heterologous proteins on the surface of microorganisms: from the screening of combinatorial libraries to live recombinant vaccines. *Nat Biotechnol* 1997;**15**:29–34.

200. Chen H, Schifferli DM. Mucosal and systemic immune responses to chimeric fimbriae expressed by *Salmonella enterica serovar typhimurium* vaccine strains. *Infect Immun* 2000;**68**:3129–39.

201. Ren T, Zamboni DS, Roy CR, Dietrich WF, Vance RE. Flagellin-deficient Legionella mutants evade caspase-1- and Naip5-mediated macrophage immunity. *PLoS Pathog* 2006;**2**:e18.

202. Alaniz RC, Cummings LA, Bergman MA, Rassoulian-Barrett SL, Cookson BT. *Salmonella typhimurium* coordinately regulates FliC location and reduces dendritic cell activation and antigen presentation to CD4+ T cells. *J Immunol* 2006;**177**:3983–93.
203. Sutterwala FS, Flavell RA. NLRC4/IPAF: a CARD carrying member of the NLR family. *Clin Immunol* 2009;**130**:2–6.
204. Miao EA, Andersen-Nissen E, Warren SE, Aderem A. TLR5 and Ipaf: dual sensors of bacterial flagellin in the innate immune system. *Semin Immunopathol* 2007;**29**:275–88.
205. Samuelson P, Gunneriusson E, Nygren PA, Stahl S. Display of proteins on bacteria. *J Biotechnol* 2002;**96**:129–54.
206. Lu Z, Murray KS, Van CV, LaVallie ER, Stahl ML, McCoy JM. Expression of thioredoxin random peptide libraries on the *Escherichia coli* cell surface as functional fusions to flagellin: a system designed for exploring protein-protein interactions. *Biotechnology* 1995;**13**:366–72.
207. Westerlund-Wikstrom B. Peptide display on bacterial flagella: principles and applications. *Int J Med Microbiol* 2000;**290**:223–30.
208. Munera D, Hultgren S, Fernandez LA. Recognition of the N-terminal lectin domain of FimH adhesin by the usher FimD is required for type 1 pilus biogenesis. *Mol Microbiol* 2007;**64**:333–46.
209. Ofek I, Hasty DL, Sharon N. Anti-adhesion therapy of bacterial diseases: prospects and problems. *FEMS Immunol Med Microbiol* 2003;**38**:181–91.
210. Wizemann TM, Adamou JE, Langermann S. Adhesins as targets for vaccine development. *Emerg Infect Dis* 1999;**5**:395–403.
211. Morgan RL, Isaacson RE, Moon HW, Brinton CC, To CC. Immunization of suckling pigs against enterotoxigenic *Escherichia coli*-induced diarrheal disease by vaccinating dams with purified 987 or K99 pili: protection correlates with pilus homology of vaccine and challenge. *Infect Immun* 1978;**22**:771–7.
212. Nagy B, Moon HW, Isaacson RE, To CC, Brinton CC. Immunization of suckling pigs against enteric enterotoxigenic *Escherichia coli* infection by vaccinating dams with purified pili. *Infect Immun* 1978;**21**:269–74.
213. Yokoyama H, Peralta RC, Diaz R, Sendo S, Ikemori Y, Kodama Y. Passive protective effect of chicken egg yolk immunoglobulins against experimental enterotoxigenic *Escherichia coli* infection in neonatal piglets. *Infect Immun* 1992;**60**:998–1007.
214. Marquardt RR, Jin LZ, Kim JW, Fang L, Frohlich AA, Baidoo SK. Passive protective effect of egg-yolk antibodies against enterotoxigenic *Escherichia coli* K88+ infection in neonatal and early-weaned piglets. *FEMS Immunol Med Microbiol* 1999;**23**:283–8.
215. Moon HW, Bunn TO. Vaccines for preventing enterotoxigenic *Escherichia coli* infections in farm animals. *Vaccine* 1993;**11**:213–20.
216. Langermann S, Mollby R, Burlein JE, Palaszynski SR, Auguste CG, DeFusco A, et al. Vaccination with FimH adhesin protects cynomolgus monkeys from colonization and infection by uropathogenic *Escherichia coli*. *J Infect Dis* 2000;**181**:774–8.
217. Strindelius L, Folkesson A, Normark S, Sjoholm I. Immunogenic properties of the Salmonella atypical fimbriae in BALB/c mice. *Vaccine* 2004;**22**:1448–56.
218. Stewart DJ, Clark BL, Emery DL, Peterson JE, Jarrett RG, O'Donnell IJ. Cross-protection from Bacteroides nodosus vaccines and the interaction of pili and adjuvants. *Aust Vet J* 1986;**63**:101–6.
219. Dhungyel OP, Whittington RJ. Modulation of inter-vaccination interval to avoid antigenic competition in multivalent footrot (Dichelobacter nodosus) vaccines in sheep. *Vaccine* 2009;**28**:470–3.
220. Lepper AW, Moore LJ, Atwell JL, Tennent JM. The protective efficacy of pili from different strains of *Moraxella bovis* within the same serogroup against infectious bovine keratoconjunctivitis. *Vet Microbiol* 1992;**32**:177–87.

221. Angelos JA, Bonifacio RG, Ball LM, Hess JF. Prevention of naturally occurring infectious bovine keratoconjunctivitis with a recombinant Moraxella bovis pilin-Moraxella bovis cytotoxin-ISCOM matrix adjuvanted vaccine. *Vet Microbiol* 2007;**125**:274–83.

222. Kao DJ, Churchill ME, Irvin RT, Hodges RS. Animal protection and structural studies of a consensus sequence vaccine targeting the receptor binding domain of the type IV pilus of *Pseudomonas aeruginosa*. *J Mol Biol* 2007;**374**:426–42.

223. Strindelius L, Degling WL, Sjoholm I. Extracellular antigens from *Salmonella enteritidis* induce effective immune response in mice after oral vaccination. *Infect Immun* 2002;**70**:1434–42.

224. Strindelius L, Filler M, Sjoholm I. Mucosal immunization with purified flagellin from Salmonella induces systemic and mucosal immune responses in C3H/HeJ mice. *Vaccine* 2004;**22**:3797–808.

225. Campodonico VL, Llosa NJ, Grout M, Doring G, Maira-Litran T, Pier GB. Evaluation of flagella and flagellin of *Pseudomonas aeruginosa* as vaccines. *Infect Immun* 2010;**78**:746–55.

226. Resnick IG, Ford CW, Shackleford GM, Berry LJ. Improved protection against cholera in adult rabbits with a combined flagellar-toxoid vaccine. *Infect Immun* 1980;**30**:375–80.

227. Holder IA, Wheeler R, Montie TC. Flagellar preparations from *Pseudomonas aeruginosa*: animal protection studies. *Infect Immun* 1982;**35**:276–80.

228. Holder IA, Naglich JG. Experimental studies of the pathogenesis of infections due to *Pseudomonas aeruginosa*: immunization using divalent flagella preparations. *J Trauma* 1986;**26**:118–22.

229. Drake D, Montie TC. Protection against *Pseudomonas aeruginosa* infection by passive transfer of anti-flagellar serum. *Can J Microbiol* 1987;**33**:755–63.

230. Rudner XL, Hazlett LD, Berk RS. Systemic and topical protection studies using *Pseudomonas aeruginosa* flagella in an ocular model of infection. *Curr Eye Res* 1992;**11**:727–38.

231. Bouckaert J, Berglund J, Schembri M, De Genst E, Cools L, Wuhrer M, et al. Receptor binding studies disclose a novel class of high-affinity inhibitors of the *Escherichia coli* FimH adhesin. *Mol Microbiol* 2005;**55**:441–55.

232. Klein T, Abgottspon D, Wittwer M, Rabbani S, Herold J, Jiang X, et al. FimH antagonists for the oral treatment of urinary tract infections: from design and synthesis to in vitro and in vivo evaluation. *J Med Chem* 2010;**53**:8627–41.

233. Ohlsson J, Jass J, Uhlin BE, Kihlberg J, Nilsson UJ. Discovery of potent inhibitors of PapG adhesins from uropathogenic *Escherichia coli* through synthesis and evaluation of galabiose derivatives. *Chembiochem* 2002;**3**:772–9.

234. Svensson A, Larsson A, Emtenas H, Hedenstrom M, Fex T, Hultgren SJ, et al. Design and evaluation of pilicides: potential novel antibacterial agents directed against uropathogenic *Escherichia coli*. *Chembiochem* 2001;**2**:915–8.

235. Pinkner JS, Remaut H, Buelens F, Miller E, Aberg V, Pemberton N, et al. Rationally designed small compounds inhibit pilus biogenesis in uropathogenic bacteria. *Proc Natl Acad Sci USA* 2006;**103**:17897–902.

236. Tramont EC, Boslego JW. Pilus vaccines. *Vaccine* 1985;**3**:3–10.

237. Cehovin A, Winterbotham M, Lucidarme J, Borrow R, Tang CM, Exley RM, et al. Sequence conservation of pilus subunits in *Neisseria meningitidis*. *Vaccine* 2010;**28**:4817–26.

238. Rudin A, Olbe L, Svennerholm AM. Monoclonal antibodies against fimbrial subunits of colonization factor antigen I (CFA/I) inhibit binding to human enterocytes and protect against enterotoxigenic *Escherichia coli* expressing heterologous colonization factors. *Microb Pathog* 1996;**21**:35–45.

239. Cegelski L, Marshall GR, Eldridge GR, Hultgren SJ. The biology and future prospects of antivirulence therapies. *Nat Rev Microbiol* 2008;**6**:17–27.

240. Clatworthy AE, Pierson E, Hung DT. Targeting virulence: a new paradigm for antimicrobial therapy. *Nat Chem Biol* 2007;**3**:541–8.
241. Kauppi AM, Nordfelth R, Uvell H, Wolf-Watz H, Elofsson M. Targeting bacterial virulence: inhibitors of type III secretion in Yersinia. *Chem Biol* 2003;**10**:241–9.
242. Nordfelth R, Kauppi AM, Norberg HA, Wolf-Watz H, Elofsson M. Small-molecule inhibitors specifically targeting type III secretion. *Infect Immun* 2005;**73**:3104–14.
243. Muschiol S, Bailey L, Gylfe A, Sundin C, Hultenby K, Bergstrom S, et al. A small-molecule inhibitor of type III secretion inhibits different stages of the infectious cycle of *Chlamydia trachomatis*. *Proc Natl Acad Sci USA* 2006;**103**:14566–71.
244. Lindberg AA. Bacteriophage receptors. *Annu Rev Microbiol* 1973;**27**:205–41.
245. Clark JR, March JB. Bacteriophages and biotechnology: vaccines, gene therapy and antibacterials. *Trends Biotechnol* 2006;**24**:212–8.
246. Evans TJ, Crow MA, Williamson NR, Orme W, Thomson NR, Komitopoulou E, et al. Characterization of a broad-host-range flagellum-dependent phage that mediates high-efficiency generalized transduction in, and between, Serratia and Pantoea. *Microbiology* 2010;**156**:240–7.
247. Alvarez MA, Rodriguez A, Suarez JE. Stable expression of the *Lactobacillus casei* bacteriophage A2 repressor blocks phage propagation during milk fermentation. *J Appl Microbiol* 1999;**86**:812–6.
248. Hyman P, Abedon ST. Bacteriophage host range and bacterial resistance. *Adv Appl Microbiol* 2010;**70**:217–48.
249. Brussow H. Phages of dairy bacteria. *Annu Rev Microbiol* 2001;**55**:283–303.
250. McCarron PA, Olwill SA, Marouf WM, Buick RJ, Walker B, Scott CJ. Antibody conjugates and therapeutic strategies. *Mol Interv* 2005;**5**:368–80.
251. Zhang L, Abbott JJ, Dong LX, Kratochvil BE, Bell D, Nelson BJ. Artificial bacterial flagella: fabrication and magnetic control. *Appl Phys Lett* 2009;**94**.
252. Zhang L, Abbott JJ, Dong L, Peyer KE, Kratochvil BE, Zhang H, et al. Characterizing the swimming properties of artificial bacterial flagella. *Nano Lett* 2009;**9**:3663–7.
253. Zhang L, Peyer KE, Nelson BJ. Artificial bacterial flagella for micromanipulation. *Lab Chip* 2010;**10**:2203–15.
254. Sitti M. Miniature devices: voyage of the microrobots. *Nature* 2009;**458**:1121–2.

The Structure of Bacterial S-Layer Proteins

Tea Pavkov-Keller,[*]
Stefan Howorka,[†] and
Walter Keller[*]

[*]Institute of Molecular Biosciences,
Structural Biology, University of Graz,
Graz, Austria

[†]Department of Chemistry, Institute of
Structural and Molecular Biology,
University College London, London,
United Kingdom

S-layers are self-assembled paracrystalline protein lattices that cover many bacteria and almost all archaea. As an important component of the bacterial cell envelope, S-layers can fulfill various biological functions and are usually the most abundantly expressed protein species in a cell. Here we review the structures of the best characterized S-layer proteins from Gram-positive and Gram-negative bacteria, as well as methods to determine their molecular architecture.

Progress in Molecular Biology
and Translational Science, Vol. 103
DOI: 10.1016/B978-0-12-415906-8.00004-2

I. Introduction

Many bacteria and archaea exhibit a regularly ordered protein layer as a component of the cell envelope, called the S-layer or surface layer (Fig. 1). S-layers are widespread and occur on almost all archaea where they constitute the only cell wall component.[2–4] S-layers are also found on pathogenic as well as biotechnologically relevant Gram-positive and Gram-negative bacteria. In Gram-positive bacteria and in some archaea, the S-layer is noncovalently bound to cell wall components such as peptidoglycan, secondary cell wall polymers (SCWPs), or pseudomurein, while in Gram-negative bacteria, the S-layer is attached to the lipopolysaccharide (LPS) component of the outer membrane.[1,5,6]

A number of recent reviews on archaeal and bacterial S-layers, their occurrence, molecular biology, biophysical and ultrastructural properties, and applications have been published.[1,5,7–14] Indeed, a separate chapter in this

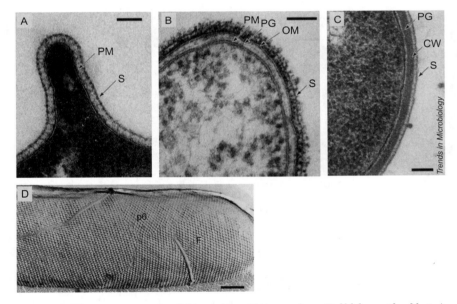

FIG. 1. Electron micrographs of thin sections of (A) an archeon (*Sulfolobus acidocaldarius*), (B) a Gram-negative bacterium (*Aeromonas salmonicida*), and (C) a Gram-positive bacterium (*Bacillus thuringiensis*). Abbreviations: CW, cell wall; OM, outer membrane; PG, peptidoglycan layer; PM, plasma membrane; S, S-layer. Scale bar = 50 nm. (D) Electron micrograph of freeze-etched preparations of intact cell of *Thermoanaerobacter thermohydrosulfuricus* (formerly *Clostridium thermohydrosulfuricus*) L111 showing a hexagonal (*p6*) surface lattice. F—flagellae. Scale bar = 100 nm. Figure adapted from Ref. 1.

issue entitled "Nanobiotechnology with S-Layer Proteins as Building Blocks" covers the topics of S-layer biology and applications. In this chapter, we focus on the current status of the structural characterization of bacterial S-layers.

A. General Biochemical Characteristics and Classification of Bacterial S-Layers

S-layers are mostly composed of one protein or glycoprotein species. If glycosylated, the degree of glycosylation and glycan composition can vary greatly.[10,15–20] Some organisms form complex S-layer lattices consisting of different protein species.[21–26] Furthermore, single strains may express different S-layer proteins depending on growth conditions as shown for *Geobacillus stearothermophilus*.[27–30]

On the sequence level, the S-layer proteins mostly show low homology,[5,31] and their molecular masses, as determined by SDS-PAGE mobility, range from 25 to 200 kDa. While differing in size, S-layer proteins usually share a high content of hydrophobic and acidic amino acids and lysine leading to an overall net negative charge and pI values in the acidic range (Ref. 31 and references therein). An exception are the S-layer proteins from *Lactobacilli*, whose pI can range between 9.4 and 10.4.[7] Another characteristic feature of S-layer proteins is the general lack or low content of sulfur-containing amino acids (Ref. 31 and references therein).

On an ultrastructural level, S-layer proteins form regular crystalline lattices, which may have an oblique $(p1, p2)$, square $(p4)$, or hexagonal $(p3, p6)$ symmetry. Lattice constants can vary from 5 to 30 nm and the S-layer thickness from 5 to 20 nm.[31] Detailed atomic resolution structures of lattices are not available, but several low-resolution three-dimensional (3D) structures have been obtained by electron microscopy (EM) of negatively stained samples.[2,32–34] Based on the reported 3D structures, Saxton and Baumeister proposed a classification scheme in which the S-layer proteins possess a heavier domain that forms the core of the morphological unit cell, and a lighter domain that provides connectivity between the units.[35] This classification was further elaborated by Hovmöller *et al.*, suggesting the most common type of subunit arrangement for $p4$ and for $p6$ symmetry (Fig. 2).[32]

A striking similarity even among quite unrelated species is the central core-forming region, which is usually oriented toward the cell envelope, giving rise to an overall corrugated inner surface. By contrast, the outer surface appears smooth despite highly variable and species-specific ultrastructure.[32,36–38] Between 30% and 70% of one unit cell is occupied by the protein, which leads to the formation of identical and well-defined pores with a diameter of 2–8 nm.[37]

F<small>IG.</small> 2. Schematic view of the monomer interactions and symmetry elements in a tetragonal and hexagonal packing arrangement of the S-layer lattice.[32]

B. Binding to the Cell Wall and Self-Assembly

S-layer proteins exhibit mostly two separated morphological regions, responsible for cell wall binding and for self-assembly,[34,39] as shown by mutagenesis studies (Ref. 40 and references therein). The position of the cell wall anchoring region within the protein, as well as its sequence composition, can vary among the bacterial species, similar to the molecular binding partner within the underlying cell wall. In several Gram-positive and Gram-negative bacteria, S-layers bind via the N-terminal region to the underlying peptidoglycan sacculus by recognizing SCWPs.[12] This specific molecular interaction with the polysaccharide is mediated by a recurring structural motif termed the "surface layer homology" (SLH) motif, which is described further below. S-layer proteins devoid of SLH motifs are anchored to different types of SCWPs through their N- or C-terminal domains.[41–44] In Gram-negative bacteria, the S-layer is attached with its N- or C-terminus to the component of the outer membrane.[45–47]

The SLH motif usually consists of about 55 amino acids of which 10–15 are conserved (Fig. 3). To date, more than 6000 sequences belonging to 285 different species possess one or more SLH domains as identified by Pfam.[50] The first characterization of SLH motifs as a putative cell wall binding domain was based on sequence comparison.[51–53] This was confirmed in mutagenesis studies by deleting the SLH motifs from the OlpB S-layer of *Clostridium thermocellum*[54] and showing that the truncated version did not bind to cell walls.[24,55,56] The corresponding binding partner of the SLH motif in the cell wall of *C. thermocellum* and other Gram-positive bacteria is a pyruvylated SCWP that is covalently linked to the peptidoglycan.[37,56–63] While the composition of the SCWPs differs between bacteria, there is nevertheless a strong

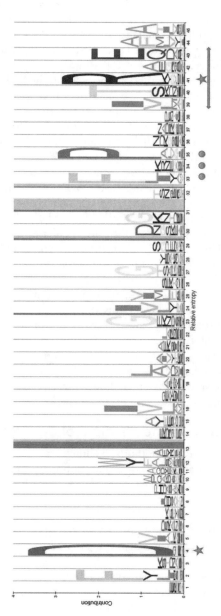

Fig. 3. HMM logo,[48,49] displaying the conservation of residues in the SLH motif. More than 6000 available sequences comprising an SLH motif were included in this analysis. The two most conserved residues (D4, R41) are marked with asterisks, the conserved "FXP" residues are marked with dots, and the "ITRAE" motif is highlighted with a horizontal arrow. (For color version of this figure, the reader is referred to the web version of this chapter.)

correlation between the existence of SLH motif(s) and the presence of the gene for the putative pyruvyltransferase CsaB, which is thought to be involved in the addition of a pyruvyl group to the SCWP.[60]

Sequence analysis has predicted that the SLH motif consists of two α-helices, which are connected by a short β-strand and a loop.[52] Additional mutational studies of conserved charged amino acids in the SLH motifs present in the xylanase XynA of *Thermoanaerobacterium thermosulfurigenes* EM1 indicate that the ITRAE motif is important for the binding function.[64] The recently determined structure of the SLH domain of the S-layer protein from *Bacillus anthracis* confirms the proposed α-helical elements and the importance of the conserved residues (see Section V).[65]

The self-assembly of S-layers can be studied *in vitro* as the recrystallized S-layers have the same lattices as those observed on the intact cells. The S-layers can be isolated from a bacterial cell by a disrupting agent such as LiCl, urea, Guanidinium hydrochloride (GdHCl), or SDS. After removal of the disrupting agent, spontaneous self-assembly into regular arrays occurs, thereby demonstrating that the individual S-layer subunits contain the information required for assembly and growth into a regular array. Small metal cations such as Ca^{2+} and Mg^{2+} can facilitate the assembly of some S-layers.[39,66] Important insights into the 2D assembly of SbpA protein from *Lysinibacillus sphaericus* on supported lipid bilayers was derived by *in situ* atomic force microscopy (AFM).[67] Protein crystallization was found to occur in several stages which were correlated to conformational changes that direct the pathway of assembly. It is suggested that, first, the extended SbpA monomers adsorb onto the surface, where they condense into amorphous clusters. The next step is a phase transition into crystalline clusters composed of compact tetramers, followed by the autocatalytic crystal growth.

C. Methods for Analyzing the Structure of S-Layers

EM-based techniques are the methods of choice for detecting the presence of an S-layer on the bacterial cell. The preparation techniques include metal shadowing,[68] freeze-fracturing and etching,[69,70] freeze drying, immunogold labeling, negative staining, thin-sectioning, as well as a cryo-electron microscopy (cryo-EM) of vitreous sections (CEMOVIS).[71,72] In order to obtain information about the 3D structure and surface topology of S-layer assemblies, various methods have been applied:

- EM: 3D reconstruction of negatively stained samples,[6,73] cryo-EM,[74,75] scanning transmission electron microscopy (STEM),[76] electron tomography of the whole cells,[77] and electron holography[78,79];
- Scanning probe microscopy: AFM *in vitro* and *in vivo* (Refs. 80–84 and references therein), scanning tunneling microscopy (STM),[36,85–87] and scanning ion conductance microscopy (SICM)[88];

- X-ray reflectivity and grazing incidence diffraction (GIXD)[89];
- 3D X-ray crystallography of N- and C-terminally truncated proteins [65,90–93];
- Small-angle X-ray scattering (SAXS).[90,91,94,95]

An enormous amount of ultrastructural as well as biochemical, biophysical, and genetic data is available for different S-layers. However, little information is available regarding the atomic structure of S-layer proteins and the spatial arrangement of domains and secondary structure elements within the S-layer lattice. One of the major hurdles toward this aim is that conventional methods for high-resolution structural analysis such as X-ray crystallography cannot be easily applied to S-layers. In particular, the growth of 3D crystals from full-length S-layer proteins is inhibited by their natural tendency to form 2D lattices and by the obvious lack of regular contacts in the third dimension. Similarly, the use of nuclear magnetic resonance (NMR) requires soluble proteins, whereas S-layer proteins assemble into lattices at low concentration. Sequence homology modeling is not a viable option for structural analysis, as S-layer proteins of different bacterial species, and even among most S-layer proteins of the same species, share little homology. In addition, the capability for prediction of the 3D atomic structure of full-length S-layer proteins is very limited, as in most cases no structures of homologous proteins are available to date.

The other technique with proven capability to solve structures at atomic resolution is electron crystallography: By recording tilt series at low dose of 2D crystals and employing 3D reconstruction using Fourier-transform techniques, high-resolution structures have been solved for proteins that form high quality 2D lattices. Currently, the structures of several non-S-layer proteins with a resolution of 3.2 Å or better are accessible in the protein data bank (PDB:2ZZ9[96]; PDB:2AT9[97]; PDB:1TVK[98]; PDB:2D57[99]; PDB: 2HA8[100]; PDB:2B60[101]; PDB:3M9I[102]). These entries are, however, exceptions as electron crystallography usually yields a resolution of 7 Å or lower, which means that secondary structure elements may be assigned in the most favorable cases, but the placement of residues including side-chain conformation is impossible. For S-layers, only two reports have used cryo-EM techniques with unstained samples to obtain comparable resolution[74,75] (see Sections VI and VIII.A). However, both these reports describe projection maps rather than 3D reconstructions. Currently, the highest resolution obtained for S-layer 3D reconstructions, from negatively stained or unstained samples, is in the order of 16 Å. Therefore, the best chance of reaching the goal of determining the atomic resolution structure of an assembled S-layer appears to be a combination of cryo-EM techniques with crystal structure determination of soluble domains or proteins that are assembly-negative due to selective mutations which interfere with the subunit–subunit interface.

Here we summarize the currently available data on the 3D structure of various bacterial S-layers, including partial crystal structures, and the methods that have been employed to elucidate them.

II. *Geobacillus stearothermophilus*

Strains belonging to the species *G. stearothermophilus* are Gram-positive organisms which frequently display an S-layer as the outermost cell envelope component.[38] The rigid cell wall layer is composed of an A1-γ chemotype peptidoglycan and an SCWP, which in wild-type strains is composed of 2,3-diacetamido mannosamine uronic acid, *N*-acetyl glucosamine, and glucose.[41,103] Variant strains that were formed in response to altered growth conditions, such as increased oxygen supply, express alternative S-layer genes and synthesize SCWPs composed of *N*-acetyl glucosamine, *N*-acetyl mannosamine, and pyruvic acid.[61] To date, the sequences of five S-layer proteins from *G. stearothermophilus* strains have been determined. SbsA is the S-layer protein expressed by the wild-type strain PV72/p6.[104] SbsB, the S-layer protein of an oxygen-induced strain variant, termed *G. stearothermophilus* PV72/p2, is expressed under altered growth conditions after the event of a genetic switch that results in the DNA rearrangement of the coding sequences *sbsA* and *sbsB*.[29] SbsC is the S-layer protein from another wild-type strain, *G. stearothermophilus* ATCC 12980, which binds to the same SCWP as SbsA.[41,42] At elevated temperatures (67°C), the same strain expresses the glycosylated S-layer protein SbsD.[30] The strain NRS 2004/3a possesses an oblique S-layer composed of the glycoprotein SgsE, which exhibits 93.9% sequence identity to SbsD at the protein level.[105] In addition, the S-layer proteins of other *G. stearothermophilus* strains have been detected and investigated by EM (see Table I), but were not investigated further.[106,107]

A. SbsA/C/D/SgsE

The structure–function relationship of the S-layer proteins produced by *G. stearothermophilus* wild-type strains has been investigated with various methods and to different degrees of complexity. Sequence comparison has shown that the N-terminal part of the four proteins is highly homologous, meaning that the different S-layer proteins are able to bind to the same SCWP. By contrast, the C-terminal crystallization regions are quite different in terms of length and sequence homology. The only exceptions are the two glycosylated S-layer proteins SbsD and SgsE, which exhibit a sequence identity of 94% over the entire amino acid sequence (903 residues) and even show a very similar glycosylation pattern.[30,105] Apart from this case, the different crystallization domains give rise to S-layer assemblies with different

TABLE I
S-LAYER PROTEINS OF *G. STEAROTHERMOPHILUS* AND THEIR PROPERTIES

Strain	Gene product	aa Residues[a]	Glycosylation	Lattice	Reference
PV72/p6	SbsA	1228	−	Hexagonal	28
PV72/p2	SbsB	920	−	Oblique	29
ATCC 12980	SbsC	1099	−	Oblique	41,42
ATCC 12980/G+	SbsD	903	+	Oblique	30
NRS 2004/3a	SgsE	903	+	Oblique	30,105
NRS 2004/3a/V2	−	n.d./120 kDa[b]	(−)[c]	Square	27
NRS 1526/3c	−	n.d./121 kDa[b]	(−)[c]	Square	106

[a]Full-length protein, including the signal peptide.
[b]n.d. not determined, molecular mass values derived from SDS-PAGE.
[c]Absence of glycosylation suggested.

FIG. 4. Electron micrograph of the cell surface of *G. stearothermophilus* NRS 2004/3a, which is undergoing oxygen-induced S-layer formation.[1]

morphological properties such as unit cell spacing, symmetry, and pore size. For example, SbsA forms a hexagonal lattice with a center-to-center spacing of 22.5 nm,[28] while SbsC yields an oblique lattice[42] with lattice constants of $a = 11 \pm 6$ nm, $b = 7 \pm 2$ nm, and $\gamma = 81°$.[108] An oblique lattice is also produced by SbsD[30] and SgsE, the S-layer protein expressed by the strain NRS 2004/3, with lattice constants of $a = 11.6$ nm, $b = 9.4$ nm, and $\gamma = 78°$.[105,109] This strain can also form different oblique and also square lattice forms under varying growth conditions[3,27] (Figs. 4 and 5).

Among the group of the S-layer proteins SbsA/C/D/SgsE, SbsC is biochemically and structurally the best characterized. Using N- and C-terminal deletion mutants,[108] it was demonstrated that the arginine- and lysine-rich N-terminal part (aa 31–257) is exclusively responsible for anchoring the S-layer subunits via the SCWP to the rigid cell wall layer.[41,42] This interaction must be independent of an SLH domain, as none was detected in sequence homology searches. The positively charged N-terminal segment is not required for self-assembly of

SbsC or for generating the wild-type-like oblique lattice structure. A similar tolerance on correct crystallization and assembly was also found for mutants that lacked the C-terminal 219 and 179 amino acids, respectively.[108]

Additional deletions of SbsC were created to obtain assembly-deficient mutants for crystallization trials. A total of three C-terminal deletion mutants, all containing the N-terminal SCWP-binding domain, and five N-terminal deletion mutants were crystallized[92,93] (Fig. 6). Two of the recombinant

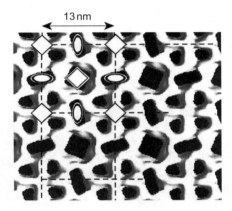

FIG. 5. Three-dimensional reconstruction of a negatively stained sample of the S-layer of *G. stearothermophilus* NRS 2004/3a/V2 exhibiting a square lattice.[110]

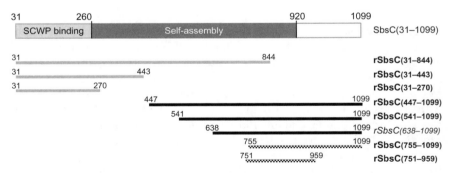

FIG. 6. Graphical representation of the SbsC protein and recombinant truncation fragments. The native mature S-layer protein SbsC$_{(31-1099)}$ is shown, with the secondary cell wall polymer (SCWP) binding and self-assembly regions indicated. C-terminal truncation fragments are shown in gray, N-terminal truncation fragments are shown in black, and the fragments obtained by unintentional proteolysis are shown as hatched lines. Start and end positions are indicated above the individual fragments. Fragments for which crystals were obtained are indicated in bold. (Adapted from Ref. 93).

C-terminal mutants, rSbsC$_{(31-844)}$ and rSbsC$_{(31-443)}$, led to the first structure determination of a bacterial S-layer protein.[91] The crystal structure of rSbsC$_{(31-844)}$ revealed a novel fold, consisting of six separate domains that are connected by short, flexible linkers (Fig. 7A and B). The first N-terminal domain from aa 31–254 is rich in α-helices and comprises the SCWP-binding domain. By comparison, the following five crystallization domains are compact and contain mainly antiparallel β-strands. Evidence for the high interdomain mobility of the rSbsC protein was obtained from the large structural differences between the crystal structures of mutants rSbsC$_{(31-844)}$ and rSbsC$_{(31-443)}$ (Fig. 7C) as well as the differences between crystal structure and the solution structure determined by SAXS.

The first three domains could be determined to high resolution from the second crystal form rSbsC$_{(31-443)}$, which revealed two new folds (Fig. 7D). As the first previously unknown structure, the SCWP-binding domain 1 consists of

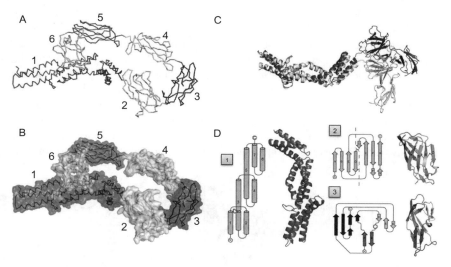

Fig. 7. (A) Structure of rSbsC$_{(31-844)}$—ribbon diagram of the domains 1 to 6. (B) Space filling presentation of domains 1 to 6, showing the ring-like conformation and the intramolecular interaction between domains 1 and 6. (C) Superposition of domain 1 (aa 32–158) indicating flexibility between the second and the third triple-helix bundle of domain 1 and between domains 1 and 2. rSbsC$_{(31-443)}$ is colored light blue and the first three domains of rSbsC$_{(31-844)}$ are colored violet. (D) Domain topology of rSbsC$_{(31-443)}$: Domain structure (right panel) and topology diagram (left panel) of domains 1, 2, and 3 in the crystal structure of rSbsC$_{(31-443)}$. Domain 1 represents a novel, all alpha-helical fold consisting of three triple-helical bundles that are connected by two continuous helices a3 and a5. Domain 2 forms beta sandwich resembling Ig-like h-type fold. The dashed line in the topology diagram indicates the separation of the two sheets. Domain 3 represents a novel topology not resembling any of the structures deposited in the PDB. Adapted from Ref. 91. (See Color Insert.)

three triple-helical bundles that are connected by two continuous helices and exhibits positively charged residues and tyrosines along the putative SCWP-binding groove. Domain 2 adopts an Ig-like fold resembling closely the h-type topology. Domain 3 also represents a novel fold, not resembling any other experimentally determined structure (Fig. 7D). Domains 4, 5, and 6 also adopt an Ig-like fold, as predicted by the Pfam server[111] but, due to the discontinuous electron density in this region, only a Cα trace could be determined.[91]

To obtain the atomic resolution structure of the C-terminal part of the SbsC protein, which is responsible for the S-layer assembly, the five N-terminal deletion mutants were crystallized (Fig. 6). The longest construct rSbsC$_{(447-1099)}$ yielded crystals with rather anisotropic diffraction, which were not suitable for structure solution. However, *in situ* proteolysis led to two shorter truncated derivatives, both of which yielded crystals. The best diffracting crystals were obtained from the deletion mutant rSbsC$_{(755-1099)}$, comprising the last three domains of the S-layer protein.[93] Their fold distantly resembles Ig-like domains, but with considerable variation in the length of β-strands as well as connecting loops (Pavkov-Keller *et al.*, unpublished data).

All available X-ray data of SbsC are compatible with the solution structure of the N-terminal deletion mutant rSbsC$_{(447-1099)}$ as determined by SAXS. The observed molecular mass is approximately 60 kDa, which is in line with a monomer. The $p(r)$ function as well as the *ab initio* models calculated from the SAXS data hinted at an elongated rod-like molecule with a cross-section of 30–40 Å, compatible with a string of Ig-like domains[91] (Fig. 8). Interestingly, the monomeric status of the N-terminal deletion mutant implies that S-layer assembly has to involve interactions between domains contained in this

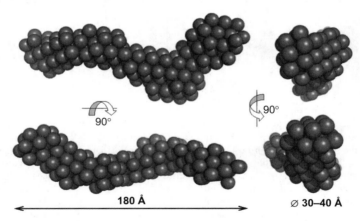

Fig. 8. Solution structure of the N-terminal deletion mutant rSbsC$_{(447-1099)}$ obtained by SAXS. The most probable bead model of rSbsC$_{(447-1099)}$ was obtained by averaging 10 *ab initio* models. (Adapted from Ref. 91).

construct with domains 2 or 3 from the missing N-terminal part of SbsC. As has been shown previously, domain 1 is essential for SCWP binding, but not for S-layer assembly.[108,112]

B. SbsB

The other S-layer protein found in *G. stearothermophilus* is SbsB. This S-layer protein shows little sequence homology to the previously described first group comprising SbsA, SbsC, SbsD, and SgsE. It shares, however, the oblique lattice type of the latter three S-layer proteins.

SbsB is produced by the variant *G. stearothermophilus* PV72, which either expresses this S-layer or protein SbsA depending on the bacterial fermentation conditions. In particular, an increase in the level of oxygen causes the switch from SbsA to SbsB.[27,28] Reflecting the two different lattice symmetries of SbsA and SbsB assemblies, the corresponding bacterial strains are referred to as PV72/p6 and PV72/p2, respectively. It is currently not known in which way the bacterial cells benefit from the change in the expression preference. Nevertheless, the switch can be followed at the genetic, protein, and ultrastructural level. At the genetic level, PCR and hybridization studies have shown that the coding region of the S-layer gene *sbsB* is unusually not located on the chromosome but on a natural megaplasmid of the bacterial strain.[29] In this location, *sbsB* is transcriptionally inactive, as the regulatory region of the gene is on the chromosome. Upon switching, *sbsB* integrates into the chromosomal expression site. The change is accompanied by movement of the *sbsA* coding region to the megaplasmid while the *sbsA* upstream regulatory region remains on the chromosome. The switch in production from SbsA to SbsB is apparent also by protein analysis and electron microscopic analysis. Indeed, it has been possible to trap cells displaying both lattice types.[28]

SbsB is expressed as a 920-amino acid long preprotein and contains a 31-residue-long N-terminal signal peptide.[113] The uncleaved signal peptide does not inhibit self-assembly as shown by the expression and lattice formation of SbsB within the heterologous host *Escherichia coli*.[113] The amino acid composition of SbsB is typical for extracellular bacterial S-layer proteins and comprises 35.4% polar, 11.8% acidic, 10.8% basic, and 49% hydrophobic residues. Valine, threonine, and alanine are the most frequently occurring residues. The isoelectric point is 5.7, but charged amino acids are not homogeneously distributed along the primary sequence. For instance, the N-terminal part of SbsB has a more basic p*I*.

Like several other S-layer proteins from *G. stearothermophilus*, SbsB comprises two main regions: an N-terminal cell wall anchoring domain and the C-terminal crystallization region. The anchoring domain was initially proposed in the course of sequence comparison, which identified SLH motifs at the N-terminus of SbsB.[113] SLH motifs are generally known to help attach bacterial exoproteins to the cell wall.[51–53] Three SLH domains were mapped to residues 32–208 of the preprotein sequence of SbsB. Experimental studies revealed that the anchoring domain's cognate binding partner in the bacterial cell wall is not the peptidoglycan but a SCWP.[56,114] The polysaccharide was initially character-ized to be composed mainly of N-acetylglucosamine (GlcNAc) and N-acetylman-nosamine (ManNAc)[56] but was later found to contain also N-acetylmannuronic acid and pyruvic acid.[115] The SCWP with a molecular weight of 24,000 constitutes about 20% of the dry weight of peptidoglycan-containing sacculi.[56] In a detailed structural analysis by NMR and mass spectrometry, the SCWP was proven to consist of the tetrasaccharide repeating unit 4-β-D-ManNAc-(1–3)-β-D-GlcNAc-(1–6)-α-D-GlcNAc featuring a (4–1)-α branching point from the last GlcNAc unit to D-ManNAc. The latter residue carries a pyruvic acid acetal at positions 4 and 6.[115] The SCWP is covalently linked to the peptidoglycan possibly via phospho-diester bonds, which can be cleaved under strongly acidic conditions.[44,114]

The specific molecular recognition between the N-terminal anchoring do-main of SbsB and the SCWP is supported by several experimental findings. (i) Only proteolytic fragments containing the N-terminus bound to the SCWP.[56] The SCWP furthermore enhanced the stability of the S-layer protein against an endoproteinase attack and specifically protected a potential cleavage site in position 160.[44] (ii) In genetic truncation studies, the N-terminus comprising the three SLH motifs but not the C-terminal remainder bound to SCWP.[61,116] The SbsB truncations are functionally and structurally intact, as confirmed by comparative thermal as well as equilibrium unfolding profiles monitored by intrinsic fluorescence.[116] The affinity of the recognition interaction is in the low nanomolar range with K_d values of 7×10^{-7} and 3×10^{-9} M as determined by surface plasmon resonance (SPR) and fluorescence anisotropy of the material, respectively. (iii) Adding increasing amounts of SCWP to monomeric full-length SbsB in recrystallization trials inhibits lattice formation, most likely as the binary complex is kept in a water-soluble state.[44] (iv) Neither the peptidoglycan nor SCWPs from other organisms, nor other polysaccharides recognize SbsB.[61] One dissenting study mapped the SCWP-binding site of SbsB between residues 272 and 362 and proposed a peptidoglycan recognition region within position 32 and 170.[114] But these data are in contradiction with all other reports and can possibly be attributed to impurities in the preparation of the SCWP and peptidoglycan. No high-resolution structure is available for the anchoring do-main of SbsB but, due to the sequence homology to the N-terminal domain of *Bacillus anthracis* S-layer protein, it can be assumed that it is similar to this recently reported structure (see Section V).

The C-terminal crystallization region of SbsB can functionally and struc-turally be separated from the cell wall anchoring domain. The crystallization region is assigned to amino acid position 208–920 of the preprotein (position 177–889 of the mature protein) as shown by genetic truncation studies in combination with EM.[116] Furthermore, the C-terminal part is rich in β-sheets (38–41%) and contains very few α-helices (4–6%).[116] SbsB exhibits multistep unfolding profiles when monitored by circular dichroism (CD) and intrinsic fluorescence [116]. Most likely, the multistep unfolding reflects the presence of several β-sheet rich domains analogous to the domain structure of SbsC.[91]

The ultrastructure of SbsB lattices has been analyzed using EM. Figure 9 displays a TEM image of a crystalline sheet that was formed in suspension and negatively stained with uranyl acetate. The morphological unit cell has lattice constants of $a = 10.4$ nm, $b = 7.9$ nm, and base angle $= 81°$. The 2D image reconstruction of the TEM micrographs revealed several depressions, which can be interpreted as pores ranging in diameter from approximately 2.5 to 6 nm. The same lattice constants were also found when the lattice was analyzed using AFM.[118]

The ultrastructural data of the SbsB lattice have been correlated with tertiary structure predictions for the protein and lattice. The molecular dynamic simulations used thermodynamic aspects of protein folding as well as steric constraints of the amino acids.[118] After performing sequence similarity searches and predicting secondary structure and domains, the obtained tertiary structure was analyzed using the mean force method by calculating the global free energy minimum, which corresponds to the thermodynamically most favorable conformation of the protein. This procedure yielded an L-shaped monomer which can virtually be assembled into an oblique lattice with the experimentally found unit cell constants.[118] A follow-up study using SAXS found that the modeled shape of the monomer resembled the calculated structure.[95]

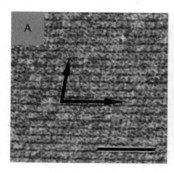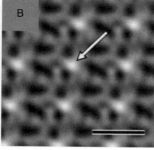

FIG. 9. Self-assembly products and digital image reconstructions of a SbsB lattice[117]: (A) The crystalline sheets were formed in suspension and negatively stained with uranyl acetate for TEM. The arrows indicate the base vectors of the oblique *p1* lattice. Bar, 50 nm. (B) The digital image reconstruction was made by Fourier processing of electron micrographs (not identical to the one shown in A). The region of highest protein mass in the SbsB lattice is the SLH domain (arrow) as inferred from fusion of a high-mass protein to the N-terminus[117] (Bars, 10 nm).

In the absence of atomic resolution data, attempts have been made to identify the position of individual residues. One approach has utilized a surface-accessibility screen to pinpoint amino acid positions that are located at the surface of the protein. Site-directed mutagenesis was applied to replace 75 residues spread along the primary sequence of SbsB into cysteine, which is not found in the wild-type protein. After confirming that the mutations did not compromise the ability to form self-assembly products, monomeric mutants, and mutants assembled into S-layers on cell wall sacculi were subjected to the surface-accessibility screen using targeted chemical modification with a bulky hydrophilic cysteine-reactive polyethylene glycol conjugate. Out of the 75 residues examined, 23 were classified as very accessible in the monomeric form of SbsB, whereas six were very accessible in the assembled form,[119] implying that these positions are at the outer surface of the lattice that faces the ambient. Further work with the 23 cyteine mutants using a photoactivatable, sulfhydryl-reactive cross-linker led to the identification of eight positions at the subunit–subunit interface.[120] As interfacial positions play an important role in the formation of regular assemblies, a follow-up study generated assembly-negative protein variants of high solubility that would facilitate high-resolution structure determination.[121] The variants were obtained using an insertion mutagenesis screen in which an epitope tag was inserted at the 23 positions at the monomer protein surface selected in the previous cysteine accessibility screen. The assembly-compromised mutants with native fold may be used for X-ray crystallography or NMR.[121]

III. *Clostridia*

The genus *Clostridium* represents a very diverse group of obligatory anaerobic, Gram-positive staining and spore-forming bacteria. Their members exhibit an extreme metabolic diversity, which is being exploited as biocatalysts for biotechnological applications. However, some members of this genus cause devastating human and animal diseases.

3D structure determination by EM methods has been performed for S-layers of two *Clostridium* species, namely, *C. aceticum* (Wieringa), a microorganism producing acetic acid from molecular hydrogen and carbon dioxide[122]; and the thermophilic bacterium *C. thermohydrosulfuricum*. The S-layer of the former has been isolated and its structure determined to 2.0 nm from negatively stained preparations.[123] The lattice exhibits tetragonal symmetry, with a cell constant of 12 nm and a thickness of 6 nm (Fig. 10). The main portion of the protein mass is positioned around the major fourfold axis, forming protrusions on the inner surface of the S-layer. Thin arms extend towards the minor fourfold axes, where the secondary contacts are established

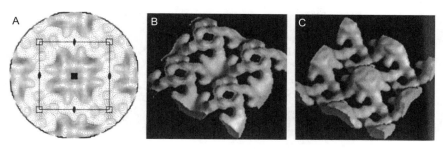

FIG. 10. 3D reconstruction of a negatively stained S-layer from *C. aceticum*. (A) Contour plot of the untilted S-layer with the unit cell and symmetry elements indicated. (B, C) View of the (B) outer and (C) inner surface. [123]

near the outer surface (Fig. 10B and C). This structure has been compared with the 3D reconstruction of the S-layer from *C. thermohydrosulfuricum*, which was determined to 2.5 nm resolution and forms a hexagonal lattice.[124] Interestingly, the overall pattern of mass distribution, as well as the protomer structure, was very similar between *C. aceticum* and *C. thermohydrosulfuricum* in spite of their different symmetries. This similarity led the authors to speculate about the possible evolutionary steps leading to the change in symmetry, although it has to be mentioned that *C. thermohydrosulfuricum* has recently been reassigned to the genus *Thermoanaerobacter*.[125]

Another *Clostridium* species that expresses an S-layer is *C. difficile*. This species is a leading cause of nosocomial infection, and in Europe it has already exceeded the number of infections caused by MRSA (methicillin-resistant *Staphylococcus aureus*). Importantly, the S-layer of *C. difficile* mediates host–pathogen interactions.[126,127] It has been shown that the proteinaceous S-layer consists of two structural components, namely, the high- and the low-molecular weight components (called HMW and LMW, respectively), which are produced by the catalytic cleavage of the precursor protein SlpA by the cysteine protease Cwp84.[128] After translocation to the surface and cleavage, the products reassemble on the surface to form the paracrystalline S-layer.

Recently, the crystal structure of a truncated form (aa 1-262) of the LMW has been determined at 2.4 Å resolution.[90] The structure consists of two domains, where the N-terminal domain exhibits a two-layer sandwich architecture while the middle domain contains a novel fold (Fig. 11). The third C-terminal domain, which is missing in the crystal structure, has been proven to be responsible for the "end-to-end" interaction of the LMW component with the HMW component.[90] Indeed, the solution structure of the LMW/HMW complex has been determined by SAXS methods, suggesting an elongated molecule with a D_{max} of 250 Å. Combined with the solution structures of the

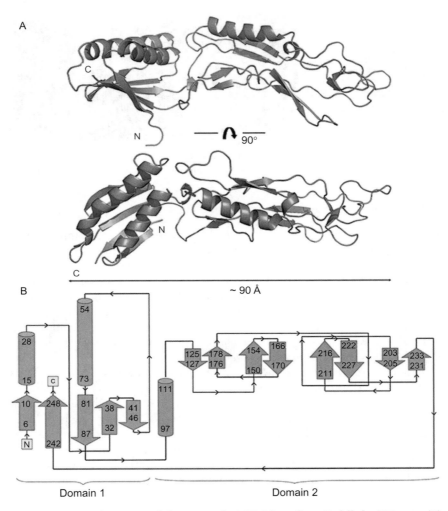

Fɪɢ. 11. Crystal structure of the truncated LMW S-layer from *C. difficile*, SLP$_{(1-262)}$. (A) Ribbon representation of the crystal structure in two different orientations related by a 90° rotation. Domain 1 is shown in green and domain 2 is shown in orange. (B) Topology diagram of domains 1 and 2 illustrating the arrangement of secondary structure elements. Domain 1 contains both the N- and C-termini and adopts a two-layer sandwich architecture, while domain 2 adopts a novel fold.[90] (See Color Insert.)

LMW alone ($D_{max} = 145$ Å) and the truncated LMW ($D_{max} = 90$ Å) used for crystallization, a plausible model for the shape of the LMW/HMW complex and its orientation on the surface of *C. difficile* could be devised[90] (Fig. 12).

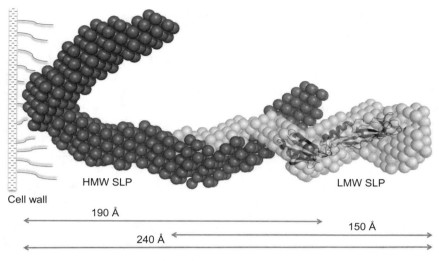

FIG. 12. SAXS structure of the native H/L complex isolated from *C. difficile* CD630.[90] Model of the orientation of the HMW and LMW SLPs on the surface of *C. difficile*. The HMW SLP is also shown interacting with the cell wall through one region of the protein; however, the extent and exact nature of this interaction is currently unknown. The cyan bead model represents the full-length LMW SLP, and the magenta bead model the HMW SLP. The LMW SLP1-262 crystal structure is shown in blue ribbon. The approximate lengths of the HMW and LMW SLPs as well as of the HMW/LMW complex were obtained by SAXS. (For color version of this figure, the reader is referred to the web version of this chapter.)

IV. Lactobacilli

Lactic acid bacteria are a heterogeneous group of Gram-positive bacteria used in the food industry. Many *Lactobacilli* have generally been regarded as safe (GRAS status).

Several *Lactobacillus* strains possess an S-layer such as *L. acidophilus*, *L. brevis*, *L. crispatus*, *L. buchneri*, *L. amylovorus*, *L. gallinarum*, *L. helveticus*, *L. johnsonii*, *L. fermentum*, and *L. gasseri*.[7,129–136] Multiple promoter structures and silent S-layer genes have been reported for several *Lactobacilli*, such as *L. acidophilus* ATCC 4356, *L. brevis* ATCC 8287, *L. crispatus* ZJ001, *L. crispatus* JCM 5810, and *L. gallinarum*.[129,133,137–142] For the SlpA protein of *L. brevis* ATCC 8287, it was demonstrated that both consecutive promoters were active at different growth stages,[140] whereas in *L. acidophilus* ATCC 4356 only the downstream promoter was used under all tested growth conditions.[129]

The mature S-layer proteins from *Lactobacilli* have a molecular mass in the range of 25–71 kDa and a p*I* of 9.4–10.4.[7] The signal peptide is typical for the general Sec-pathway and includes 25–30 amino acids at the N-terminus.[38]

Most of the *Lactobacillus* S-layer proteins appear to be nonglycosylated[7,143] and the primary structures show the highest similarity for the C-terminal region.[7] The N-termini of the S-layer proteins from *L. acidophilus* ATCC 4356 (SlpA), *L. helveticus* ATCC 12046 (SlpH), and *L. crispatus* JCM 5810 (CbsA) exhibit a low similarity, except for the high identity of the signal peptide; the remaining C-terminal sequence of *L. acidophilus* SlpA and *L. helveticus* SlpH proteins show the greatest similarity (Fig. 13). In contrast, the S-layer protein SlpA from *L. brevis* ATCC 8287 is not related in sequence to the SlpA protein from *L. acidophilus* and *L. crispatus*, or with SlpH from *L. helveticus* (Fig. 13). Furthermore, different *L. brevis* S-layer proteins exhibit high similarity at their N-terminal region whereas their C-termini are quite divergent.[145]

As indicated earlier, the regions important for the cell wall binding and self-assembly are different between the S-layer proteins from *Lactobacilli*.[43,146,147] The C-terminal region of the SlpA protein from *L. acidophilus* ATCC 4356 was shown to be responsible for the cell wall binding and the N-terminal region for self-assembly.[43] For example, a C-terminal fragment containing one-third of the SlpA, which was fused to the green fluorescent protein, had the ability to bind to *L. acidophilus*, *L. helveticus*, and *L. crispatus* cells from which the S-layer was removed by treatment with 5M LiCl. Correspondingly, the N-terminal fragment containing two-thirds of SlpA could be recrystallized into mono- and multilayered 2D crystals.[43] The same arrangement of domains was reported for CbsA from *L. crispatus* JCM 5810.[148] Truncation studies revealed that the region consisting of aa 32–271 carries the information for the self-assembly of CbsA into a periodic structure.

In contrast, for SlpA S-layer protein from *L. brevis* ATCC 8287, the cell wall binding region was mapped to the first 145 amino acids of SlpA, as shown by cell wall binding assays using genetic truncation mutants.[146] It is currently not known to which cell wall component the N-terminal domain binds. One possible candidate, teichoic acids, can be ruled out. This SCWP is a major constituent of the cell wall of Gram-positive bacteria,[149] but when *L. brevis* was treated with an acid that usually hydrolyzes the linkage between the teichoic acids and the peptidoglycan, the SlpA lattice on the cell walls was not affected.[146] Truncation studies further showed that the C-terminal part of SlpA protein from *L. brevis* ATCC 8287 (aa 179–435) is important for the self-assembly.[146]

No atomic resolution structure of any *Lactobacillus* S-layer protein is available to date. In one report, the positions of individual amino acid residues were mapped to the surface of the *L. brevis* ATCC 8287 SlpA protein in the monomeric and assembled state.[150] Using the combined approach of cysteine scanning mutagenesis and targeted chemical modification, the steric accessibility of 46 single cysteine residues spread along the self-assembly domain of SlpA was probed using thiol-specific reagents. The analysis led to the

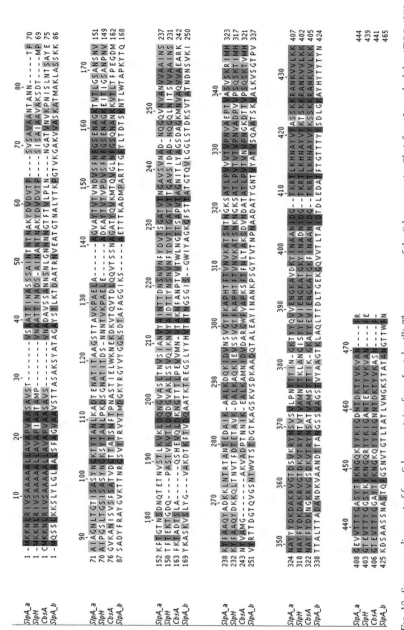

FIG. 13. Sequence alignment of four S-layer proteins from *Lactobacilli*. The sequences of the S-layer proteins SlpA from *L. acidophilus* ATCC 4356 (SlpA_a), SlpH from *L. helveticus* ATCC 12046 (SlpH), CbsA from *L. crispatus* JCM 5810 (CbsA), and SlpA from *L. brevis* ATCC 8287 (SlpA_b) aligned by ClustalW.[144] The residues are shaded based on their conservation: dark (conserved in all four proteins) to light (conserved in two sequences only).

```
              20                    40                    60                    80
MQSSLKKSLY LGLAALSFAG VAAVSTTASA KSYATAGAYS TLKTDAATRN VEATGTNALY TKPGTVKGAK VVASKATMAK
              100                   120                   140                   160
LASSKKSADY FRAYGVKTTN RGSVYYRVVT MDGKYRGYVY GGKSDTAFAG GIKSAETTTK ADMPARTTGF YLTDTSKNTL
              180                   200                   220                   240
WTAPKYTQYK ASKVSLYGVA KDTKFTVDQA ATKTREGSLY YHVTATNGSG ISGWIYAGKG FSTTATGTQV LGGLSTDKSV
              260                   280                   300                   320
TANNDNSVKI VYRTTDGTQV GSNTWVTSTD GTKAGSKVSD KAADQTALEA YINANKPSGY TVNNPNAADA TYGNTVYATV
              340                   360                   380                   400
SQAATSKVAL KVSGTPVTTA LTTADANDKV AANDTTANGS SVAGSTVYAA GTKLAQLTTD LTGEKGQVVT LTAIDTDLED
              420                   440                   460
ANFTGTTTYY SDLGKAYHYT YTYNKDSAAS SNASTQFGSN VTGTLTATLV MGKSTATANG TTWFN
```

Fig. 14. Solvent accessibility of mutated residues in the SlpA protein from *L. brevis* ATCC 8287 as reported in Ref. 150. All mutated residues are situated in the self-assembly domain and are shaded according to their assignment to four different groups: black background and white letters: outer surface; gray and white letters: pore/interface; gray and black letters: protein interior; and bold letters: inner surface. The signal peptide is underlined.

assignment of residues into four groups according to their location in the protein monomer and lattice structure: outer surface of the lattice (nine residues), inner surface of the lattice (9), protein interior (12), and protein–protein interface/pore regions (16)[150] (Fig. 14).

EM analysis of cell wall fragments or *in vitro* recrystallized S-layers from *L. brevis* 8287, *L. buchneri* ATCC 4005, *L. fermentum* F-4 (NCTC 7230), *L. helveticus* ATCC 12046, and *L. acidophilus* ATCC 4356 has been performed.[43,131,151–154] A regular array exhibiting tetragonal symmetry was reported for the S-layer of *L. fermentum* F-4 (NCTC 7230).[131] A hexagonal array with a center-to-center spacing of about 6 nm was found on the wall surface of freeze-etched *L. buchneri* ATCC 4005.[154] S-layers from *L. acidophilus* ATCC 4356 and *L. helveticus* ATCC 12046 show an oblique lattice[43,146,151] with very similar reported lattice dimensions: $a = 4.5$ nm, $b = 9.6$ nm, $\gamma = 77°$ for S-layer from *L. helveticus* ATCC 12046 and $a = 11.8$ nm, $b = 5.3$ nm, $\gamma = 102°$ for the S-layer from *L. acidophilus* ATCC 4356. The projection map of the latter could be calculated and compared with that of the C-terminally truncated SlpA protein (Fig. 15). The putative position of the C-terminal third of the protein could also be determined (Fig. 15). The lattice formed by the SlpA protein from *L. brevis* ATTC 8287 was first characterized as tetragonal.[152,153] Later, TEM studies of SlpA isolated from the wild-type ATCC 8287 cells and a recombinant SlpA showed an oblique lattice with a very similar unit cell dimensions $a = 9.39$ nm, $b = 6.10$ nm, $\gamma = 79.8°$ and $a = 10.38$ nm, $b = 6.36$ nm, $\gamma = 72.7°$.[146]

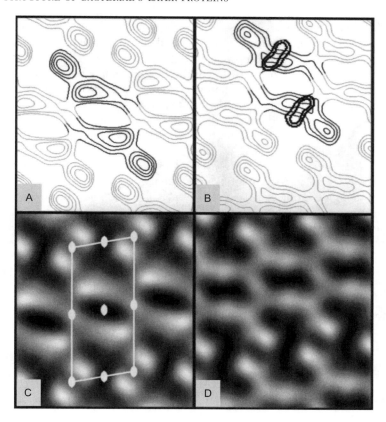

FIG. 15. Projection structures of the 2D crystals formed by full-length SlpA from *L. acidophilus ATCC 4356* (A, C) and the C-terminally truncated construct, SAN (B, D). One repeat unit containing two 2-fold related motifs is outlined (C). The difference density map (SlpA protein minus SAN) yields two peaks (in bold (B)), which indicate the positions of the C-terminal cell wall-binding domains. The unit cell $a = 118$ Å, $b = 53$ Å, $\gamma = 102°$ together with the $p2$ symmetry elements is drawn in (C). Figure adapted from Ref. 43.

V. *Bacillus anthracis*

B. anthracis is a Gram-positive, spore-forming bacterium that possesses a proteinaceous capsule.[155] It causes anthrax, a mostly lethal disease affecting both humans and animals. The cell wall of *B. anthracis* is surrounded by an S-layer that is composed of two proteins, Sap and EA1, and a poly-y-D-glutamic capsule.[24,156,157] The S-layer is found in both capsulated and non-capsulated vegetative *B. anthracis* cells. When present, the capsule is situated on the exterior of the cell and completely covers the S-layer. However, it does not require the S-layer for attachment to the cell surface.[158]

The Sap protein is composed of 814 residues, including a 29 amino acid signal peptide. The mature protein has a calculated molecular mass of 83.7 kDa and a pI of 6.[157] The EA1 protein is slightly longer, with 861 amino acids (including a 29-residue long signal peptide) and a calculated molecular mass of 88.5 kDa and a pI of 5.5 for the mature protein.[24] Both proteins possess three SLH motifs at their N-termini, whereas the rest of the sequences show only limited similarity (22% identity and 42% similarity). The SLH motifs of both proteins are sufficient for binding to the *B. anthracis* cell wall as C-terminal deletion constructs EA1$_{(32-213)}$ and Sap$_{(32-211)}$ bind to the extracted cell walls of *B. anthracis* with high affinity K_d values of 1.8×10^{-7} and 1.0×10^{-7} M, respectively.[159] The SLH motifs bind noncovalently to the pyruvylated SCWP[60] composed of galactose (Gal), N-acetylglucosamine, and N-acetylman-nose in an approximately 3:2:1 ratio.[160,161]

Recently, the structure of the SLH domain of the *B. anthracis* S-layer protein Sap has been solved to high resolution, confirming the importance of the conserved residues and giving a structure-based explanation for the mode of SCWP

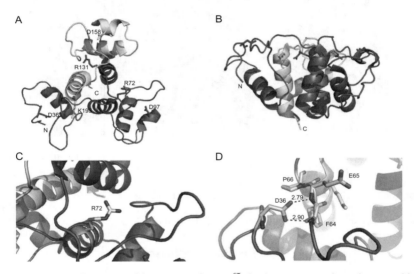

FIG. 16. Crystal structure of the Sap-SLH domain.[65] The three SLH motifs are shown as ribbon diagram and colored blue, red, and yellow, respectively. The N- and C-termini are labeled and the conserved aspartate residues (D36, D97, and D158) and basic residues (R72, R131, and K193) are shown in stick presentation. (A) View along the pseudo-threefold axis; (B) side view (A rotated by 90° around the horizontal axis); (C) R72 from the ITRAE motif of the first SLH motif is intercalated between the two lobes of the second SLH motif (red); (D) D36 of the first SLH motif represents the residue with the highest conservation in a comparison of all known SLH motifs. It forms a double hydrogen bond to the main chain of the opposite loop region (residues F64 and E65), thereby stabilizing the conformation of the interhelix loop of the SLH motif. (See Color Insert.)

binding.[65] In the crystal structure, the three SLH motifs form an internal trimer that adopts the shape of a three-pronged spindle (Fig. 16A and B). The C-terminal helices of the SLH motifs build a parallel three-helix bundle along the trimer axis, whereas the N-terminal helices form the three peripheral prongs and are positioned to interact with the C-terminal helix of the neighboring SLH motif. Long, structured loops connect the two helices, but the β-strand that has been predicted based on sequence comparison does not occur, at least not in the unbound form of the SLH structure. Interestingly, the SLH motifs are interdigitated in a way that the side chains of the essential arginine/lysine residues (R72, R131, K193) of the ITRAE motif point to the putative SCWP-binding groove of the adjacent SLH motif, where they contribute to trimer stabilization as well as to the positively charged surface patch, which is supposed to form the binding site for the negatively charged SCWP [65] (Fig. 16C). Another highly conserved residue, the aspartate at the N-terminal end of each SLH motif (D36, D97, and D158), also contributes to the trimer stability by forming a double H-bond to the main chain amides of the long interhelix loop (F64 and E65 in the first SLH motif; Fig. 16D), thus fixing the conformation of the loop that lines the putative SCWP-binding groove. The biological importance of these conserved residues has been proven by performing SCWP-binding assays with triple point mutants in which all three conserved aspartates (D36, D97, and D158) or the three conserved basic residues (R72, R131, K193) were replaced by alanines. Both triple mutants exhibited a strongly reduced SCWP-binding capacity as compared to the wild-type SLH.[65]

Construction of single and double Sap/EA1 mutants showed that Sap and EA1 are colocalized at the cell surface of *B. anthracis*. Only in the double mutant the S-layer does not form at the cell surface.[24,158] *In vitro* regulation studies of EA1 and Sap biosynthesis suggested that the proteins are produced sequentially, with Sap preceding EA1.[162] This developmental switch could not be observed *in vivo*. In fact, it was shown that, *in vivo*, the *B. anthracis* S-layer is mainly composed of EA1.[163]

As shown by EM of negatively stained S-layer fragments, both proteins are able to form a regular 2D crystalline array, but with different cell constants.[164] The calculated 20 Å resolution projection map of EA1 revealed *p1* symmetry and unit cell parameters of $a = 69$ Å, $b = 83$ Å, and $\gamma = 106°$ (Fig. 17A). The observed protein structure can be divided into four main domains: a major domain and three smaller domains, called head and wings (Fig. 17A, inset), forming a bird shape. The surface projections of a smooth and a corrugated side, obtained by combination of freeze fracture and freeze-etching, could be calculated. The corrugated side showed only one large globular feature, whereas the smooth side displayed two smaller domains (Fig. 17B and C). Comparison of the three projection maps suggested that the globular domain at the corrugated side corresponds to the major domain in the negative stain projection.[164]

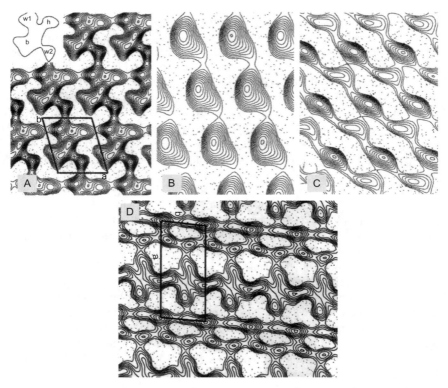

F<small>IG</small>. 17. Structural analysis of the EA1 and Sap S-layers of *B. anthracis*. Projection maps of (A) negatively stained EA1 array and (B) corrugated and (C) smooth side, obtained by freeze-etching. Inset in (A): Schematic representation of the protein mass as a flying bird: b—major domain, h—head, w1 and w2—wings. (D) 30 Å projection map of Sap array. The unit cells for both S-layers are indicated (A, D).[164]

The projection map of Sap could be calculated to 30 Å resolution, indicating a lower quality of the 2D arrays. The unit cell parameters of Sap layer were $a = 184$ Å, $b = 81$ Å, and $\gamma = 84°$ when $p1$ symmetry was applied (Fig. 17D), but the projection map also revealed a possible $p2$ symmetry. The domain boundaries of individual subunits could not be unambiguously defined.[164] The 604 amino acid C-terminal domain (Sap$_c$) obtained by digestion of Sap with proteinase K[57] starts immediately after the SLH motifs and was shown to constitute a crystallization domain.[40] Furthermore, negatively stained 2D crystals of His$_6$-Sap$_c$ attached onto a nickel-containing lipid film and diffracted to 18 Å resolution. The unit cell parameters were consistent with those obtained for full-length Sap protein. To determine the putative domains necessary for the self-assembly, a two-hybrid system was used.[40] A random library of polypeptides, derived from the Sap_c domain, was constructed and screened. Interestingly, it was shown that larger fragments with at least 155 amino acids are required to ensure efficient interaction. Based on these results, the Sap_c could be divided in an N-terminal, a central, and a C-terminal subdomain. A polypeptide containing amino acids 148–358, covering the complete central subdomain, was able to interact with itself, whereas both other domains interacted only with the whole Sap_c.[40]

VI. *Lysinibacillus sphaericus*

L. sphaericus is a spore-forming, Gram-positive, motile, rod-shaped insect pathogen. Four different *L. sphaericus* strains are known to carry an S-layer.[165–169] The properties of the S-layer proteins from the best studied pathogenic and nonpathogenic *L. sphaericus* strains are summarized in Table II.

The molecular weight of the mature S-layer proteins ranges between 122 and 129 kDa with a p*I* of approximately 5.0. The sequence comparison shows highest sequence homology between the S-layer proteins from *L. sphaericus* strains P-1 (T-layer), CCM 2177 (SbpA), NCTC 9602 (SlfA), and JG-A12 (SlfB). The S-layer proteins from the pathogenic strains *L. sphaericus* C3-41 (SlpC) and *L. sphaericus* 2362 are 100% identical but show only 30–37% sequence similarity to other *L. sphaericus* S-layer proteins (Table III). By comparison, a high sequence identity of 65% was found for the full-length proteins SbpA and SlfB. The highest sequence similarity between all listed S-layers is at the N-terminus, where three SLH motifs are located.[59,171] The central regions of SbpA, T-layer, SlfA, and SlfB are of lower homology. Only the C-terminal regions of SbpA and SlfB are more similar.[169]

The N-terminal region containing SLH motifs is responsible for attachment of the S-layer protein to the cell wall.[59,63] Similar to other Gram-positive bacteria, the SbpA S-layer protein binds to a pyruvylated SCWP, composed of GlcNAc, ManNAc, and pyruvic acid.[59] Characterization of the SCWP binding by SPR spectroscopy revealed the existence of three different binding sites for recombinant SbpA, showing high, medium, and low affinity.[63] Furthermore, in SPR analysis of C-terminally truncated SbpA, the N-terminal part containing three SLH motifs (amino acids 31–202) was not sufficient for SCWP binding. Only a construct comprising amino acids 31–318 was able to specifically bind to the SCWP.[63] The recently solved structure of the homologous SLH domain of the *B. anthracis* S-layer protein Sap accounts for this set of observations[65] (see Section V). The structure shows that each SLH motif contains two helices and that the C-terminal helices of the three SLH motifs together form a three-helix bundle. This three-helix bundle is central to the pseudo-trimeric arrangement

TABLE II
PROPERTIES OF *L. SPHAERICUS* S-LAYER PROTEINS

	Strain	Protein name	MM (kDa)	Amino acids	Signal peptide	p*I*	Accession code	Reference
Pathogenic	C3-41	SlpC	122.2	1176	1–30	5.0	ABQ00415	166
	2362	na	122.2	1176	1–30	5.0	AAA50256	167
Nonpathogenic	P-1	na	127.0	1252	1–30	4.8	CAA02847	170
	CCM 2177/ATCC 4525	SbpA	129	1268	1–30	4.7	AAF22978	168
	NCTC 9602	SlfA	126.6	1228	1–31	5.2	CAH61070	169
	JG-A12	SlfB	126.3	1238	1–31	5.1	CAH61072	169

na - not available.

TABLE III
SEQUENCE IDENTITY/SIMILARITY BETWEEN THE S-LAYER PROTEINS OF DIFFERENT *L. SPHAERICUS* STRAINS

SbpA	45%/55%				
SlfA	42%/54%	33%/44%			
SlfB	46%/57%	44%/54%	65%/71%		
SlpC	27%/37%	27%/37%	20%/30%	22%/33%	
2362[a]	27%/37%	27%/37%	20%/30%	22%/33%	100%
	T-layer[a]	SbpA	SlfA	SlfB	SlpC

Sequence alignment was performed by ClustalW [144] and the identities/similarities were calculated using the server SIAS (http://imed.med.ucm.es/Tools/sias.html).
[a]S-layer proteins from *L. sphericus* strains P-1 and 2362 are referred to as T-layer and 2362.

of the SLH motifs and appears to be essential for the stability and integrity of the SLH domain fold. Aligning the SbpA$_{(31-202)}$ sequence with the Sap-SLH sequence, which was used in the structure determination, provides a reason as to why the SbpA$_{(31-202)}$ is unable to bind to the cell wall.[63] In particular, the C-terminal helix of the third SLH motif is truncated, which most likely thwarts the folding of a stable SLH domain. Future experiments can confirm this hypothesis by incorporating the sequence up to A209 (i.e., SbpA$_{(31-209)}$) which—according to an alignment with the Sap-SLH structure—should complete the third SLH motif and hence exhibit SCWP-binding activity.

The secondary structure of the S-layer proteins from *L. sphaericus* strains NCTC 9602 and JG-A12, namely, SlfA and SlfB, was studied by FTIR (Fourier-transform infrared spectroscopy). It showed a similar secondary structure comprising of 35% β-sheets and a small amount of α-helical content.[169,172] The same was observed by circular dichrosim measurements for the full-length and truncations of SbpA.[63] A higher α-helical content was measured for the SLH motifs containing the fragment rSbpA$_{(31-202)}$, whereas the rest of the protein showed typical spectra of a β-sheet-like structure.

The self-assembly of SbpA *in vitro* was found to be strongly dependent on the presence of calcium ions and sodium chloride as well as on temperature and protein concentration.[59,173] Upon dialysis of the GdHCl-extracted SbpA against 10 mM EDTA or distilled water, mostly amorphous aggregates were observed, whereas in the presence of $CaCl_2$, SbpA reassembled in double-layered flat sheets up to 2 μm in size. It was also shown that the central region of the SbpA protein is involved in the self-assembly.[63] The native square lattice was still preserved when the last 213 amino acids and the N-terminal SLH region (aa 31–202) of SbpA were deleted. Deletion of the 350 amino acids from the C-terminus (aa 918–1268) induced a change to the oblique lattice, and after the deletion of the last 500 amino acids (aa 768–1268) only unstructured aggregates were observed.

The 3D structure of the negatively stained S-layer samples from *L. sphaericus* P-1 was determined by EM to a resolution of 2.5 nm.[165,174] The report suggested that the negative staining does not introduce major changes into the S-layer native structure, since the projections of the negatively stained and unstained (frozen-hydrated)[175] S-layers show similar features. This layer was also referred to as the T-layer,[68,165,174–176] and the *L. sphaericus* P-1 was originally identified as a *Bacillus brevis* and later as *Bacillus sphaericus* species.[177]

Recently, an EM structure with very similar features was reported for SbpA, the S-layer from *L. sphaericus* CCM 2177.[94] Furthermore, the 7 Å cryo-EM projection map could be calculated from the sugar-embedded SbpA lipid monolayer crystals.[74] A similar structure was also published for the S-layer from *S. ureae*.[178] The structure of SbpA shows *p4* symmetry with the lattice constant of ca. 13 nm and three putative domains were defined: the major (M), minor (m), and the arm (A) domain[174] (Fig. 18). Perpendicular cross-sections through the S-layer structure indicate that the structure can be divided into two halves: the first half containing the major domain and the second half containing the minor and the arm domains. The connection between the major, minor, and arm domains is still unclear. Norville and colleagues proposed two different possibilities: a compact, triangular form, or an extended, linear arrangement of the domains.[74] The inner and the outer surfaces of the S-layer show a rough and a smooth structure, respectively, [74] as generally observed for other crystalline arrays. The thickness of the S-layer was estimated to be between 6 and 9 nm. These values were obtained by different methods. Negative-stain EM reconstructions yielded values of 8[174] and 6 nm[94]; AFM 8 nm[179]; and X-ray reflectometry and grazing incidence 9 nm.[89] It was also observed that at least 25% of the area within the S-layer is occupied by water.[89]

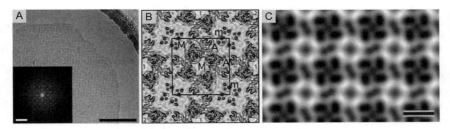

FIG. 18. Electron microscopy analysis of *L. sphaericus* CCM 2177 SbpA monolayer. (A) Frozen trehalose-embedded SbpA 2D crystals (scale bar = 100 nm) and (B) a corresponding *p4*-symmetrized 7 Å projection map. The Fourier transform is shown as inset in image (A) (scale bar = 6.5 nm⁻¹). A unit cell with a side length of $a = b = 133$ Å is outlined in black (B). The major (M), arm (A) and minor (m) domains of two SbpA subunits are labeled. Two putative domain arrangements are shown in light (compact organization) and dark (extended) gray circles. (C) Computer image reconstruction of scanning force microscopic images of the topography of the square (*p4*) S-layer lattice of *L. sphaericus* CCM 2177. Images adapted from Refs. 31,74.

In the absence of the atomic resolution structure, it is difficult to locate the secondary structure elements in the available EM structures, especially the position of the individual β-strands. The 7 Å cryo-EM projection map of SbpA indicates that the minor domain could adopt three α-helices positioned perpendicular to the plane of the lipid monolayer.[74] This suggests that the m-domain comprises essentially the SLH domain and that the C-terminal helices of the SLH motifs, which form the central three-helix bundle of the SLH domain,[65] are positioned next to the minor fourfold axis (Fig. 18B).

Recently, an "atomistic" model of the *p4* type SbpA lattice was generated by molecular modeling, SAXS, and TEM.[94] The authors of the study suggest that the tetramer is formed from monomers that are interlocked into one another and that the N-termini of SpbA are located on the inner surface responsible for anchoring to the cell wall. In this model, the C-termini are located on the outer surface of the S-layer, in line with previous AFM experiments on rSbpA-*Strep*-tag fusion proteins recrystallized on a silicon surface.[180]

VII. *Corynebacterium glutamicum*

S-layers also occur on the biotechnologically relevant *Corynebacterium glutamicum*. *C. glutamicum* belongs to a distinct group of bacteria that include pathogenic mycobacteria and nocardia which share an unusual structure of the cell wall as common feature. Despite being Gram-positive, the organization of the cell envelope resembles that of Gram-negative bacteria. In particular, the cell wall contains an outer lipid layer which is mainly composed of mycolic acid and is linked to the underlying thick arabinogalactan–peptidoglycan polymer meshwork,[181] as shown by cryo-EM of thin sections.[72]

The S-layer of *C. glutamicum* is composed of the PS2 protein which is encoded by the *cspB* gene. Genetic experiments with a deletion mutant followed by the reintroduction of the gene have unequivocally proven that PS2 forms the S-layer lattice.[182,183] The length of the PS2 protein depends on the particular bacterium. In a comparative analysis of 28 strains, the molecular masses ranged from 55 to 66 kDa and the protein length from 490 to 510 amino acids, as determined by sequencing of the corresponding genes.[184] The p*I* was between 4.11 and 4.27, whereas the content of hydrophobic amino acids at 50% was similar. Of note is a high proportion of alanine residues (18.5–19.8%), which are mainly localized in the N- and C-terminal regions of the PS2 protein. Independent of the particular S-layer variant, all PS2 proteins share a nearly identical 30 amino acid long N-terminus.[184] This signal sequence directs S-layer protein transport across the inner cell membrane via the ATP- and proton motive force-dependent Sec machinery pathway.[185] As confirmed by pulse chase experiments, a second transport step across the cell wall and the outer mycolate layer is independent of energy sources and faster than the first one.[184] The existence of a second specific translocation machinery has, however, not been proven.

Apart from the almost identical N-terminus, a C-terminal sequence of 21 mostly hydrophobic amino acids is also highly conserved as it anchors the protein into the outer membrane.[184] In some strains, the stretch of apolar amino acids encompasses the final 27 residues.[186] The anchoring of protein via the lipophilic C-terminal tail has been demonstrated in a series of experiments. For example, the S-layer could be detached from cells as organized sheets by using detergents at room temperature.[186] Furthermore, treatment of cells with a protease caused the gradual shortening of PS2 into a slightly lower molecular mass form. This was accompanied by shedding of patches of ordered surface layers from the cell into the medium,[183,186] consistent with the cleavage of the protein at the C-terminus, with the N-terminal remainder staying intact. Finally, genetically deleting the C-terminal stretch of 27 residues produced a protein that was unable to form an organized S-layer, as it did not bind to the outer membrane but was released into the medium.[186] Indeed, the position of the C-terminus could be mapped within the ultrastructure of the lattice by comparing images of full-length and C-terminally fragmented PS2 lattices[187] as described below.

The PS2 protein forms a $p6$ lattice which can be visualized by AFM imaging[184,187] or EM analysis of freeze-etched samples.[183,186] The unit cell dimensions of the best ordered lattice imaged at a lateral resolution of better than 1 nm are $a = b = 16.0 \pm 0.2$ nm and $\gamma = 60 \pm 1°$.[187] Other lattices from different strains feature unit cell dimension that range from 15.2 to 17.4 nm.[184] The thickness of the best studied S-layer was 4.6 ± 0.1 nm for the native form and 4.1 ± 0.1 nm for the proteolytically cleaved lattice lacking the C-terminus.[187] By comparing the averaged images for full (Fig. 19A and E) and fragmented S-layer (Fig. 19C and F), a difference map could be constructed that showed the position of the C-terminus (Fig. 19G). As images for both sides of the S-layer were taken, the AFM study calculated a 3D reconstruction of the S-layer.[187] AFM imaging was also combined with single-molecule force spectroscopy to examine the biomechanical properties of PS2 and understand the remarkable stability of the PS2 S-layer which requires detergent treatment at 70°C for solubilization of the protein monomer.[186] The structural resilience also helped AFM imaging of the lattice on living cells.[188]

VIII. *Deinococcus–Thermus*

Members of the *Deinococcus–Thermus* phylum include many species that are resistant to extreme radiation, as well as several thermophiles. They have been recognized solely on the basis of their branching patterns in 16S rRNA

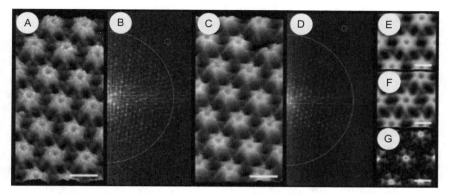

FIG. 19. High-resolution AFM topographs of native and digested S-layer from *C. glutamicum*. All images displayed are 15° tilted. (A) Cell-wall facing side of the native S-layer (scale bar = 15 nm; full gray scale: 3.5 nm). (B) Calculated power spectrum of the full image shown in (A). Large circle: 1 nm^{-1}; small circle: diffraction spot at subnanometer resolution. (C) Cell-wall facing side of the digested S-layer (scale bar = 15 nm; full gray scale: 3.0 nm). (D) Calculated power spectrum of the full image shown in (C). Large circle: 1 nm^{-1}; small circle: diffraction spot at subnanometer resolution. (E) Sixfold symmetrized cross-correlation average of image (A) (scale bar = 5 nm; full gray scale: 3.5 nm). (F) Sixfold symmetrized cross-correlation average of image (C) (scale bar = 5 nm; full gray scale 3.5 nm). (G) Sixfold symmetrized difference map calculated from averages (E) and (F) (scale bar = 5 nm; full gray scale: 1 nm).[187]

and other phylogenetic trees.[189] As characteristic feature, bacteria from the *Deinococcus–Thermus* group have thick cell walls and stain as Gram-positive bacteria, but structurally they are closer to Gram-negative bacteria given their second membrane.

A. *Deinococcus radiodurans*

The structure of the *Deinococcus radiodurans* cell envelope has been studied since the 1960s. It consists of the inner membrane, the periplasmic space, a peptidoglycan layer, an interstitial layer, the lipid-rich backing-layer, the S-layer, and a long-chain carbohydrate coat as the outermost layer.[190–193] The last four layers, together with carotenoids, form the so-called "pink envelope". Upon removal of the membrane-like backing-layer, the bacteria lose their typical shape, suggesting that this layer is responsible for the curvature of the bacteria.[194] The S-layer of the *D. radiodurans* contains at least two proteins: the Hpi protein that assembles into the hexagonally packed intermediate (HPI) layer, and SlpA, a homolog of an S-layer protein SlpA from *Thermus thermophilus*.[25]

1. Hpi

The Hpi protein consists of 1036 amino acids of which 17 belong to the potential signaling peptide. The mature protein has a molecular mass of 106 kDa and a p*I* of 4.8.[195,196] Interestingly, the Hpi protein contains seven Cys residues, which are rarely observed in S-layer proteins. These cysteine residues are possibly involved in the formation of three disulfide bonds (aa 74–86, 256–275, 642–754).[196] Another interesting feature is the existence of a Thr-rich region which lies between amino acids 71 and 247.

In addition to biochemical analysis, [197] the HPI layer has been structurally characterized by EM, [75,190,194,198–202] STM,[76,203] and AFM *in vitro* [80,204–206] and *in vivo*.[82] The isolated HPI layer exhibit *p*6 symmetry and unit cell dimensions of $a = b = 18$ nm.[76,197,199] This lattice is consistent with the hexagonal array observed on the bacteria in their native state, unperturbed by sample preparation.[82] The isolation of the HPI layer from the whole cells is achieved with 2% (w/v) SDS. [197,199] The complete dissociation of the HPI layer occurs in the presence of 2% (w/v) SDS and temperatures close to 100°C. In the presence of EDTA, the dissociation temperature is lowered to 60°C, indicating that divalent cations contribute to the stabilization of the layer.[197]

The secondary structure of the Hpi protein was estimated by infrared spectroscopy: 30% of the protein is composed of β-sheets/turns and the rest is a random coil. At temperatures between 25 and 60°C and in the presence of detergents, the secondary structure remained the same. Upon lowering the pH to 2.2, a dramatic change could be observed, suggesting a transition to a complete random coil structure. At this pH, the lattice structure is destroyed as well, indicating that the β-structures are important in maintaining the integrity of the HPI layer.[197]

3D reconstructions of HPI layer up to 1.8 nm resolution, using negatively stained,[199] rotary-shadowed, [207] as well as unstained, freeze-dried sample[208] have been reported. Furthermore, the projection structure to 0.8 nm could be obtained by cryo-EM of aurothioglucose-embedded HPI layer sheets.[75] The detailed characterization of both the inner and outer surfaces was achieved with AFM measurements.[80,204,205] Structural data obtained from all *in vitro* methods as well as *in vivo* AFM[82] agree well, confirming that the reconstructed structure of the HPI layer corresponds to that observed on the bacterial cell. According to this analysis, the HPI layer exhibits a core concentrated around the sixfold axis with a radius of 2.2 nm, a central pore of 2.5–3 nm in diameter, and fine spokes that connect adjacent units.[75,194,197,199,204] The core is surrounded by six relatively large openings showing threefold symmetry; the spokes exhibit a twofold symmetry (Fig. 20). Mass mapping experiments of the unstained, freeze-dried HPI layer by STEM have shown that about 85% of its total mass (MM = 655 kDa for a hexamer) is contained in the core.[76]

The hydrophobic inner surface of the S-layer lattice interacts with the outer membrane, while the outer surface is more hydrophilic.[199,205] When the inner surface was analyzed with AFM, pores with and without a central plug were found. By repeating the measurement after 5 min, switching from the open to the closed state was observed for individual pores[205] (Fig. 21). It is not known what induces the switching, although tip-induced conformational changes are ruled out. It was suggested that molecules up to 5 kDa could pass through the central 2.2 nm wide channel.[205]

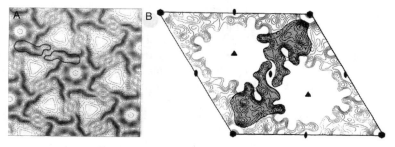

FIG. 20. Electron microscopy analysis of the HPI layer from *D. radiodurans*. (A) Central section of the 3D reconstruction from the negatively stained HPI layer. (B) The 8 Å projection map of one unit cell of the HPI layer obtained by cryo-EM. Putative subunit boundaries are outlined in both projections. Symmetry elements are marked in image (B).[75,199]

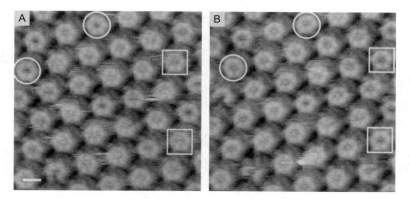

FIG. 21. Conformational changes of inner surface of HPI layer. (A) Some pores are in an open and others in an obstructed conformation. (B) Area shown in panel A imaged 5 min later. Several pores that were open in the first measurement are now closed (circles), while closed ones opened during the time interval (squares). Scale bar = 10 nm. (Figure from Ref. 205).

The thickness of the HPI layer was derived by different methods: STEM,[76] STM, [86,87] AFM, [204,206] and TEM.[199,207] The results vary between 5.7 and 8.6 nm for a single layer. The value of 6.9 ± 0.5 nm obtained by STEM data collected on unstained samples is likely compromised by radiation damage. Considering other methods and their reliability, it was concluded that the correct thickness is around 8 nm.[201,205]

2. SLPA

The existence of an additional protein involved in the maintenance and integrity of the S-layer in *D. radiodurans* was first reported by Cava *et al.* [62] SlpA (DR2577) is a proposed homolog of an S-layer protein from *T. thermophilus*, since both proteins contain one SLH motif at the N-terminus. SlpA comprises 1167 amino acids, of which the first 19 amino acids most likely constitute a signaling peptide.[209] The mature protein has a theoretical molecular mass of 121.8 kD, a theoretical p*I* of 4.9 and contains two cysteines.[209]

Genetic studies showed that deleting the *slpa* gene leads to a change in the structure of the cell envelope as well as in resistance to solvent and shear stress. By comparison deleting the *hpi* gene caused little alternations in the structure of the cell envelope or its reaction to shear- or solvent-induced stress.[25] This indicates that the SlpA protein may play an important role in the preservation of the cell envelope structure. Since the Hpi protein does not contain any SLH motifs, it was proposed that the SlpA protein could be responsible for the anchoring to the inner cell envelope.[25] The structure of the SlpA lattice has not been analyzed so far.

B. *Thermus thermophilus*

When the cell wall of *T. thermophilus* was studied by freeze-fracture EM, mostly amorphous, irregular surfaces resembling the outer membrane of other Gram-negative bacteria were uncovered along with some patches of highly ordered regular structure.[210] A treatment of the cells with EDTA before freeze-etching displayed a regular hexagonal array which was similar to the patches on the untreated cells. This array is composed of the SlpA (also called P100) protein with an apparent molecular mass of 100 kDa,[211] showing a hexagonal symmetry (*p6*) with unit cell dimensions $a = b = 24 \pm 2$ nm.[210]

SlpA is biochemically,[211,212] genetically,[213] and functionally[214] well characterized. The SlpA layer can be removed from the bacterial cell with ionic detergents such as SDS and Triton at an elevated temperature of 60 °C. [211] Purified SlpA forms Ca^{2+}-stabilized trimers, indicating that the trimers are the structural subunits forming the regular array.[215]

At the N-terminus of SlpA, one SLH motif could be identified by sequence comparison. A truncated form of the S-layer gene without the SLH motif could not attach to the bacterial envelope *in vivo*. In contrast, the assembly to regular arrays occurred in the absence of the peptidoglycan similar to the wild-type SlpA protein.[55] It was also shown that the pyruvylated SCWP is required for the binding of SlpA to the cell wall, and that the CsaB protein was required for SCWP pyruvylation.[62]

After isolating the SlpA protein from the membrane fractions, three different classes of 2D crystals were observed depending on the extraction procedure (see Table IV).[215] The SlpA protein was present in all three crystal forms, as shown by partial proteolysis and Western blotting.[215] The crystals obtained from cell envelopes of *T. thermophilus* by sequential extractions in the presence of 1% Triton X-100 and 50 mM EDTA showed a hexagonal shape. These so-called S1 crystals contained two layers with a total thickness of 35.6 nm. By comparison, the thickness of the native S-layer determined by thin sections and negative stain 3D reconstruction was around 18.2 and 11 nm, respectively. This indicates that the negatively stained sample was not completely embedded by the stain.[216] The overall structure showed that the main protein mass was concentrated around the sixfold symmetry axis and connected by arms with the smaller protein mass organized around a threefold symmetry axis (Fig. 22). The structural subunits were interpreted to be composed of three elongated domains: a 10-nm thick domain involved in the formation of the hexameric core and two other domains of 7.5 and 3 nm length which provide the connection within the subunits.[216] Comparing the negative stain 3D reconstruction with the freeze-fractured, freeze-etched, and metal-shadowed SlpA arrays indicated that the compact sixfold core is in contact with the underlying supporting membrane and the complex network domain presents itself toward the outer face of the layer.[210,216]

TABLE IV

DIFFERENT 2D CRYSTALS OBTAINED FROM THE S-LAYER PROTEIN SLPA OF *T. THERMOPHILUS*

	Extraction	Crystal description	Symmetry	Unit cell dimensions	Thickness
S1	2× Triton X-100 and EDTA	Hexagonal shape	$p6$ ($a=b$)	19.2 ± 0.2 nm[a] 24 ± 2 nm[b]	18.2 nm[c] 11 nm[a]
S2	2× Triton X-100	Rectangular crystals	$p2$	$a = 24.1 \pm 0.2$ nm $b = 21 \pm 0.2$ nm	10 nm[c] 6.3 nm[a]
pS2	(1) Triton X-100 (2) Triton X-100 and EDTA	Tubular crystals with irregular contours	$p3$ ($a=b$)	18 ± 0.3 nm	10 nm[c] 3.4 nm[a]

Values obtained from [a]negative stain 3D-EM reconstruction, [b]freeze-fracture shadowing, and [c]thin sections.

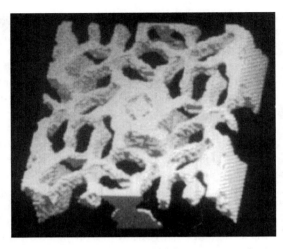

Fig. 22. Structure of the SlpA layer of *T. thermophilus* as present in the S1 crystals. The image displays the volume representation of the 3D reconstruction of the negatively stained S1 crystals, with the outer surface facing up.[216]

Rectangular crystals, or S2 crystals, with a *p2* point symmetry were obtained after two sequential extractions with 1% Triton X-100. S2 crystals showed a continuous organization without defined domain borders. The thickness of the layer was estimated to be 6.3 nm, which is less than the thickness of 10 nm estimated from the thin sections.[216]

Tubular crystals with irregular contours, pS2 crystals, were observed after solubilizing the cell envelopes with 1% Triton X-100 in the first extraction and 1% Triton X-100 and 50 mM EDTA in the second. The thickness of the reconstruction was around 3.4 nm, indicating that a large portion of pS2 crystals was not embedded by the stain. Prominent triplets of channels perpendicular to the membrane plane could be observed.[216]

It was suggested that the S1 crystals with *p6* symmetry represent the native S-layer detected by freeze-etching on whole cells.[210,216] The S2 and pS2 crystals with morphology similar to that of bacterial porins[215] could not be detected on the cell of the *T. thermophilus* grown under different osmotic conditions.[216]

IX. S-Layers from Gram-Negative Bacteria

No atomic resolution structure of S-layers from Gram-negative bacteria is reported to date. Here we discuss S-layers for which their 3D EM structures are available. The S-layers are classified based on their symmetry and belong either to the hexagonal (*p6*) or the tetragonal (*p4*) group. Within the S-layers belonging to the *p6* symmetry group we cover the S-layer from *Caulobacter crescentus* as representative example. The S-layers belonging to a *p4* symmetry group are discussed together.

A. S-Layer Proteins from Gram-Negative Bacteria with Hexagonal Symmetry

Structures of several *p6* S-layer assemblies from the following Gram-negative bacteria have been reconstructed by negative stain EM: *C. crescentus*,[73] *Aquaspirillum serpens*,[217,218] *Wollinela recta*,[219] and *Bacteroides buccae*.[21] Here we focus on the best studied system in this group, the S-layer from *C. crescentus*.

C. crescentus is a rod-shaped oligotrophic bacterium playing an important role in the carbon cycle.[220,221] The S-layer of *C. crescentus* is composed of a single protein, RsaA, which self-assembles into a hexagonally arranged lattice.[73,222] RsaA contains 1026 amino acids (98 kDa), has no N-terminal signal leader peptide[223] and is secreted by a type I mechanism to the cell surface relying on the uncleaved C-terminal secretion signal.[45,224–227] The same mechanism is reported for the S-layers from *Campylobacter fetus*[228] and *Serratia marcescens*.[229] Deletion of the N-terminal part of RsaA or linker-peptide insertions of 4–6 amino acids in the N-terminus of RsaA disabled the binding of the protein to the cell surface and its crystallization into regular arrays.[45,224] Binding of RsaA to the outer membrane occurs via noncovalent interactions with a specific smooth LPS that is modified with an *O*-polysaccharide.[230,231] Mutant strains missing smooth LPS were unable to attach the S-layer to the cell wall. Instead, the crystalline S-layer arrays were found adjacent to cells.[230] The region of the first 225 amino acids was found to be responsible for binding to the smooth LPS. This was shown by an assay evaluating the reattachment of small RsaA fragments onto the S-layer negative cells.[232] Following the initial observation that another S-layer protein, SapA from *C. fetus*, attaches specifically with its N-terminus to the smooth LPS only in the presence of divalent cations,[233,234] a related requirement for Ca^{2+} ions to achieve self-assembly and recrystallization was also described for the RsaA protein under different extraction conditions such as low pH or addition of EDTA.[230,235] High quality long-range-ordered self-assemblies could be obtained only in the presence of Ca^{2+} ions in combination with lipid vesicles containing smooth LPS. Omitting the latter yielded crystals of lower quality.[236] The relevance of calcium suggests that the cation is involved in interactions between the monomers, possibly within the calcium-binding RTX (repeat in toxin) motifs located mostly at the C-termini.[223,236] These glycine- and aspartate-rich nonapeptide repeats are involved in the binding of calcium in some enzymes and enable the proper folding and biological activity.[223] The contribution of the further domain(s) in subunit–subunit interactions remains unknown.

The periodic S-layer structure on the cell surface of *C. crescentus* CB15 was studied by different EM methods: negative staining, thin-sectioning, freeze-etching,[73,222] and electron tomography.[77] A 3D reconstruction was calculated from the natural crystal patches containing S-layer and material originating from the outer membrane.[73] Removing the membrane vesicles by detergent (deoxycholate) treatment rearranged the S-layer to double-layer sheets. Since the resolution was close to 2.1 nm for both, the native patches and the purified material, the untreated sample was preferred for the reconstruction.[73]

FIG. 23. Horizontal sections through the 3D reconstruction of RsaA, the S-layer protein from *C. crescentus*, obtained from a complete, continuous tilt series, sampled at 0.15-nm spacings. The sections, starting from top left (outer surface) to bottom right (inner surface), are displayed with a distance of 0.6 nm. Figure from Ref. 73.

The 3D structure of RsaA exhibits a hexagonal core consisting of six monomers forming a pore with a diameter of approximately 2.5–3.5 nm. The distance between the core regions is 22 nm. Arms connect one monomer with another one from the adjacent unit cell. The horizontal sections suggest that the main contact between the neighboring subunits is around the threefold axis and a minor one around the twofold axis (Fig. 23). In the direction perpendicular to the plane of the 2D array, the structure can be divided into three parts: the first containing only the arms around the threefold axis; the second containing both arms and the core region; and the third part containing only the core. The thickness of the reconstruction is close to 7 nm.[73] The authors of the study speculated that the putative monomer of the S-layer is an elongated and twisted molecule.

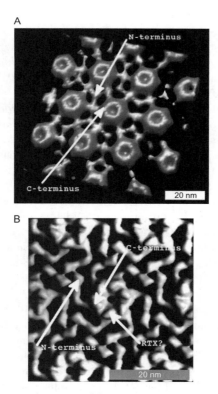

FIG. 24. Three-dimensional structure of the RsaA array of *C. crescentus*. (A) Isosurface of the RsaA hexagon obtained by electron tomography. The N-termini of RsaA monomers are arranged around the threefold and the C-termini around the sixfold symmetry axes. (B) 3D reconstruction of the negatively stained RsaA 2D crystals. The putative location of the RTX-motifs within the U-shaped RsaA monomers is indicated. Figure from Ref. 77.

The structure and organization of the RsaA protein within the lattice was assessed directly on the growing cells by cryo-electron tomography.[77] A short-range order of hexagonally organized subunits and areas of stacked double layers, especially when RsaA was overexpressed, were observed. Combining the information obtained from high-resolution amino acid residue-specific Nanogold labeling and the subtomogram averaging with the 3D EM reconstruction obtained by negative stain technique, positions of the N- and C-termini could be determined. It was proposed that the U-shaped RsaA monomer proceeds from the N-terminus, forming the threefold axis, to the C-terminus forming the sixfold symmetry axis[77] (Fig. 24).

B. S-Layer Proteins from Gram-Negative Bacteria with Tetragonal Symmetry

3D EM structures of tetragonal S-layer-arrays from Gram-negative bacteria to a maximum resolution of 1.3 nm were reported from the fish pathogens [11,237] *Aeromonas salmonicida A449-TM5* [238–240] and *Aeromonas hydrophila TF7* [241,242] as well as from the nonpathogenic *Azobacter vinelandii UW1* [243] and *Delftia acidovorans* ATCC 15668.[244,245]

On the sequence level, only S-layers from *A. salmonicida* and *A. hydrophila* have been characterized. The S-layer protein from *A. salmonicida*, VapA, is composed of 502 amino acid residues with the first 21 comprising the signal peptide. The mature protein has a theoretical molecular mass of 50.8 kDa and a p*I* of 5.1.[246] The S-layer protein AhsA from *A. hydrophila* consists of 472 amino acids (signal peptide 1–22 aa), and has a molecular mass of 45.4 kDa and a p*I* of 6.6.[247] SDS-PAGE analysis of AhsA suggested that the protein is larger (52 kDa) with a lower p*I* of 4.60.[248] This discrepancy was later explained by the posttrans-lational modification of the AhsA protein via phosphorylation of tyrosine.[247] Only 17% sequence identity and 29% similarity were found between VapA and AhsA, respectively. This reflects the general feature of S-layers as a very diverse group of proteins, even between related bacterial species.

Circular dichroism measurements suggest that AhsA, VapA, and the S-layer protein from *A. vinelandii* consist mainly of β-sheets (44%, 51%, 35%) and a smaller amount of α-helices (19%, 14%, and 4%), respectively.[242,248–250]

The reported 3D reconstructions of VapA, AhsA, and the S-layer protein from *A. vinelandii* show a similar structural arrangement, as suggested by comparison of the projection maps (Fig. 25). It has to be mentioned that, for all reconstructions compared here, S-layer samples devoid of the outer membrane material were used. A similar architecture is also observed for the S-layer of the *Pseudomonas*-like strain EU2 (Fig. 25A).[251] The major protein mass is situated at one of the fourfold axes, and a minor part at the second fourfold symmetry axis. The connection between the monomers occurs at the major

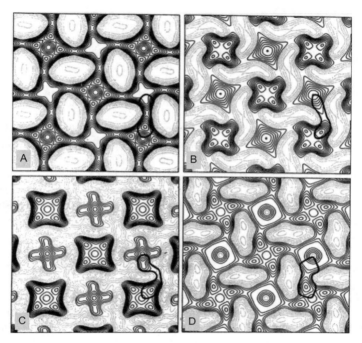

FIG. 25. Comparison of the projected structures obtained by EM of different Gram-negative S-layers with tetragonal symmetry. (A) *Pseudomonas*-like strain EU2, (B) *A. hydrophila*, (C) *A. vinelandii*, and (D) *A. salmonicida*. From Ref. 251.

fourfold axis of the S-layer. Two tetrameric units are related by a twofold symmetry axis.[240–243,251] The lattice constants of these S-layers are between 11.3 and 13.6 nm and the thickness is 5–5.6 nm.

Stewart *et al.* observed an interesting feature of VapA arrays.[238] Two different conformations, defined as type I (closed) and type II (open), were detected in the fragments dissociated from the cell surface. Open conformation of the S-layer was found in the double-layered sheets, mostly in the S-layer material released by the mutant cells lacking LPS.[240] The closed conformation was observed mostly on intact cells, suggesting that the presence/absence of the LPS is important for the conformational change and attachment of the S-layer to the bacterial surface. Furthermore, a difference in lattice constants of 1 nm between the two conformations was reported.[238] A similar observation was made for AhsA.[243]

In another study, it was shown that the C-terminal truncated AhsA (38.6 kDa) is exported to the cell surface but does not adhere to it, nor does it form a tetragonal array. Instead, cup-like structures were observed.[46] Similar results were found for the 39.4 kDa N-terminal tryptic fragment of VapA.[47] This indicates that the C-terminus is important for both self-assembly and attachment to the cell surface. It was also suggested that the N-terminal part of both S-layer proteins is important for tetramer formation within the major core of the surface array.[46,47]

Presence of divalent cations, especially Ca^{2+}, has been shown to be an important factor in the assembly process of S-layers. Detailed analysis confirmed the influence of the divalent cations on assembly of VapA and the S-layer from *A. vinelandii*.[252,253] Ring-shaped units were observed in the absence of Ca^{2+}, representing a tetrameric subunit. Only in the presence of Ca^{2+}, native crystalline arrays were observed, indicating that Ca^{2+} is involved in the interactions between the two tetrameric subunits.[254] The same was shown for the assembly of S-layer protein subunits of *A. vinelandii in vivo*.[253] Furthermore, the presence of Ca^{2+} enhances the reattachment of VapA protein to LPS O-polysaccharide.[254]

The S-layer from *Delftia acidovorans* ATCC 15668 differs structurally from other tetragonal S-layers of Gram-negative bacteria (Fig. 26). Indeed, it was also speculated whether the arrays should be classified as a true S-layer or rather as a regularly arrayed outer membrane protein.[244] 3D reconstructions of S-layer with and without native outer membrane material show the same features in the central region and some minor differences near the inner and outer surfaces.[244,245] The pattern consists of massive identical subunits, composed presumably of two monomers tightly connected through the twofold axis (Fig. 26). These dimeric subunits are arranged in a so-called cobblestone pattern with one pore sitting on the first fourfold axis and another pore sitting

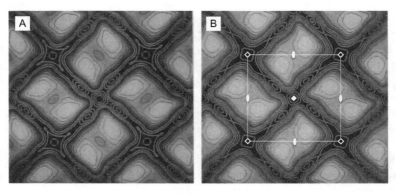

FIG. 26. Projection structure of the tetragonal S-layer from *Delftia acidovorans* ATCC 15668. Similar structures were obtained from the negatively stained samples obtained from (A) the outer membrane and (B) from the recrystallized purified protein. The unit cell and the symmetry elements are indicated in (B). Figure adapted from Ref. 245.

on the second fourfold axis.[244,245,255] The horizontal sections of the 3D reconstruction show little interaction between the dimeric subunits, which is in accordance with the observation that the S-layer tends to disintegrate easily.[245] The square unit cell has a lattice constant of 10.5–11 nm and a thickness of 3.5–5 nm.[244,245,255]

X. Concluding Remarks

To date, the 3D structures of S-layers from a variety of bacterial species have been determined by EM techniques to a resolution range of 1.6–3 nm, providing information on their shape, symmetry, and thickness. In addition, AFM methods have been developed and applied to visualize the S-layers in their *in vivo* environment, that is, attached to the intact bacterial cell. What is still missing is the high-resolution structure of assembled S-layers that can provide information on the spatial arrangement of residues, which is required to understand the assembly process, the structure of the inner and outer surface, and the architecture and the possible functionality of the pores. Progress toward this end has been made very recently by elucidating the crystal structures of soluble truncated versions of S-layer proteins. In future, it will be necessary to determine the structures of the missing parts or domains of the respective S-layer proteins in order to complete the high-resolution structures. The major breakthrough in the field of S-layer research will be the combination of the high-resolution partial structures with the 3D reconstruction of assembled S-layers.

ACKNOWLEDGMENTS

We profusely thank Dr. David Albesa-Jove for providing the SAXS data for the *C. difficile* S-layer and Dr. Monika Oberer for reading and critically discussing the chapter. We greatfully acknowledge the funding by the Austrian Science Fund (FWF): P17885-N11.

REFERENCES

1. Sleytr UB, Beveridge TJ. Bacterial S-layers. *Trends Microbiol* 1999;**7**:253–60.
2. Baumeister W, Lembcke G. Structural features of archaebacterial cell envelopes. *J Bioenerg Biomembr* 1992;**24**:567–75.
3. Sleytr UB, Messner P, Pum D, Sára M. Crystalline bacterial cell surface layers (S layers): from supramolecular cell structure to biomimetics and nanotechnology. *Angew Chem Int Ed Engl* 1999;**38**:1035–54.
4. Engelhardt H. Are S-layers exoskeletons? The basic function of protein surface layers revisited. *J Struct Biol* 2007;**160**:115–24.

5. Claus H, Akca E, Debaerdemaeker T, Evrard C, Declercq JP, Harris JR, et al. Molecular organization of selected prokaryotic S-layer proteins. *Can J Microbiol* 2005;**51**:731–43.

6. König H. Archaeobacterial cell envelopes. *Can J Microbiol* 1988;**34**:395–406.

7. Avall-Jaaskelainen S, Palva A. Lactobacillus surface layers and their applications. *FEMS Microbiol Rev* 2005;**29**:511–29.

8. Bahl H, Scholz H, Bayan N, Chami M, Leblon G, Gulik-Krzywicki T, et al. Molecular biology of S-layers. *FEMS Microbiol Rev* 1997;**20**:47–98.

9. Engelhardt H. Mechanism of osmoprotection by archaeal S-layers: a theoretical study. *J Struct Biol* 2007;**160**:190–9.

10. Messner P, Steiner K, Zarschler K, Schäffer C. S-layer nanoglycobiology of bacteria. *Carbohydr Res* 2008;**343**:1934–51.

11. Noonan B, Trust TJ. The synthesis, secretion and role in virulence of the paracrystalline surface protein layers of *Aeromonas salmonicida* and *A. hydrophila*. *FEMS Microbiol Lett* 1997;**154**:1–7.

12. Schäffer C, Messner P. The structure of secondary cell wall polymers: how Gram-positive bacteria stick their cell walls together. *Microbiology* 2005;**151**:643–51.

13. Sleytr UB, Egelseer EM, Ilk N, Pum D, Schuster B. S-Layers as a basic building block in a molecular construction kit. *FEBS J* 2007;**274**:323–34.

14. Thompson SA. Campylobacter surface-layers (S-layers) and immune evasion. *Ann Periodontol* 2002;**7**:43–53.

15. Messner P, Sleytr UB. Bacterial surface layer glycoproteins. *Glycobiology* 1991;**1**:545–51.

16. Messner P, Allmaier G, Schäffer C, Wugeditsch T, Lortal S, König H, et al. Biochemistry of S-layers. *FEMS Microbiol Rev* 1997;**20**:25–46.

17. Moens S, Vanderleyden J. Glycoproteins in prokaryotes. *Arch Microbiol* 1997;**168**:169–75.

18. Schäffer C, Messner P. Glycobiology of surface layer proteins. *Biochimie* 2001;**83**:591–9.

19. Upreti RK, Kumar M, Shankar V. Bacterial glycoproteins: functions, biosynthesis and applications. *Proteomics* 2003;**3**:363–79.

20. Schäffer C, Messner P. Surface-layer glycoproteins: an example for the diversity of bacterial glycosylation with promising impacts on nanobiotechnology. *Glycobiology* 2004;**14**:31R–42R.

21. Sjogren A, Hovmoller S, Farrants G, Ranta H, Haapasalo M, Ranta K, et al. Structures of two different surface layers found in six Bacteroides strains. *J Bacteriol* 1985;**164**:1278–82.

22. Messner P, Sleytr UB. Crystalline bacterial cell-surface layers. *Adv Microb Physiol* 1992;**33**:213–75.

23. Smith SH, Murray RG, Hall M. The surface structure of *Leptotrichia buccalis*. *Can J Microbiol* 1994;**40**:90–8.

24. Mesnage S, Tosi-Couture E, Mock M, Gounon P, Fouet A. Molecular characterization of the *Bacillus anthracis* main S-layer component: evidence that it is the major cell-associated antigen. *Mol Microbiol* 1997;**23**:1147–55.

25. Rothfuss H, Lara JC, Schmid AK, Lidstrom ME. Involvement of the S-layer proteins Hpi and SlpA in the maintenance of cell envelope integrity in *Deinococcus radiodurans* R1. *Microbiology* 2006;**152**:2779–87.

26. Calabi E, Ward S, Wren B, Paxton T, Panico M, Morris H, et al. Molecular characterization of the surface layer proteins from *Clostridium difficile*. *Mol Microbiol* 2001;**40**:1187–99.

27. Sára M, Pum D, Küpcü S, Messner P, Sleytr UB. Isolation of two physiologically induced variant strains of *Bacillus stearothermophilus* NRS 2004/3a and characterization of their S-layer lattices. *J Bacteriol* 1994;**176**:848–60.

28. Sára M, Kuen B, Mayer HF, Mandl F, Schuster KC, Sleytr UB. Dynamics in oxygen-induced changes in S-layer protein synthesis from *Bacillus stearothermophilus* PV72 and the S-layer-deficient variant T5 in continuous culture and studies of the cell wall composition. *J Bacteriol* 1996;**178**:2108–17.

29. Scholz HC, Riedmann E, Witte A, Lubitz W, Kuen B. S-layer variation in *Bacillus stearothermophilus* PV72 is based on DNA rearrangements between the chromosome and the naturally occurring megaplasmids. *J Bacteriol* 2001;**183**:1672–9.

30. Egelseer EM, Danhorn T, Pleschberger M, Hotzy C, Sleytr UB, Sára M. Characterization of an S-layer glycoprotein produced in the course of S-layer variation of *Bacillus stearothermophilus* ATCC 12980 and sequencing and cloning of the sbsD gene encoding the protein moiety. *Arch Microbiol* 2001;**177**:70–80.

31. Sleytr UB, Huber C, Ilk N, Pum D, Schuster B, Egelseer EM. S-layers as a tool kit for nanobiotechnological applications. *FEMS Microbiol Lett* 2007;**267**:131–44.

32. Hovmöller S, Sjogren A, Wang DN. The structure of crystalline bacterial surface layers. *Prog Biophys Mol Biol* 1988;**51**:131–63.

33. Baumeister W, Wildhaber I, Engelhardt H. Bacterial surface proteins. Some structural, functional and evolutionary aspects. *Biophys Chem* 1988;**29**:39–49.

34. Baumeister W, Wildhaber I, Phipps BM. Principles of organization in eubacterial and archaebacterial surface proteins. *Can J Microbiol* 1989;**35**:215–27.

35. Saxton WO, Baumeister W. Principles of organization in S layers. *J Mol Biol* 1986;**187**:251–3.

36. Beveridge TJ. Bacterial S-Layers. *Curr Opin Struct Biol* 1994;**4**:204–12.

37. Sára M, Sleytr UB. Crystalline bacterial cell surface layers (S-layers): from cell structure to biomimetics. *Prog Biophys Mol Biol* 1996;**65**:83–111.

38. Sára M, Sleytr UB. S-Layer proteins. *J Bacteriol* 2000;**182**:859–68.

39. Engelhardt H, Peters J. Structural research on surface layers: a focus on stability, surface layer homology domains, and surface layer-cell wall interactions. *J Struct Biol* 1998;**124**:276–302.

40. Candela T, Mignot T, Hagnerelle X, Haustant M, Fouet A. Genetic analysis of Bacillus anthracis Sap S-layer protein crystallization domain. *Microbiology* 2005;**151**:1485–90.

41. Egelseer EM, Leitner K, Jarosch M, Hotzy C, Zayni S, Sleytr UB, et al. The S-layer proteins of two *Bacillus stearothermophilus* wild-type strains are bound via their N-terminal region to a secondary cell wall polymer of identical chemical composition. *J Bacteriol* 1998;**180**:1488–95.

42. Jarosch M, Egelseer EM, Mattanovich D, Sleytr UB, Sára M. S-layer gene sbsC of *Bacillus stearothermophilus* ATCC 12980: molecular characterization and heterologous expression in *Escherichia coli*. *Microbiology* 2000;**146**:273–81.

43. Smit E, Oling F, Demel R, Martinez B, Pouwels PH. The S-layer protein of *Lactobacillus acidophilus* ATCC 4356: identification and characterisation of domains responsible for S-protein assembly and cell wall binding. *J Mol Biol* 2001;**305**:245–57.

44. Sára M, Dekitsch C, Mayer HF, Egelseer EM, Sleytr UB. Influence of the secondary cell wall polymer on the reassembly, recrystallization, and stability properties of the S-layer protein from *Bacillus stearothermophilus* PV72/p2. *J Bacteriol* 1998;**180**:4146–53.

45. Bingle WH, Nomellini JF, Smit J. Linker mutagenesis of the *Caulobacter crescentus* S-layer protein: toward a definition of an N-terminal anchoring region and a C-terminal secretion signal and the potential for heterologous protein secretion. *J Bacteriol* 1997;**179**:601–11.

46. Thomas S, Austin JW, McCubbin WD, Kay CM, Trust TJ. Roles of structural domains in the morphology and surface anchoring of the tetragonal paracrystalline array of *Aeromonas hydrophila*. Biochemical characterization of the major structural domain. *J Mol Biol* 1992;**228**:652–61.

47. Doig P, McCubbin WD, Kay CM, Trust TJ. Distribution of surface-exposed and non-accessible amino acid sequences among the two major structural domains of the S-layer protein of *Aeromonas salmonicida*. *J Mol Biol* 1993;**233**:753–65.

48. Schneider TD, Stephens RM. Sequence logos: a new way to display consensus sequences. *Nucleic Acids Res* 1990;**18**:6097–100.

49. Schuster-Bockler B, Schultz J, Rahmann S. HMM Logos for visualization of protein families. *BMC Bioinformatics* 2004;**5**:7.

50. Finn RD, Mistry J, Tate J, Coggill P, Heger A, Pollington JE, et al. The Pfam protein families database. *Nucleic Acids Res* 2010;**38**:D211–22.
51. Fujino T, Beguin P, Aubert JP. Organization of a *Clostridium thermocellum* gene cluster encoding the cellulosomal scaffolding protein CipA and a protein possibly involved in attachment of the cellulosome to the cell surface. *J Bacteriol* 1993;**175**:1891–9.
52. Lupas A, Engelhardt H, Peters J, Santarius U, Volker S, Baumeister W. Domain structure of the *Acetogenium kivui* surface layer revealed by electron crystallography and sequence analysis. *J Bacteriol* 1994;**176**:1224–33.
53. Matuschek M, Burchhardt G, Sahm K, Bahl H. Pullulanase of *Thermoanaerobacterium thermosulfurigenes* EM1 (*Clostridium thermosulfurogenes*): molecular analysis of the gene, composite structure of the enzyme, and a common model for its attachment to the cell surface. *J Bacteriol* 1994;**176**:3295–302.
54. Lemaire M, Ohayon H, Gounon P, Fujino T, Beguin P. OlpB, a new outer layer protein of *Clostridium thermocellum*, and binding of its S-layer-like domains to components of the cell envelope. *J Bacteriol* 1995;**177**:2451–9.
55. Olabarria G, Carrascosa JL, de Pedro MA, Berenguer J. A conserved motif in S-layer proteins is involved in peptidoglycan binding in *Thermus thermophilus*. *J Bacteriol* 1996;**178**:4765–72.
56. Ries W, Hotzy C, Schocher I, Sleytr UB, Sára M. Evidence that the N-terminal part of the S-layer protein from *Bacillus stearothermophilus* PV72/p2 recognizes a secondary cell wall polymer. *J Bacteriol* 1997;**179**:3892–8.
57. Mesnage S, Tosi-Couture E, Fouet A. Production and cell surface anchoring of functional fusions between the SLH motifs of the *Bacillus anthracis* S-layer proteins and the *Bacillus subtilis* levansucrase. *Mol Microbiol* 1999;**31**:927–36.
58. Brechtel E, Bahl H. In *Thermoanaerobacterium thermosulfurigenes* EM1 S-layer homology domains do not attach to peptidoglycan. *J Bacteriol* 1999;**181**:5017–23.
59. Ilk N, Kosma P, Puchberger M, Egelseer EM, Mayer HF, Sleytr UB, et al. Structural and functional analyses of the secondary cell wall polymer of *Bacillus sphaericus* CCM 2177 that serves as an S-layer-specific anchor. *J Bacteriol* 1999;**181**:7643–6.
60. Mesnage S, Fontaine T, Mignot T, Delepierre M, Mock M, Fouet A. Bacterial SLH domain proteins are non-covalently anchored to the cell surface via a conserved mechanism involving wall polysaccharide pyruvylation. *EMBO J* 2000;**19**:4473–84.
61. Mader C, Huber C, Moll D, Sleytr UB, Sára M. Interaction of the crystalline bacterial cell surface layer protein SbsB and the secondary cell wall polymer of *Geobacillus stearothermophilus* PV72 assessed by real-time surface plasmon resonance biosensor technology. *J Bacteriol* 2004;**186**:1758–68.
62. Cava F, de Pedro MA, Schwarz H, Henne A, Berenguer J. Binding to pyruvylated compounds as an ancestral mechanism to anchor the outer envelope in primitive bacteria. *Mol Microbiol* 2004;**52**:677–90.
63. Huber C, Ilk N, Runzler D, Egelseer EM, Weigert S, Sleytr UB, et al. The three S-layer-like homology motifs of the S-layer protein SbpA of *Bacillus sphaericus* CCM 2177 are not sufficient for binding to the pyruvylated secondary cell wall polymer. *Mol Microbiol* 2005;**55**:197–205.
64. May A, Pusztahelyi T, Hoffmann N, Fischer RJ, Bahl H. Mutagenesis of conserved charged amino acids in SLH domains of *Thermoanaerobacterium thermosulfurigenes* EM1 affects attachment to cell wall sacculi. *Arch Microbiol* 2006;**185**:263–9.
65. Kern J, Wilton R, Zhang R, Binkowski TA, Joachimiak A, Schneewind O. Structure of the SLH domains from *Bacillus anthracis* surface array protein. *J Biol Chem* 2011;**286**:26042–9.
66. Sotiropoulou S, Mark SS, Angert ER, Batt CA. Nanoporous S-layer protein lattices. A biological ion gate with calcium selectivity. *J Phys Chem* 2007;**111**:13232–7.

67. Chung S, Shin SH, Bertozzi CR, De Yoreo JJ. Self-catalyzed growth of S layers via an amorphous-to-crystalline transition limited by folding kinetics. *Proc Natl Acad Sci USA* 2010;**107**:16536–41.

68. Smith PR, Kistler J. Surface reliefs computed from micrographs of heavy metal-shadowed specimens. *J Ultrastruct Res* 1977;**61**:124–33.

69. Moor H, Muhlethaler K. Fine structure in frozen-etched yeast cells. *J Cell Biol* 1963;**17**:609–28.

70. Remsen CC, Watson SW. Freeze-etching of bacteria. *Int Rev Cytol* 1972;**33**:253–96.

71. Eltsov M, Dubochet J. Fine structure of the *Deinococcus radiodurans* nucleoid revealed by cryoelectron microscopy of vitreous sections. *J Bacteriol* 2005;**187**:8047–54.

72. Zuber B, Chami M, Houssin C, Dubochet J, Griffiths G, Daffe M. Direct visualization of the outer membrane of mycobacteria and corynebacteria in their native state. *J Bacteriol* 2008;**190**:5672–80.

73. Smit J, Engelhardt H, Volker S, Smith SH, Baumeister W. The S-layer of *Caulobacter crescentus*: three-dimensional image reconstruction and structure analysis by electron micros-copy. *J Bacteriol* 1992;**174**:6527–38.

74. Norville JE, Kelly DF, Knight Jr. TF, Belcher AM, Walz T. 7 Å projection map of the S-layer protein SbpA obtained with trehalose-embedded monolayer crystals. *J Struct Biol* 2007;**160**:313–23.

75. Rachel R, Jakubowski U, Tietz H, Hegerl R, Baumeister W. Projected structure of the surface protein of *Deinococcus radiodurans* determined to 8 Å resolution by cryomicroscopy. *Ultra-microscopy* 1986;**20**:305–16.

76. Engel A, Baumeister W, Saxton WO. Mass mapping of a protein complex with the scanning transmission electron microscope. *Proc Natl Acad Sci USA* 1982;**79**:4050–4.

77. Amat F, Comolli LR, Nomellini JF, Moussavi F, Downing KH, Smit J, et al. Analysis of the intact surface layer of *Caulobacter crescentus* by cryo-electron tomography. *J Bacteriol* 2010;**192**:5855–65.

78. Simon P, Lichte H, Formanek P, Lehmann M, Huhle R, Carrillo-Cabrera W, et al. Electron holography of biological samples. *Micron* 2008;**39**:229–56.

79. Simon P, Lichte H, Wahl R, Mertig M, Pompe W. Electron holography of non-stained bacterial surface layer proteins. *Biochim Biophys Acta* 2004;**1663**:178–87.

80. Wiegrabe W, Nonnenmacher M, Guckenberger R, Wolter O. Atomic force microscopy of a hydrated bacterial surface protein. *J Microsc* 1991;**163**:79–84.

81. Ohnesorge F, Heckl WM, Haberle W, Pum D, Sára M, Schindler H, et al. Scanning force microscopy studies of the S-layers from *Bacillus coagulans* E38-66, *Bacillus sphaericus* CCM2177 and of an antibody binding process. *Ultramicroscopy* 1992;**42–44**:1236–42.

82. Lister TE, Pinhero PJ. In vivo atomic force microscopy of surface proteins on *Deinococcus radiodurans*. *Langmuir* 2001;**17**:2624–8.

83. Dufrene YF. Application of atomic force microscopy to microbial surfaces: from reconstituted cell surface layers to living cells. *Micron* 2001;**32**:153–65.

84. Allison DP, Mortensen NP, Sullivan CJ, Doktycz MJ. Atomic force microscopy of biological samples. *Wiley Interdiscip Rev Nanomed Nanobiotechnol* 2010;**2**:618–34.

85. Beveridge TJ, Graham LL. Surface-layers of bacteria. *Microbiol Rev* 1991;**55**:684–705.

86. Wang ZH, Hartmann T, Baumeister W, Guckenberger R. Thickness determination of biological samples with a zeta-calibrated scanning tunneling microscope. *Proc Natl Acad Sci USA* 1990;**87**:9343–7.

87. Stemmer A, Reichelt R, Wyss R, Engel A. Biological structures imaged in a hybrid scanning transmission electron microscope and scanning tunneling microscope. *Ultramicroscopy* 1991;**35**:255–64.

88. Shevchuk AI, Frolenkov GI, Sanchez D, James PS, Freedman N, Lab MJ, et al. Imaging proteins in membranes of living cells by high-resolution scanning ion conductance microscopy. *Angew Chem Int Ed Engl* 2006;**45**:2212–6.

89. Weygand M, Wetzer B, Pum D, Sleytr UB, Cuvillier N, Kjaer K, et al. Bacterial S-layer protein coupling to lipids: x-ray reflectivity and grazing incidence diffraction studies. *Biophys J* 1999;**76**:458–68.

90. Fagan RP, Albesa-Jove D, Qazi O, Svergun DI, Brown KA, Fairweather NF. Structural insights into the molecular organization of the S-layer from *Clostridium difficile*. *Mol Microbiol* 2009;**71**:1308–22.

91. Pavkov T, Egelseer EM, Tesarz M, Svergun DI, Sleytr UB, Keller W. The structure and binding behavior of the bacterial cell surface layer protein SbsC. *Structure* 2008;**16**:1226–37.

92. Pavkov T, Oberer M, Egelseer EM, Sára M, Sleytr UB, Keller W. Crystallization and preliminary structure determination of the C-terminal truncated domain of the S-layer protein SbsC. *Acta Crystallogr D Biol Crystallogr* 2003;**59**:1466–8.

93. Kroutil M, Pavkov T, Birner-Gruenberger R, Tesarz M, Sleytr UB, Egelseer EM, et al. Towards the structure of the C-terminal part of the S-layer protein SbsC. *Acta Crystallogr Sect F Struct Biol Cryst Commun* 2009;**65**:1042–7.

94. Horejs C, Gollner H, Pum D, Sleytr UB, Peterlik H, Jungbauer A, et al. Atomistic structure of monomolecular surface layer self-assemblies: toward functionalized nanostructures. *ACS Nano* 2011;**5**:228–97.

95. Horejs C, Pum D, Sleytr UB, Peterlik H, Jungbauer A, Tscheliessnig R. Surface layer protein characterization by small angle x-ray scattering and a fractal mean force concept: from protein structure to nanodisk assemblies. *J Chem Phys* 2010;**133**:175102.

96. Tani K, Mitsuma T, Hiroaki Y, Kamegawa A, Nishikawa K, Tanimura Y, et al. Mechanism of aquaporin-4's fast and highly selective water conduction and proton exclusion. *J Mol Biol* 2009;**389**:694–706.

97. Mitsuoka K, Hirai T, Murata K, Miyazawa A, Kidera A, Kimura Y, et al. The structure of bacteriorhodopsin at 3.0 A resolution based on electron crystallography: implication of the charge distribution. *J Mol Biol* 1999;**286**:861–82.

98. Nettles JH, Li H, Cornett B, Krahn JM, Snyder JP, Downing KH. The binding mode of epothilone A on alpha, beta-tubulin by electron crystallography. *Science* 2004;**305**:866–9.

99. Hiroaki Y, Tani K, Kamegawa A, Gyobu N, Nishikawa K, Suzuki H, et al. Implications of the aquaporin-4 structure on array formation and cell adhesion. *J Mol Biol* 2006;**355**:628–39.

100. Holm PJ, Bhakat P, Jegerschold C, Gyobu N, Mitsuoka K, Fujiyoshi Y, et al. Structural basis for detoxification and oxidative stress protection in membranes. *J Mol Biol* 2006;**360**:934–45.

101. Gonen T, Cheng Y, Sliz P, Hiroaki Y, Fujiyoshi Y, Harrison SC, et al. Lipid-protein interactions in double-layered two-dimensional AQP0 crystals. *Nature* 2005;**438**:633–8.

102. Hite RK, Li Z, Walz T. Principles of membrane protein interactions with annular lipids deduced from aquaporin-0 2D crystals. *EMBO J* 2010;**29**:1652–8.

103. Schäffer C, Kahlig H, Christian R, Schulz G, Zayni S, Messner P. The diacetamidodideoxyuronic-acid-containing glycan chain of *Bacillus stearothermophilus* NRS 2004/3a represents the secondary cell-wall polymer of wild-type B. stearothermophilus strains. *Microbiology* 1999;**145**:1575–83.

104. Kuen B, Sleytr UB, Lubitz W. Sequence analysis of the sbsA gene encoding the 130-kDa surface-layer protein of *Bacillus stearothermophilus* strain PV72. *Gene* 1994;**145**:115–20.

105. Schäffer C, Wugeditsch T, Kahlig H, Scheberl A, Zayni S, Messner P. The surface layer (S-layer) glycoprotein of *Geobacillus stearothermophilus* NRS 2004/3a. Analysis of its glycosylation. *J Biol Chem* 2002;**277**:6230–9.

106. Sleytr UB, Sára M, Kupcu Z, Messner P. Structural and chemical characterization of S-layers of selected strains of *Bacillus stearothermophilus* and *Desulfotomaculum nigrificans*. *Arch Microbiol* 1986;**146**:19–24.

107. Messner P, Hollaus F, Sleytr UB. Paracrystalline cell wall surface layers of different *Bacillus stearothermophilus* strains. *Int J Syst Evol Microbiol* 1984;**34**:202–10.

108. Jarosch M, Egelseer EM, Huber C, Moll D, Mattanovich D, Sleytr UB, et al. Analysis of the structure-function relationship of the S-layer protein SbsC of *Bacillus stearothermophilus* ATCC 12980 by producing truncated forms. *Microbiology* 2001;**147**:1353–63.

109. Messner P, Pum D, Sleytr UB. Characterization of the ultrastructure and the self-assembly of the surface layer of *Bacillus stearothermophilus* strain NRS 2004/3a. *J Ultrastruct Mol Struct Res* 1986;**97**:73–88.

110. Pum D, Sleytr UB. The application of bacterial S-layers in molecular nanotechnology. *Trends Biotechnol* 1999;**17**:8–12.

111. Bateman A, Coin L, Durbin R, Finn RD, Hollich V, Griffiths-Jones S, et al. The Pfam protein families database. *Nucleic Acids Res* 2004;**32**:D138–41.

112. Ferner-Ortner J, Mader C, Ilk N, Sleytr UB, Egelseer EM. High-affinity interaction between the S-layer protein SbsC and the secondary cell wall polymer of *Geobacillus stearothermophilus* ATCC 12980 determined by surface plasmon resonance technology. *J Bacteriol* 2007;**189**:7154–8.

113. Kuen B, Koch A, Asenbauer E, Sára M, Lubitz W. Molecular characterization of the *Bacillus stearothermophilus* PV72 S-layer gene sbsB induced by oxidative stress. *J Bacteriol* 1997;**179**:1664–70.

114. Sára M, Egelseer EM, Dekitsch C, Sleytr UB. Identification of two binding domains, one for peptidoglycan and another for a secondary cell wall polymer, on the N-terminal part of the S-layer protein SbsB from *Bacillus stearothermophilus* PV72/p2. *J Bacteriol* 1998;**180**:6780–3.

115. Petersen BO, Sára M, Mader C, Mayer HF, Sleytr UB, Pabst M, et al. Structural characterization of the acid-degraded secondary cell wall polymer of *Geobacillus stearothermophilus* PV72/p2. *Carbohydr Res* 2008;**343**:1346–58.

116. Runzler D, Huber C, Moll D, Kohler G, Sára M. Biophysical characterization of the entire bacterial surface layer protein SbsB and its two distinct functional domains. *J Biol Chem* 2004;**279**:5207–15.

117. Moll D, Huber C, Schlegel B, Pum D, Sleytr UB, Sára M. S-layer-streptavidin fusion proteins as template for nanopatterned molecular arrays. *Proc Natl Acad Sci USA* 2002;**99**:14646–51.

118. Horejs C, Pum D, Sleytr UB, Tscheliessnig R. Structure prediction of an S-layer protein by the mean force method. *J Chem Phys* 2008;**128**:065106.

119. Howorka S, Sára M, Wang Y, Kuen B, Sleytr UB, Lubitz W, et al. Surface-accessible residues in the monomeric and assembled forms of a bacterial surface layer protein. *J Biol Chem* 2000;**275**:37876–86.

120. Kinns H, Howorka S. The surface location of individual residues in a bacterial S-layer protein. *J Mol Biol* 2008;**377**:589–604.

121. Kinns H, Badelt-Lichtblau H, Egelseer EM, Sleytr UB, Howorka S. Identifying assembly-inhibiting and assembly-tolerant sites in the SbsB S-layer protein from *Geobacillus stearothermophilus*. *J Mol Biol* 2010;**395**:742–53.

122. Braun M, Mayer F, Gottschalk G. *Clostridium aceticum* (Wieringa), a microorganism producing acetic acid from molecular hydrogen and carbon dioxide. *Arch Microbiol* 1981;**128**:288–93.

123. Woodcock CL, Engelhardt H, Baumeister W. The tetragonal surface layer of *Clostridium aceticum*: three-dimensional structure and comparison with the hexagonal layer of *Clostridium thermohydrosulfuricum*. *Eur J Cell Biol* 1986;**42**:211–7.

124. Cejka Z, Hegerl R, Baumeister W. Three-dimensional structure of the surface layer protein of *Clostridium thermohydrosulfuricum*. *J Ultrastruct Mol Struct Res* 1986;**96**:1–11.
125. Onyenwoke RU, Kevbrin VV, Lysenko AM, Wiegel J. *Thermoanaerobacter pseudethanolicus* sp. nov., a thermophilic heterotrophic anaerobe from Yellowstone National Park. *Int J Syst Evol Microbiol* 2007;**57**:2191–3.
126. Calabi E, Fairweather N. Patterns of sequence conservation in the S-Layer proteins and related sequences in *Clostridium difficile*. *J Bacteriol* 2002;**184**:3886–97.
127. Masuda K, Itoh M, Kawata T. Characterization and reassembly of a regular array in the cell wall of *Clostridium difficile* GAI 4131. *Microbiol Immunol* 1989;**33**:287–98.
128. Dang TH, de la Riva L, Fagan RP, Storck EM, Heal WP, Janoir C, et al. Chemical probes of surface layer biogenesis in Clostridium difficile. *ACS Chem Biol* 2010;**5**:279–85.
129. Boot HJ, Kolen CP, Andreadaki FJ, Leer RJ, Pouwels PH. The Lactobacillus acidophilus S-layer protein gene expression site comprises two consensus promoter sequences, one of which directs transcription of stable mRNA. *J Bacteriol* 1996;**178**:5388–94.
130. Masuda K, Kawata T. Reassembly of a regularly arranged protein in the cell wall of *Lactobacillus buchneri* and its reattachment to cell walls: chemical modification studies. *Microbiol Immunol* 1985;**29**:927–38.
131. Kawata T, Masuda K, Yoshino K, Fujimoto M. Regular array in cell wall of *Lactobacillus fermenti* as revealed by freeze-etching and negative staining. *Jpn J Microbiol* 1974;**18**:469–76.
132. Masuda K. Heterogeneity of S-layer proteins of *Lactobacillus acidophilus* strains. *Microbiol Immunol* 1992;**36**:297–301.
133. Hagen KE, Guan LL, Tannock GW, Korver DR, Allison GE. Detection, characterization, and in vitro and in vivo expression of genes encoding S-proteins in *Lactobacillus gallinarum* strains isolated from chicken crops. *Appl Environ Microbiol* 2005;**71**:6633–43.
134. Kant R, Paulin L, Alatalo E, de Vos WM, Palva A. Genome sequence of *Lactobacillus amylovorus* GRL1112. *J Bacteriol* 2011;**193**:789–90.
135. Ventura M, Jankovic I, Walker DC, Pridmore RD, Zink R. Identification and characterization of novel surface proteins in *Lactobacillus johnsonii* and *Lactobacillus gasseri*. *Appl Environ Microbiol* 2002;**68**:6172–81.
136. Yasui T, Yoda K, Kamiya T. Analysis of S-layer proteins of *Lactobacillus brevis*. *FEMS Microbiol Lett* 1995;**133**:181–6.
137. Boot HJ, Kolen CP, Pouwels PH. Identification, cloning, and nucleotide sequence of a silent S-layer protein gene of *Lactobacillus acidophilus* ATCC 4356 which has extensive similarity with the S-layer protein gene of this species. *J Bacteriol* 1995;**177**:7222–30.
138. Boot HJ, Kolen CP, Pot B, Kersters K, Pouwels PH. The presence of two S-layer-protein-encoding genes is conserved among species related to *Lactobacillus acidophilus*. *Microbiology* 1996;**142**:2375–84.
139. Vidgren G, Palva I, Pakkanen R, Lounatmaa K, Palva A. S-layer protein gene of *Lactobacillus brevis*: cloning by polymerase chain reaction and determination of the nucleotide sequence. *J Bacteriol* 1992;**174**:7419–27.
140. Hynonen U, Avall-Jaaskelainen S, Palva A. Characterization and separate activities of the two promoters of the *Lactobacillus brevis* S-layer protein gene. *Appl Microbiol Biotechnol* 2010;**87**:657–68.
141. Chen X, Chen Y, Li X, Chen N, Fang W. Characterization of surface layer proteins in *Lactobacillus crispatus* isolate ZJ001. *J Microbiol Biotechnol* 2009;**19**:1176–83.
142. Sillanpaa J, Martinez B, Antikainen J, Toba T, Kalkkinen N, Tankka S, et al. Characterization of the collagen-binding S-layer protein CbsA of *Lactobacillus crispatus*. *J Bacteriol* 2000;**182**:6440–50.
143. Masuda K, Kawata T. Distribution and chemical characterization of regular arrays in the cell-walls of strains of the genus Lactobacillus. *FEMS Microbiol Lett* 1983;**20**:145–50.

144. Thompson JD, Higgins DG, Gibson TJ. CLUSTAL W: improving the sensitivity of progressive multiple sequence alignment through sequence weighting, position-specific gap penalties and weight matrix choice. *Nucleic Acids Res* 1994;**22**:4673–80.

145. Jakava-Viljanen M, Avall-Jaaskelainen S, Messner P, Sleytr UB, Palva A. Isolation of three new surface layer protein genes (slp) from *Lactobacillus brevis* ATCC 14869 and characterization of the change in their expression under aerated and anaerobic conditions. *J Bacteriol* 2002;**184**:6786–95.

146. Avall-Jaaskelainen S, Hynonen U, Ilk N, Pum D, Sleytr UB, Palva A. Identification and characterization of domains responsible for self-assembly and cell wall binding of the surface layer protein of *Lactobacillus brevis* ATCC 8287. *BMC Microbiol* 2008;**8**:165.

147. Smit E, Pouwels PH. One repeat of the cell wall binding domain is sufficient for anchoring the *Lactobacillus acidophilus* surface layer protein. *J Bacteriol* 2002;**184**:4617–9.

148. Antikainen J, Anton L, Sillanpaa J, Korhonen TK. Domains in the S-layer protein CbsA of *Lactobacillus crispatus* involved in adherence to collagens, laminin and lipoteichoic acids and in self-assembly. *Mol Microbiol* 2002;**46**:381–94.

149. Sanchez Carballo PM, Vilen H, Palva A, Holst O. Structural characterization of teichoic acids from *Lactobacillus brevis*. *Carbohydr Res* 2010;**345**:538–42.

150. Vilen H, Hynonen U, Badelt-Lichtblau H, Ilk N, Jaaskelainen P, Torkkeli M, et al. Surface location of individual residues of SlpA provides insight into the *Lactobacillus brevis* S-layer. *J Bacteriol* 2009;**191**:3339–49.

151. Lortal S, Vanheijenoort J, Gruber K, Sleytr UB. S-Layer of *Lactobacillus helveticus* ATCC 12046: isolation, chemical characterization and re-formation after extraction with lithium-chloride. *J Gen Microbiol* 1992;**138**:611–8.

152. Masuda K, Kawata T. Ultrastructure and partial characterization of a regular array in the cell wall of *Lactobacillus brevis*. *Microbiol Immunol* 1979;**23**:941–53.

153. Masuda K, Kawata T. Reassembly of the regularly arranged subunits in the cell-wall of *Lactobacillus brevis* and their reattachment to cell-walls. *Microbiol Immunol* 1980;**24**:299–308.

154. Masuda K, Kawata T. Characterization of a regular array in the wall of *Lactobacillus buchneri* and its reattachment to the other wall components. *J Gen Microbiol* 1981;**124**:81–90.

155. Hanby WE, Rydon HN. The capsular substance of *Bacillus anthracis*. *Biochem J* 1946;**40**:297–309.

156. Holt SC, Leadbett ER. Comparative ultrastructure of selected aerobic spore-forming bacteria - a freeze-etching study. *Bacteriol Rev* 1969;**33**:346–78.

157. Etienne-Toumelin I, Sirard JC, Duflot E, Mock M, Fouet A. Characterization of the *Bacillus anthracis* S-layer: cloning and sequencing of the structural gene. *J Bacteriol* 1995;**177**:614–20.

158. Mesnage S, Tosi-Couture E, Gounon P, Mock M, Fouet A. The capsule and S-layer: two independent and yet compatible macromolecular structures in *Bacillus anthracis*. *J Bacteriol* 1998;**180**:52–8.

159. Chauvaux S, Matuschek M, Beguin P. Distinct affinity of binding sites for S-layer homologous domains in *Clostridium thermocellum* and *Bacillus anthracis* cell envelopes. *J Bacteriol* 1999;**181**:2455–8.

160. Choudhury B, Leoff C, Saile E, Wilkins P, Quinn CP, Kannenberg EL, et al. The structure of the major cell wall polysaccharide of *Bacillus anthracis* is species-specific. *J Biol Chem* 2006;**281**:27932–41.

161. Leoff C, Choudhury B, Saile E, Quinn CP, Carlson RW, Kannenberg EL. Structural elucidation of the nonclassical secondary cell wall polysaccharide from *Bacillus cereus* ATCC 10987. *J Biol Chem* 2008;**283**:29812–21.

162. Mignot T, Mesnage S, Couture-Tosi E, Mock M, Fouet A. Developmental switch of S-layer protein synthesis in *Bacillus anthracis*. *Mol Microbiol* 2002;**43**:1615–27.

163. Mignot T, Mock M, Fouet A. A plasmid-encoded regulator couples the synthesis of toxins and surface structures in *Bacillus anthracis*. *Mol Microbiol* 2003;**47**:917–27.

164. Couture-Tosi E, Delacroix H, Mignot T, Mesnage S, Chami M, Fouet A, et al. Structural analysis and evidence for dynamic emergence of *Bacillus anthracis* S-layer networks. *J Bacteriol* 2002;**184**:6448–56.

165. Aebi U, Smith PR, Dubochet J, Henry C, Kellenberger E. A study of the structure of the T-layer of *Bacillus brevis*. *J Supramol Struct* 1973;**1**:498–522.

166. Hu X, Li J, Hansen BM, Yuan Z. Phylogenetic analysis and heterologous expression of surface layer protein SlpC of *Bacillus sphaericus* C3-41. *Biosci Biotechnol Biochem* 2008;**72**:1257–63.

167. Bowditch RD, Baumann P, Yousten AA. Cloning and sequencing of the gene encoding a 125-kilodalton surface-layer protein from *Bacillus sphaericus* 2362 and of a related cryptic gene. *J Bacteriol* 1989;**171**:4178–88.

168. Ilk N, Vollenkle C, Egelseer EM, Breitwieser A, Sleytr UB, Sára M. Molecular characterization of the S-layer gene, sbpA, of *Bacillus sphaericus* CCM 2177 and production of a functional S-layer fusion protein with the ability to recrystallize in a defined orientation while presenting the fused allergen. *Appl Environ Microbiol* 2002;**68**:3251–60.

169. Pollmann K, Raff J, Schnorpfeil M, Radeva G, Selenska-Pobell S. Novel surface layer protein genes in *Bacillus sphaericus* associated with unusual insertion elements. *Microbiology* 2005;**151**:2961–73.

170. Deblaere RY, Desomer J, Dhaese P. *Expression of surface layer proteins*. Vol. WO 9519371-A2.

171. Pollmann K, Matys S. Construction of an S-layer protein exhibiting modified self-assembling properties and enhanced metal binding capacities. *Appl Microbiol Biotechnol* 2007;**75**:1079–85.

172. Fahmy K, Merroun M, Pollmann K, Raff J, Savchuk O, Hennig C, et al. Secondary structure and Pd(II) coordination in S-layer proteins from *Bacillus sphaericus* studied by infrared and X-ray absorption spectroscopy. *Biophys J* 2006;**91**:996–1007.

173. Teixeira LM, Strickland A, Mark SS, Bergkvist M, Sierra-Sastre Y, Batt CA. Entropically driven self-assembly of *Lysinibacillus sphaericus* S-layer proteins analyzed under various environmental conditions. *Macromol Biosci* 2010;**10**:147–55.

174. Lepault J, Martin N, Leonard K. Three-dimensional structure of the T-layer of *Bacillus sphaericus* P-1. *J Bacteriol* 1986;**168**:303–8.

175. Lepault J, Pitt T. Projected structure of unstained, frozen-hydrated T-layer of *Bacillus brevis*. *EMBO J* 1984;**3**:101–5.

176. Kistler J, Aebi U, Kellenberger E. Freeze drying and shadowing a two-dimensional periodic specimen. *J Ultrastruct Res* 1977;**59**:76–86.

177. Howard L, Tipper DJ. A polypeptide bacteriophage receptor: modified cell wall protein subunits in bacteriophage-resistant mutants of *Bacillus sphaericus* strain P-1. *J Bacteriol* 1973;**113**:1491–504.

178. Engelhardt H, Saxton WO, Baumeister W. Three-dimensional structure of the tetragonal surface layer of *Sporosarcina ureae*. *J Bacteriol* 1986;**168**:309–17.

179. Wetzer B, Pum D, Sleytr UB. S-layer stabilized solid supported lipid bilayers. *J Struct Biol* 1997;**119**:123–8.

180. Tang J, Ebner A, Ilk N, Lichtblau H, Huber C, Zhu R, et al. High-affinity tags fused to s-layer proteins probed by atomic force microscopy. *Langmuir* 2008;**24**:1324–9.

181. Bayan N, Houssin C, Chami M, Leblon G. Mycomembrane and S-layer: two important structures of *Corynebacterium glutamicum* cell envelope with promising biotechnology applications. *J Biotechnol* 2003;**104**:55–67.

182. Peyret JL, Bayan N, Joliff G, Gulik-Krzywicki T, Mathieu L, Schechter E, et al. Characterization of the cspB gene encoding PS2, an ordered surface-layer protein in *Corynebacterium glutamicum*. *Mol Microbiol* 1993;**9**:97–109.

183. Chami M, Bayan N, Dedieu J, Leblon G, Shechter E, Gulik-Krzywicki T. Organization of the outer layers of the cell envelope of *Corynebacterium glutamicum*: a combined freeze-etch electron microscopy and biochemical study. *Biol Cell* 1995;**83**:219–29.

184. Hansmeier N, Bartels FW, Ros R, Anselmetti D, Tauch A, Puhler A, et al. Classification of hyper-variable *Corynebacterium glutamicum* surface-layer proteins by sequence analyses and atomic force microscopy. *J Biotechnol* 2004;**112**:177–93.

185. Houssin C, Nguyen DT, Leblon G, Bayan N. S-layer protein transport across the cell wall of *Corynebacterium glutamicum*: in vivo kinetics and energy requirements. *FEMS Microbiol Lett* 2002;**217**:71–9.

186. Chami M, Bayan N, Peyret JL, Gulik-Krzywicki T, Leblon G, Shechter E. The S-layer protein of *Corynebacterium glutamicum* is anchored to the cell wall by its C-terminal hydrophobic domain. *Mol Microbiol* 1997;**23**:483–92.

187. Scheuring S, Stahlberg H, Chami M, Houssin C, Rigaud JL, Engel A. Charting and unzipping the surface layer of *Corynebacterium glutamicum* with the atomic force microscope. *Mol Microbiol* 2002;**44**:675–84.

188. Dupres V, Alsteens D, Pauwels K, Dufrene YF. In vivo imaging of S-layer nanoarrays on *Corynebacterium glutamicum*. *Langmuir* 2009;**25**:9653–5.

189. Griffiths E, Gupta RS. Identification of signature proteins that are distinctive of the Deino-coccus-Thermus phylum. *Int Microbiol* 2007;**10**:201–8.

190. Emde B, Heisse W, Karrenberg F, Baumeister W. Structural investigations on the cell-wall of Micrococcus-Radiodurans. *Eur J Cell Biol* 1980;**22**:461–3.

191. Kubler O, Baumeister W. The structure of a periodic cell wall component (HPI-layer of *Micrococcus radiodurans*). *Cytobiologie* 1978;**17**:1–9.

192. Thompson BG, Murray RGE, Boyce JF. The association of the surface array and the outer-membrane of *Deinococcus radiodurans*. *Can J Microbiol* 1982;**28**:1081–8.

193. Work E, Griffiths H. Morphology and chemistry of cell walls of *Micrococcus radiodurans*. *J Bacteriol* 1968;**95**:641–57.

194. Baumeister W, Kubler O. Topographic study of the cell surface of *Micrococcus radiodurans*. *Proc Natl Acad Sci USA* 1978;**75**:5525–8.

195. Peters J, Baumeister W. Molecular cloning, expression, and characterization of the gene for the surface (HPI)-layer protein of *Deinococcus radiodurans* in *Escherichia coli*. *J Bacteriol* 1986;**167**:1048–54.

196. Peters J, Peters M, Lottspeich F, Schafer W, Baumeister W. Nucleotide sequence analysis of the gene encoding the *Deinococcus radiodurans* surface protein, derived amino acid sequence, and complementary protein chemical studies. *J Bacteriol* 1987;**169**:5216–23.

197. Baumeister W, Karrenberg F, Rachel R, Engel A, ten Heggeler B, Saxton WO. The major cell envelope protein of *Micrococcus radiodurans* (R1). Structural and chemical characterization. *Eur J Biochem* 1982;**125**:535–44.

198. Sleytr UB, Kocur M, Glauert AM, Thornley MJ. A study by freeze-etching of the fine structure of *Micrococcus radiodurans*. *Arch Mikrobiol* 1973;**94**:77–87.

199. Baumeister W, Barth M, Hegerl R, Guckenberger R, Hahn M, Saxton WO. Three-dimensional structure of the regular surface layer (HPI layer) of *Deinococcus radiodurans*. *J Mol Biol* 1986;**187**:241–50.

200. Baumeister W, Kubler O, Zingsheim HP. The structure of the cell envelope of *Micrococcus radiodurans* as revealed by metal shadowing and decoration. *J Ultrastruct Res* 1981;**75**:60–71.

201. Wildhaber I, Gross H, Engel A, Baumeister W. The effects of air-drying and freeze-drying on the structure of a regular protein layer. *Ultramicroscopy* 1985;**16**:411–22.

202. Karrenberg FH, Wildhaber I, Baumeister W. Surface structure variants in *Deinococcus radiodurans*. *Curr Microbiol* 1987;**16**:15–20.

203. Guckenberger R, Wiegrabe W, Baumeister W. Scanning tunnelling microscopy of biomacro-molecules. *J Microsc* 1988;**152**:795–802.
204. Karrasch S, Hegerl R, Hoh JH, Baumeister W, Engel A. Atomic force microscopy produces faithful high-resolution images of protein surfaces in an aqueous environment. *Proc Natl Acad Sci USA* 1994;**91**:836–8.
205. Muller DJ, Baumeister W, Engel A. Conformational change of the hexagonally packed intermediate layer of *Deinococcus radiodurans* monitored by atomic force microscopy. *J Bacteriol* 1996;**178**:3025–30.
206. Muller DJ, Baumeister W, Engel A. Controlled unzipping of a bacterial surface layer with atomic force microscopy. *Proc Natl Acad Sci USA* 1999;**96**:13170–4.
207. Wildhaber I, Hegerl R, Barth M, Gross H, Baumeister W. Three-dimensional reconstruction of a freeze-dried and metal-shadowed bacterial surface-layer. *Ultramicroscopy* 1986;**19**:57–67.
208. Rachel R, Jakubowski U, Baumeister W. Electron microscopy of unstained, freeze-dried macromolecular assemblies. *J Microsc* 1986;**141**:179–91.
209. White O, Eisen JA, Heidelberg JF, Hickey EK, Peterson JD, Dodson RJ, et al. Genome sequence of the radioresistant bacterium *Deinococcus radiodurans* R1. *Science* 1999;**286**:1571–7.
210. Caston JR, Carrascosa JL, Depedro MA, Berenguer J. Identification of a crystalline surface-layer on the cell-envelope of the thermophilic eubacterium *Thermus-Thermophilus*. *FEMS Microbiol Lett* 1988;**51**:225–30.
211. Berenguer J, Faraldo ML, de Pedro MA. Ca2+-stabilized oligomeric protein complexes are major components of the cell envelope of "*Thermus thermophilus*" HB8. *J Bacteriol* 1988;**170**:2441–7.
212. Faraldo MLM, Depedro MA, Berenguer J. Purification, Composition and Ca-2+-Binding Properties of the Monomeric Protein of the S-Layer of *Thermus-Thermophilus*. *FEBS Lett* 1988;**235**:117–21.
213. Faraldo MM, de Pedro MA, Berenguer J. Sequence of the S-layer gene of *Thermus thermophilus* HB8 and functionality of its promoter in *Escherichia coli*. *J Bacteriol* 1992;**174**:7458–62.
214. Lasa I, Caston JR, Fernandez-Herrero LA, de Pedro MA, Berenguer J. Insertional mutagenesis in the extreme thermophilic eubacteria *Thermus thermophilus* HB8. *Mol Microbiol* 1992;**6**:1555–64.
215. Caston JR, Berenguer J, de Pedro MA, Carrascosa JL. S-layer protein from *Thermus thermophilus* HB8 assembles into porin-like structures. *Mol Microbiol* 1993;**9**:65–75.
216. Caston JR, Berenguer J, Kocsis E, Carrascosa JL. Three-dimensional structure of different aggregates built up by the S-layer protein of *Thermus thermophilus*. *J Struct Biol* 1994;**113**:164–76.
217. Dickson MR, Downing KH, Wu WH, Glaeser RM. Three-dimensional structure of the surface layer protein of *Aquaspirillum serpens* VHA determined by electron crystallography. *J Bacteriol* 1986;**167**:1025–34.
218. Kist ML, Murray RG. Components of the regular surface array of *Aquaspirillum serpens* MW5 and their assembly in vitro. *J Bacteriol* 1984;**157**:599–606.
219. Dokland T, Olsen I, Farrants G, Johansen BV. Three-dimensional structure of the surface layer of *Wolinella recta*. *Oral Microbiol Immunol* 1990;**5**:162–5.
220. Poindexter JLS, Cohenbazire G. Fine structure of stalked bacteria belonging to family Caulobacteraceae. *J Cell Biol* 1964;**23**:587–607.
221. Poindexter JS. Biological properties and classification of the Caulobacter group. *Bacteriol Rev* 1964;**28**:231–95.
222. Smit J, Grano DA, Glaeser RM, Agabian N. Periodic surface array in *Caulobacter crescentus*: fine structure and chemical analysis. *J Bacteriol* 1981;**146**:1135–50.

223. Gilchrist A, Fisher JA, Smit J. Nucleotide sequence analysis of the gene encoding the *Caulobacter crescentus* paracrystalline surface layer protein. *Can J Microbiol* 1992;**38**:193–202.
224. Bingle WH, Le KD, Smit J. The extreme N-terminus of the *Caulobacter crescentus* surface-layer protein directs export of passenger proteins from the cytoplasm but is not required for secretion of the native protein. *Can J Microbiol* 1996;**42**:672–84.
225. Awram P, Smit J. The *Caulobacter crescentus* paracrystalline S-layer protein is secreted by an ABC transporter (type I) secretion apparatus. *J Bacteriol* 1998;**180**:3062–9.
226. Fernandez LA, Berenguer J. Secretion and assembly of regular surface structures in Gram-negative bacteria. *FEMS Microbiol Rev* 2000;**24**:21–44.
227. Toporowski MC, Nomellini JF, Awram P, Smit J. Two outer membrane proteins are required for maximal type I secretion of the *Caulobacter crescentus* S-layer protein. *J Bacteriol* 2004;**186**:8000–9.
228. Thompson SA, Shedd OL, Ray KC, Beins MH, Jorgensen JP, Blaser MJ. *Campylobacter fetus* surface layer proteins are transported by a type I secretion system. *J Bacteriol* 1998;**180**:6450–8.
229. Kawai E, Akatsuka H, Idei A, Shibatani T, Omori K. Serratia marcescens S-layer protein is secreted extracellularly via an ATP-binding cassette exporter, the Lip system. *Mol Microbiol* 1998;**27**:941–52.
230. Walker SG, Karunaratne DN, Ravenscroft N, Smit J. Characterization of mutants of *Caulobacter crescentus* defective in surface attachment of the paracrystalline surface layer. *J Bacteriol* 1994;**176**:6312–23.
231. Awram P, Smit J. Identification of lipopolysaccharide O antigen synthesis genes required for attachment of the S-layer of *Caulobacter crescentus*. *Microbiology* 2001;**147**:1451–60.
232. Ford MJ, Nomellini JF, Smit J. S-layer anchoring and localization of an S-layer-associated protease in *Caulobacter crescentus*. *J Bacteriol* 2007;**189**:2226–37.
233. Yang LY, Pei ZH, Fujimoto S, Blaser MJ. Reattachment of surface array proteins to *Campylobacter* fetus cells. *J Bacteriol* 1992;**174**:1258–67.
234. Dworkin J, Tummuru MK, Blaser MJ. A lipopolysaccharide-binding domain of the *Campylobacter fetus* S-layer protein resides within the conserved N terminus of a family of silent and divergent homologs. *J Bacteriol* 1995;**177**:1734–41.
235. Walker SG, Smith SH, Smit J. Isolation and comparison of the paracrystalline surface layer proteins of freshwater caulobacters. *J Bacteriol* 1992;**174**:1783–92.
236. Nomellini JF, Küpcü S, Sleytr UB, Smit J. Factors controlling in vitro recrystallization of the *Caulobacter crescentus* paracrystalline S-layer. *J Bacteriol* 1997;**179**:6349–54.
237. Trust TJ, Klotz FW, Miller LH. Pathogenesis of infectious diseases of fish. *Annu Rev Microbiol* 1986;**40**:479–502.
238. Stewart M, Beveridge TJ, Trust TJ. Two patterns in the *Aeromonas salmonicida* A-Layer may reflect a structural transformation that alters permeability. *J Bacteriol* 1986;**166**:120–7.
239. Dooley JS, Trust TJ. Surface protein composition of *Aeromonas hydrophila* strains virulent for fish: identification of a surface array protein. *J Bacteriol* 1988;**170**:499–506.
240. Dooley JS, Engelhardt H, Baumeister W, Kay WW, Trust TJ. Three-dimensional structure of an open form of the surface layer from the fish pathogen *Aeromonas salmonicida*. *J Bacteriol* 1989;**171**:190–7.
241. Murray RG, Dooley JS, Whippey PW, Trust TJ. Structure of an S layer on a pathogenic strain of *Aeromonas hydrophila*. *J Bacteriol* 1988;**170**:2625–30.
242. Al-Karadaghi S, Wang DN, Hovmoller S. Three-dimensional structure of the crystalline surface layer from *Aeromonas hydrophila*. *J Ultrastruct Mol Struct Res* 1988;**101**:92–7.
243. Bingle WH, Engelhardt H, Page WJ, Baumeister W. Three-dimensional structure of the regular tetragonal surface layer of *Azotobacter vinelandii*. *J Bacteriol* 1987;**169**:5008–15.

244. Chalcroft JP, Engelhardt H, Baumeister W. Three-dimensional structure of a regular surface layer from *Pseudomonas acidovorans*. *Arch Microbiol* 1986;**144**:196–200.

245. Engelhardt H, Gerbl-Rieger S, Santarius U, Baumeister W. The three-dimensional structure of the regular surface protein of *Comamonas acidovorans* derived from native outer membranes and reconstituted two-dimensional crystals. *Mol Microbiol* 1991;**5**:1695–702.

246. Chu S, Cavaignac S, Feutrier J, Phipps BM, Kostrzynska M, Kay WW, et al. Structure of the tetragonal surface virulence array protein and gene of *Aeromonas salmonicida*. *J Biol Chem* 1991;**266**:15258–65.

247. Thomas SR, Trust TJ. Tyrosine phosphorylation of the tetragonal paracrystalline array of *Aeromonas hydrophila*: molecular cloning and high-level expression of the S-layer protein gene. *J Mol Biol* 1995;**245**:568–81.

248. Dooley JS, McCubbin WD, Kay CM, Trust TJ. Isolation and biochemical characterization of the S-layer protein from a pathogenic *Aeromonas hydrophila* strain. *J Bacteriol* 1988;**170**:2631–8.

249. Phipps BM, Trust TJ, Ishiguro EE, Kay WW. Purification and characterization of the cell surface virulent A protein from *Aeromonas salmonicida*. *Biochemistry* 1983;**22**:2934–9.

250. Bingle WH, Whippey PW, Doran JL, Murray RG, Page WJ. Structure of the *Azotobacter vinelandii* surface layer. *J Bacteriol* 1987;**169**:802–10.

251. Austin JW, Stewart M, Murray RG. Structural and chemical characterization of the S layer of a Pseudomonas-like bacterium. *J Bacteriol* 1990;**172**:808–17.

252. Garduno RA, Phipps BM, Baumeister W, Kay WW. Novel structural patterns in divalent cation-depleted surface layers of *Aeromonas salmonicida*. *J Struct Biol* 1992;**109**:184–95.

253. Doran JL, Bingle WH, Page WJ. Role of calcium in assembly of the *Azotobacter vinelandii* surface array. *J Gen Microbiol* 1987;**133**:399–413.

254. Garduno RA, Phipps BM, Kay WW. Physical and functional S-layer reconstitution in *Aeromonas salmonicida*. *J Bacteriol* 1995;**177**:2684–94.

255. Paul A, Engelhardt H, Jakubowski U, Baumeister W. Two-dimensional crystallization of a bacterial surface protein on lipid vesicles under controlled conditions. *Biophys J* 1992;**61**:172–88.

FURTHER READING

Crowther RA, Sleytr UB. An analysis of the fine structure of the surface layers from two strains of Clostridia, including correction for distorted images. *J Ultrastruct Res* 1977;**58**:41–9.

Spider Silk: Understanding the Structure–Function Relationship of a Natural Fiber

MARTIN HUMENIK, THOMAS
SCHEIBEL, AND ANDREW SMITH

*Lehrstuhl Biomaterialien,
Universität Bayreuth, Bayreuth,
Germany*

Spider silk is of great interest because of its extraordinary physical properties, such as strength and toughness. Here we discuss how these physical properties relate to the way in which spiders have utilized this material in prey capture, forcing its evolution to a high-performance fiber. Female spiders can produce up to seven different types of silk, and all these have different physical properties, which relate to their various functions.

Progress in Molecular Biology
and Translational Science, Vol. 103
DOI: 10.1016/B978-0-12-415906-8.00007-8

The variation in properties are due to underlying differences in the proteins making up these silks. As our understanding of spider silk has increased in the recent years, it has been possible to produce recombinant versions of the respective proteins. Recombinant proteins open up the potential to produce synthetic silk fibers with properties similar to those of the natural spider silk threads.

I. Synopsis

Nature has produced a wonderful array of structures and materials over the past billion years. The evolution of life has created fantastic materials from the iridescent shells of mollusks and diatoms, to the chitin armor of arthropods, to the cellulose and lignin structures of massive trees. Many of these materials have been exploited by humans throughout the ages. Now, in the technological era, we are starting to understand the molecular structure of these materials. Based on this understanding, potential human applications and also the diversity of bioinspired materials used by man will increase.

One such material that has been exploited by humans for millennia is silk,[1] a material which appears to have evolved multiple times in arthropods. A number of insects extrude silk from glands in various locations, while spiders have evolved a system where silk is extruded from the abdomen.[2] Spider silk has caught the attention of researchers because of the wonderful structures that spiders produce to capture flying insects. However, to understand this material it is necessary to first look at the varied properties of spider silk. Here, we briefly describe the breadth of spider species, indicating that research concentrates on a relatively few species. We discuss how spiders use silk and describe in more detail the webs used to capture flying insects and why this has resulted in the evolution of silk with specific properties that are comparable with or superior to man-made materials. A detailed description of the most well-studied silk, namely, dragline silk, is given, detailing its production, structure, function, and properties. This will act as a basis to then discuss the various other silks that are produced by spiders, and how differences in sequence relate to the different properties and hence functions that these silks have been employed for. The discussion then leads on to the attempts to produce recombinant versions of spider silks. We highlight the methods by which spider silk proteins have been assembled into threads. Finally, we give a future outlook of this field of research, looking at how the natural silk protein sequences relate to their properties and functions and how this understanding provides the potential to produce materials with a much wider range of properties than currently being tested.

II. Spiders

Spiders (Aranea) belong to the class Arachnida which includes the scorpions, ticks, and mites, and are part of the Arthropod phylum. Spiders can basically be defined as air-breathing Arthropods with two body parts, the cephalothorax and the abdomen. They have eight legs attached to the cephalothorax, and chelicerae which are venom-injecting fangs. Most spiders have eight simple eyes but some species have more or fewer eyes. Spiders are able to ingest only liquids, as they do not possess mandibles, and thus their venom liquidizes their prey prior to the spider ingesting it. The most important feature that has captured the interest of materials scientists is the fact that all spiders produce silk from spinnerets on their lower abdomen, which vary between species from two to eight. These spinnerets have a selection of spigots through which the silk is extruded from internal glands. The arrangement and number of spigots vary between the species, sex, and age of the spider.[3–6] It is likely that the use of silk evolved once in spiders, and then over time has evolved into multiple different forms to fulfill different applications or functions for the spider.

The Aranea order is the seventh most diverse order of all organisms and consists of over 42,000 species in 109 families.[7] All except one known species of spider[8] are predators and primarily feed on insects. The Aranea can be broken down into three suborders, the Mesothelae being the most ancient, consisting of only one family. The other two suborders are larger sister groups Mygalomorphae and Araneomorphae, which diverged around 240 million years ago.[9]

The Mygalomorphae have downward-pointing chelicerae and incorporate the large bodied spiders that include the spiders more popularly known as "tarantulas" (Theraphosidae) and "funnel web spiders" (Hexathelidae). However, some species are only a centimeter long and the smallest less than a millimeter.[10] Some of the larger Mygalomorphae prey on the young of larger animals utilizing their venomous bite to kill or anesthetize them. This suborder of spiders tends to live in burrows or retreats and use silk threads to wrap up their prey, for the production of egg cases, and to extend their sensory range outside of their burrows. Thus they have not evolved any obvious specialized applications for their silks that would generate highly engineered threads.

The Araneomorphae is the largest suborder consisting of about 90 families, 2700 genera, and over 32,000 species. They have chelicerae that point diagonally downward and cross over one another. Many spiders use silk just for egg-wrapping and the lining of burrows, but do not use it for hunting. However, there are a number of spiders that utilize prey-capture devices made of silk. The prey-capture devices produced by Araneomorphae are varied, ranging from substrate sheets, aerial sheets, orb webs, cob webs, and bola, all suited to different preys (Fig. 1). As the capture of prey is an important driving force

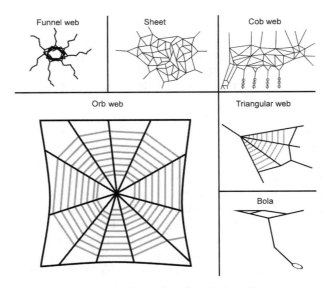

Fig. 1. Various web structures used by spiders, from the basic burrow arrangement to sheets and then the more complex and familiar orb web. The three structures on the right are developments from the orb web using the same materials. Black lines represent MA silk, gray lines flagelliform silk covered in aggregate silk, and circles represent aggregate silk added to MA silk in cob webs and bola.

in the evolution of predators, some of these spiders have evolved to have silk with properties that make it capable of stopping and holding the prey even in flight (Fig. 2). As a consequence of the evolution of prey-capture devices and hence of the properties of spider silk, the variety of silks produced by Araneomorphae has expanded.[11] This has allowed some spider species to take a more passive role in hunting, just having to get to their captured prey and kill it before it manages to escape from the web.

The prey-capture devices that have received the most attention are the orb webs produced by some Araneomorphae of the Orbiculariae clade. The spiders of this clade produce the largest range of silks with up to seven silk types in adult female spiders.[12] The orb web was an evolutionary move away from webs based on the surrounding substrate toward an aerial frame that is more suited to capturing aerial prey. The frame of the orb web, which is made of major ampullate (MA) silk, is connected by mooring threads made from the same material to nearby structures. The frame then supports radii, also made from MA silk, which are used initially as connection points to lay down an auxiliary spiral made from minor ampullate (MI) silk. This MI silk is later replaced by the capture spiral which has sticky properties for capturing flying insects. The ancestral orb web weavers, namely, cribellate spiders, relied on a capture spiral

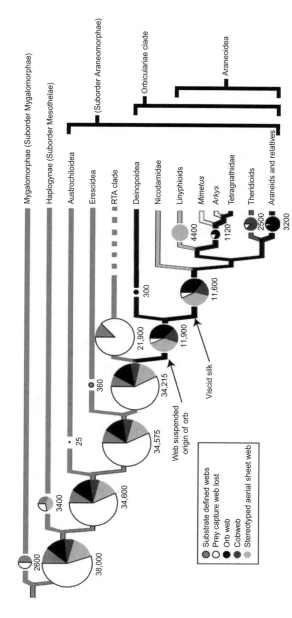

Mygalomorphae (Suborder Mygalomorphae)

Haplogynae (Suborder Mesothelae)

Austrochiloidea

Eresoidea

RTA clade

Deinopoidea

Nicodamidae

Linyphioids

Mimetus

Arkys

Tetragnathidae

Theridioids

Araneids and relatives

(Suborder Araneomorphae)

Orbiculariae clade

Araneoidea

2600

3400

25

360

34,600

34,575

34,215

21,900

11,900

Web suspended
origin of orb

Viscid silk

11,600

38,000

300

4400

1120

2500

3200

Substrate defined webs
Prey capture web lost
Orb web
Cobweb
Stereotyped aerial sheet web

Fig. 2. Association between web spinning and species diversity; the numbers represent species. Modified with permission from Ref. 11.

consisting of a dry pseudoflagelliform silk. Cribellate silk is manually combed by special devices on the hind legs of the spiders and is placed between two threads of the pseudoflagelliform silk.[13] Cribellate silk is still seen in the Deinopoidea super family of spiders. It forms a tufted surface on the capture spiral with sufficient surface area for van der Waals forces to hold onto the prey.[14,15]

The ecribellate spiders lack the cribellate silk gland. Ecribellate spiders including the Araneoidea, a sister group to the Deinopoidea, have replaced the pseudoflagelliform and cribellate silk capture thread with a more elastic flagelliform silk coated with an adhesive viscid glue (aggregate silk) that is laid down as the orb web is spun, meaning that a similar sized web can be spun in one-sixth of the time of cribellate webs.[16]

The orb web, whether produced by an ecribellate or a cribellate spider, is an impressive piece of engineering with varying elasticities and strengths in the component threads. In the case of orb webs made by *Araneus sericatus*, they are capable of trapping objects with a kinetic energy of up to about 2×10^{-4} J, sufficient to capture a house fly, but, interestingly, not strong enough to capture larger insects. This is most likely to prevent the spider from having to deal with a prey that can potentially injure the spider while it struggles. The various structural components of a web are all held under tension, with each component of the structure having different ranges of tension, and this is balanced out throughout the web, even though there may be varying numbers of radii on the top and bottom halves of the orb web.[17] Spiders do not need a specific orientation to spin orb webs, as it has been shown that it is possible for spiders to spin orb webs in microgravity with little difference in structure. The only major apparent difference is that in microgravity there was no difference between the top and bottom half of an orb web. This is most likely due to the lack of gravity supplying a reference point and its effect on the weight of the structure.[18] Additionally, it has been shown that orb-weaving spiders can sense the amount of the silk components that they have available. Spiders that are starved and prevented from consuming their own webs, which is a common occurrence in orb weavers, can ration out there remaining stocks to ensure that a complete web is spun but with smaller radii and a decrease in spiral turns in the web as well as spacing of the aggregate droplets that make the capture thread sticky.[19]

The species within the Orbiculariae clade account for roughly one-fourth of current spider diversity; however, most of these species no longer spin orb webs but have evolved to spin cob webs (Theridiidae) and sheet webs (Linyphiidae), or bola (Mastophoreae), or abandoned using silk as a prey-capture technique altogether, indicating that the orb web is just a stepping stone in the evolution of one method of prey capture. The largest spider clade, the retrolateral tibial apophysis (RTA) clade, mostly do not use silk for prey capture at all, highlighting the specialist nature of the orb weavers. It appears that spiders have evolved toward improving and simplifying the methods of prey capture to reduce the required energy expenditure for feeding, with the later cob webs of the Theridiidae relying on less adhesive viscid glue and some webs of Linyphiidae lacking sticky silk altogether, while the bola of the Mastophoreae requiring a small silk fiber and just a droplet of sticky silk at the end of the thread. However, these other structures are often used to capture different categories of prey such as terrestrial prey, or specific species of prey. For example, Mastophoreae tend to target specific moth species, but are still opportunistic enough to try capturing nearby prey with their sticky bola. As such, it seems that the development of the orb web resulted in the evolution and expansion of the repertoire of spider silks and their properties, which led to the development of new methods of prey capture that would not have been possible with the earlier silks.

Researchers have concentrated their efforts on the orb-weaving spiders, specifically the ecribellates, as their silk appears to have the best properties. Thus the main species that are discussed here are those belonging to the Araneidae family with most research concentrating on species from the genii *Argiope*, *Araneus* (specifically *Araneus diadematus*), and *Nephila* (specifically *Nephila clavipes*). Additionally, much research has been undertaken on the *Lactrodectus* genus, more commonly known as "black widow spiders," from the Theridiidae which spin cob webs but still utilize the same range of silks as the orb weavers. Early research has concentrated on the MA silk for two reasons. Firstly, this is the standard silk produced by spiders, which acts as a lifeline or dragline, the more common name for this silk, and thus is easy to collect. Secondly, this silk appears to be the strongest silk, as it is used for the frame and radii of the orb web. Since this is the best characterized spider silk, we start by describing it in detail.

III. Nomenclature of Silk Types

Spiders produce a wide range of silks that have a variety of functions. As a consequence, the different silks have evolved to have different properties.[20] Orb-weaving spiders have the largest selection of silk glands, with adult female spiders having up to seven different glands. Here we describe the details of the different silk proteins that are produced by spiders in relation to the glands that they come from as well as the function of the different silks. Therefore, we highlight the primary sequence of the various silk proteins and the physical properties of silk threads and how these two features are linked to each other. It is worth noting the naming convention for spider silks: the name consists of the first two letters of the gland where it is produced, followed by "Sp" for spidroin. Sequences with no orthologs or where the gland of origin is unknown are classed as fibroin and named after the species they were found in; for example, ADF-1 is the first fibroin from *A. diadematus*. As a consequence, silks from the same gland may have multiple names, sometimes because there are multiple proteins, and at other times because analogous proteins are given different names after how they were discovered. Additionally, some silks can be found in glands other than their "main" gland.

The silk that is best characterized is produced by the MA gland and, as previously mentioned, is called MA silk. This silk is present in all spiders, as one of its primary functions is that of a dragline, designed to take the weight of the spider should it fall. As such, this silk is used to describe the general process by which silk fibers are produced. Additionally, MI silk is described at the same time, since, although typically produced as a different silk from a different gland, it has physical properties similar to those of MA silk. For all the other silks, less is known about their material properties and their production, and therefore only limited information is available thereon.

IV. Set-up of Silk Proteins

A. General Set-up

It appears that the fundamental protein sequences of spidroins are very similar to one another. All tend to be very large proteins, mostly in the order of 200–350 kDa or larger. They consist of three main components: a central repetitive core with nonrepetitive N- and C-termini.

B. Repetitive Domains of Spidroins

The core repetitive regions of the spidroins make up the vast majority of the protein and these are highly specialized to the different silks. The general structure is composed of short sequences that are repeated a certain number of times to create a modular unit. This modular unit is then copied a number of times, creating a sequence with multiple levels of hierarchy. The exact sequences and details of the repeats of these regions are discussed in the relevant sections.

C. Nonrepetitive Domains of Spidroins

The nonrepetitive domains within a spidroin are homologous to one another. It appears that they have a highly conserved nature among the different silk types and even among phylogenetically distant spider species.[21–27] These domains comprise approximately 100–150 amino acids[25,26,28] and, in context of the more than 3000 amino acids of the fully sequenced black widow MA spidroins (MaSps),[26] they would seem to be almost negligible structural or functional moieties.

Polyclonal antibodies derived against fusion proteins containing the conserved C-terminal regions of both spidroin 1 and 2 from *N. clavipes* provided the first evidence that its C-terminal domain is not cleaved off in the gland and is retained in the dragline fiber.[29] Moreover, the stabilization of MaSp dimers[29,30] seems to result from a conserved cysteine residue present in the C-terminal domain. Predictions of the secondary structure,[25,31] as well as a recently published NMR solution structure of the ADF-3 C-terminal domain at atomic resolution,[32] have shown that the C-terminal domains adopt a predominantly α-helical conformation *in vitro*. The solved structure is a new protein fold composed of a parallel-oriented dimeric five-helix bundle in which the longest helix contains the single cysteine residue, thus being the main dimerization site (Fig. 3A).[32] In the folded state, the hydrophobic residues are buried and the hydrophilic amino acids such as glutamine and serine are exposed to the surface.

Because of the repetitive nature of the spidroin's messenger RNA, resulting in secondary structure formation of the mRNA, and of large transcript sizes that exceed 10 kb, it has been quite difficult to obtain sequence data encoding N-terminal sequences. The first sequence came from flagelliform spidroins.[33] Sequences of genomic clones and complementary DNA allowed the derivation of several dragline silk N-terminal coding regions and their comparison with flagelliform or tubuliform N-terminal regions.[23,24] The analysis of these data revealed the presence of a secretion signal, a high conservation between spider species, and the hydrophilic nature of spider silk N-terminal domains, but further functions had remained unassigned. Recently, more insight into its structure and function has been brought about by the X-ray structure of purified *Euprosthenops australis* N-terminal domain of MA dragline silk, as well as related experiments performed on recombinant miniature MA-like spidroins.[34,35] Similar to the aforementioned *A. diadematus* C-terminal domain,[32] the X-ray structure of *E. australis* N-terminal domain at pH 6 shows a homodimer of dipolar, five-helix bundle subunits that lack homologs (Fig. 3B). However, in contrast to the C-terminus, this dimer possesses an antiparallel orientation, and, additionally at pH 7, it is monomeric.

In the following we discuss MA and MI ampullate silks, including the spinning process of MA silk. However, we do not discuss the differences in the various silk glands and spinning processes. It is known that shape and structure of the various spider silk glands are different, with most being less complex than the MA silk gland and often being just an elongated, round lumen leading to a spigot. Additionally, the spigots of the different glands vary in morphology, number, and position, in addition to also varying between sexes,

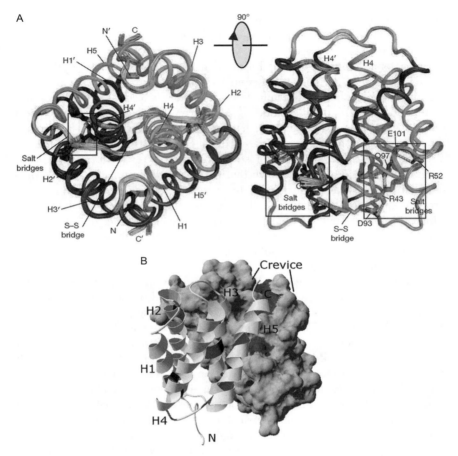

FIG. 3. NMR and X-ray crystal structures of the N- and C-termini, respectively. (A) NMR structure of a dimer of the C-terminus, monomers in dark and light gray. Highlighted regions are the residues involved in the charge interface.[32] (B) X-ray crystal structure of the N-terminus, one monomer shown as a space-filling model and the other as a ribbon structure. Conserved residues are shown in dark gray, structure based on the PDB file 3LR2.

age, and species. The size of the spigot tip obviously has an influence on the width of the thread formed, and the number of spigots determines the number of silk fibers that form a thread. Further details of the arrangement of the silk spigots are beyond this review but have been characterized in a number of papers and books for a wide number of species.[4,6,36] In the following, we discuss the spidroin sequences and their properties; in this sense, we compare all the silks to one another.

V. Major Ampullate Silk (aka Dragline Silk, Lifeline Silk) and Minor Ampullate Silk

A. Major and Minor Ampullate Gland

The MA gland of orb-weaving spiders produces spidroins (MaSps) that form the dragline silk, which is the thread with the most characterized mechanical properties. Dragline silk competes with man-made, high-technology fibers especially in terms of toughness[37] (Table I). Most studies have focused on the solid dragline silk, partly because of its relatively simple collection by "milking" [39] and also of its unique mechanical properties.

Generally, large-scale spider silk production is hampered by the fact that most spider species are cannibals and therefore farming is not feasible. The molecular composition of the fiber and the spinning process both contribute to the observed mechanical properties. The MA gland represents a highly specialized secretory system and is anatomically subdivided into a long tail (A-zone) and a wider sac (B-zone; Fig. 4). The A-zone fulfills the main secretory function of the gland, possessing one type of endothelial columnar cells. These cells are capable of extensive silk production, since they harbor a large endoplasmic reticulum and large numbers of secretory vesicles filled by short, narrow filaments. The cells of the B-zone morphologically resemble those of the A-zone, but their secretory vesicles contain hexagonal columnar liquid crystals, instead. These secretory vesicles are believed to produce glycoproteins[12] covering the A-zone secretion flowing toward the funnel (Fig. 5B, Section 1), a thickened cuticular structure that connects the ampullate gland to a S-shaped spinning duct. The silk solution, or spinning dope, flows continuously under intra-abdominal body pressure from the B-zone down the first and second limb of the duct. The drawdown process starts in the third limb, where the protein solution is abruptly pulled away from the cuticule. The epithelium progressively increases in height (Fig. 5B, Section 2) facilitating ion exchange, acidification, and extraction of water between the first and third limb of the duct. An additional coat of lipids and glycoproteins is proposed to be added to the thread in the drawdown taper.[43,44] The spinning duct connects to a structure called the "valve" (Fig. 5B, Section 3), which is surrounded by tall columnar cells specialized for rapid water recovery. The valve serves as a clamp for gripping the thread, and it may also be used as a pump to push out a broken thread, thus renewing the spinning process. The whole spinning apparatus terminates at the spigot (Fig. 5B, Section 4) which possesses flexible lips that are thought to assist in water retention by the final stripping off of the surface water.[12,28]

TABLE I

COMPARISON OF THE MECHANICAL PROPERTIES OF *A. DIADEMATUS* DRAGLINE SILK AND OTHER NATURAL AND SYNTHETIC FIBERS IN THE ORDER OF TOUGHNESS

Material	Density (g cm^{-3})	Strength (GPa) (breaking point)	Stiffness (GPa) (Young's modulus)	Extensibility (%) (elongation to break)	Toughness (MJ m^{-3}) (work to fracture)
A. *diadematus* silk (dragline)	1.3	1.1	10	27	180
A. *diadematus* silk (flagelliform)	1.3	0.5	0.003	270	150
Nylon fiber	1.1	0.95	5	18	80
Bombyx mori silk (cocoon)	1.3	0.6	7	18	70
Wool (at 100% RH)	1.3	0.2	0.5	5	60
Kevlar 49™ fiber	1.4	3.6	130	2.7	50
Carbon fiber	1.8	4	300	1.3	25
High-tensile steel	7.8	1.5	200	0.8	6
Elastin	1.3	0.002	0.001	15	2

RH, relative humidity. Data taken from Refs. 37,38.

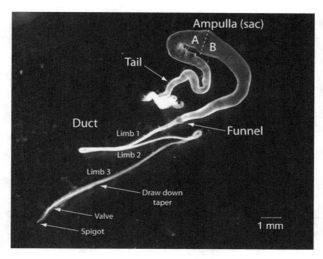

FIG. 4. Typical image of the MA gland of *N. clavipes*. All tissues surrounding the gland and the duct were removed, conserving only the anterior tissue attaching the tail to the body wall. The different parts of the glands are indicated. Adapted with permission from Ref. 40. Copyright 2008 American Chemical Society.

B. Spinning Silk

The spinning process of dragline silk is based on the conversion of a highly viscous and concentrated aqueous protein solution into an insoluble fiber at ambient temperature and pressure. The protein solution secreted into the tail of the gland comprises up to 50% (w/v) protein.[12,45] Because of the amphiphilic character of spidroins, they are thought to form a micelle-like structure.[32,44–46] Small droplets were observed in the gland, which first coalesce as the solution flows distally along the gland and subsequently become elongated in the duct into thin structures called canaliculi.[46]

The high concentration of spider silk protein and consequently the high viscosity of the solution in the dope is a prerequisite for the production of fine streams that do not undergo capillary breakup. Low-concentration aqueous solutions do not readily produce continuous streams/threads of material, because the surface tension of water results in the stream breaking up into droplets.[2] Additionally, the spinning dope in this high protein concentration condition has properties of a crystalline liquid within the first and second limbs of the duct. Molecules in these regions exhibit nematic ordering, where molecules all have their long axis aligned with one another.[46–48] Such behavior is observed quite often for biopolymers, which self-assemble into fibrillar structures. They can orient and pack into lyotropic mesophases in aqueous solution

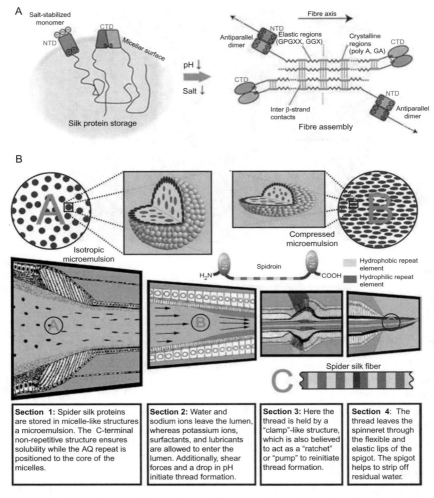

FIG. 5. (A) Mechanism of the initiation of fiber assembly including the pH- and salt-dependent role of the N-terminal domain of spider dragline silk. During silk protein storage at high
protein concentrations, both N- and C-terminal domains are most likely located at the surface of the
formed protein micelles. At fiber-forming conditions (lower pH, less salt), the N-terminal domains
are able to dimerize in an antiparallel fashion. CTD, C-terminal domain; NTD, N-terminal
domain.[41] (B) Representation of macroscopic processes in the dragline spinneret of a spider. AQ
in section 1 are modules in the repetitive region of the spidroin (A, hydrophobic poly-alanine-rich
motif: GPYGPGASA$_6$GGYGPGSGQQ; Q, hydrophilic glutamine- and glycine-rich motif:
(GPGQQ)$_4$).[42] Copyright Wiley-VCH Verlag GmbH & Co. KGaA. Reproduced with permission.

at sufficiently high concentration. Nematic, chiral nematic (cholesteric), or hexagonal columnar phases are the most commonly observed mesophases for several classes of biopolymers including DNA, peptides, polymer/peptides conjugates, and glycopolymers.[49]

The alignment of spidroin molecules in a liquid crystal fluid maintains the extensional viscosity, preventing capillary breakup, but decreases the flow viscosity by an order of magnitude. Upon moderate increase of shear rate, a phenomenon known as "shear thinning" is observed in the tapering spinning duct. Shear thinning is one factor that allows the formation of fibers under the low pulling forces of a spider just walking along a surface. The Newtonian laminar flow of a common solution, in which the shear rate determines the stress forces and not the viscosity, would require much higher pulling forces.[2,12,50,51] The long axis of the elongated proteins in the liquid crystalline solution undergoes rearrangement during spinning, from perpendicular to the fiber axis in the A-zone to parallel in the B-zone and the first limb of the spinning duct. In the narrower second limb of the duct, molecules are bent progressively forward and backward, producing an alternating pattern known as the "cellular optical texture" caused by ordering of molecules into a layered disk.[46,52] The hyperbolic geometry of the duct from the funnel to drawdown taper results in a practically constant elongation rate of the spinning dope.[46] This generates a low and uniform stress, which may prevent coagulation seeding before the molecules are arranged for spinning,[28,53] as proper molecular arrangement determines the threads' toughness.[12,53] In the third limb of the duct, the low and uniform forces of elongational flow suddenly change to a much higher stress as the forming fiber starts to be pulled away from the cuticule and is stretched. These abrupt changes bring aligned proteins close enough to form intermolecular hydrogen bonds, resulting in changes in the protein secondary structure, such as β-sheet formation.[40,54]

The geometry of the drawdown taper also changes from a hyperbolic to a much steeper exponential curve.[46] Protein aggregation and local crystallization increase the hydrophobicity of the protein surfaces through burial of hydrophilic residues,[12] initiating phase separation and exclusion of water, respectively, which is facilitated by the epithelial cells' active extraction of water. The phase separation is also supported by the ion exchange of chaotropic sodium and chloride ions for the relatively kosmotropic potassium and phosphate ions.[55] An additional drop in pH, controlled by a proton pump, may minimize repulsion forces between spidroins as a result of the protonation of the amino-acid side chains.[55–57] The final drawdown and stretching occurs in the air as the fiber leaves the spigot and may be accompanied by further evaporation of water[12] (Fig. 5B).

The effect of the nonrepetitive termini on assembly has been brought into focus by experimental studies performed with recombinant proteins containing shorter repetitive sequences of MaSps. It has been revealed that the C-terminal domain could be a prerequisite of the supramolecular self-assembly process. This was manifested in insect cells where the formation of a fiber-like structure was observed only if this domain was a part of recombinant spidroins.[25] Additional effects on higher order supramolecular assembly of the MaSps were shown in the form of fully reversible lower critical solubilization temperature behavior[58] as well as by shear-dependent aggregation where the repetitive core assembled into an aligned fibrillar structure only if furnished with the C-terminal domain.[32,59] The folded C-terminal domain seems to be stabilized by the chaotropic nature of sodium chloride, which keeps the repetitive core in solution.[32] Such findings explain why spidroins are stored in the gland in the presence of NaCl (100–150 mM).[55] Additionally, the C-terminal domain seems to function as a shear-sensitive switch, since partial unfolding of this domain, observed under shear stress, leads to exposure of more hydrophobic residues.[32]

In vitro aggregation assays performed on recombinant mini-spidroins, comprising the N-terminal domain and/or the corresponding C-terminal domain, clearly demonstrated that mini-spidroins furnished with N-terminal domains are prone to dimerization under slightly acidic condition (pH = 6.3, condition in the spinning duct). Interestingly, at higher pH values (pH 6.9 and higher, storage conditions in the MA gland) such spidroins are stabilized in solution.[35] Recent NMR and light scattering studies on the N-terminal domain of *Latrodectus hesperus* confirm that a combination of pH and salt concentration controls the dimerization. At neutral pH and high salt concentration, the N-terminal domain is a monomer. As the pH is lowered toward pH 6 and/or the salt concentration is reduced, dimerization occurs.[41] Mini-spidroins furnished with the C-terminal domain or only the "naked" repetitive core do not reveal similar pH-dependent aggregation probably due to the 4-repeat core being too short.[35] It has been proposed that changes in salt concentration and composition in the spinning duct result in a partial destabilization of the C-terminus, which thus allows the associated repeat sequences to start forming β-sheet-rich regions. Additionally, the lower pH causes the N-terminus to dimerize in an antiparallel fashion to create head-to-tail dimers of dimers.[41]

The role of MA silk C- and N-termini in storage of spidroins as well as in triggering the spidroin assembly during the spinning process could be summarized as follows:

i. Both domains seem to prevent spidroin aggregation and stabilize a solution competent state under physiological conditions, that is, the presence of chaotropic sodium and chloride ions and neutral pH in the lumen of the gland.

ii. As the conditions change significantly in the lumen of the duct, a continual increase in a shear stress and a change in pH cause partial unfolding of the C-terminal domain, triggering ordered spidroin assembly,[41,59,60] and a drop in pH leads to an increased oligomerization triggered by the titratable N-terminal domain.[35]

iii. The concomitant interplay of shear and pH triggers is plausibly enhanced by the third crucial factor, the exchange of chaotropic versus kosmotropic ions that promotes β-sheet structure formation and assembly of the amphiphilic repetitive core.[60,61]

iv. These changes triggered at the molecular level are probably further interlinked by organization of the individual spidroins into higher ordered molecular structures. On the basis of strong parallel (C-terminus) and antiparallel (N-terminus) dimerization propensities of the nonrepetitive domains, we can hypothesize the formation of oligomeric branched structures or a kind of spidroin network (Fig. 5A).

v. The presence of the C- or N-terminal domains leads to formation of vesicle-like supramolecular assemblies (Fig. 5).[35,42,59,62] Such emulsification would enable faster dehydration and polymerization of the spinning dope. Macroscopic changes into a vesicle shape with coalescence and elongation upon shear stress may hypothetically be responsible for the fibrillar appearance of the dragline fiber.

Nevertheless, the hypotheses stated in (iv) and (v) have yet to be proven by further efforts to fully understand the mechanism that leads to the fiber having properties that are "stronger than steel and lighter than air."

C. Major and Minor Ampullate Spidroins

The MA and MI spinning dope of the most studied genera, namely, *Nephila, Argiope, Latrodectus,* and *Araneus*, consists of two proteins called MA spidroin 1 and 2 (MaSp1 and 2) and MI spidroin 1 and 2 (MiSp 1 and 2). In the case of the European garden spider *A. diadematus*, the analogous proteins are named *A. diadematus* fibroin 2, 3, and 4 (ADF-2, -3, and -4) for MaSps and ADF-1 for MI spidroins. All these share several common features: (i) high molecular weight ranging from 200 to 350 kDa or larger,[30,63] (ii) dimerization and oligomerization via cysteine bridges,[64] (iii) unusual amino-acid composition having more than 50% of alanine and glycine residues of total content (Fig. 6), and (iv) a core polypeptide, comprising highly repetitive amino-acid motifs, flanked by nonrepetitive carboxy (C-) and amino (N-) terminal domains.[23,28]

The repetitive units (Fig. 7) of MaSp1 are composed of a few GA motifs linked to poly-alanine sequences $(A)_n$ ($n = 4$–12) that are followed by several GGX (X = Y, L, and Q) repeats. MaSp2 typically contains more proline residues,

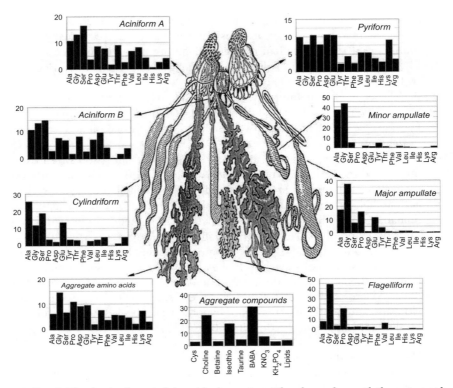

Fig. 6. The glands of a typical Araneid orb-weaving spider, shown along with the amino-acid composition of the spinning solution in the lumen of each of these glands. Reprinted by permission from Macmillan Publishers Ltd: Ref. 12, copyright 2001.

reflecting the fact that GGX motifs are alternated with GPGXX repeats (X = Q, G, Y).[28,66] These repetitive motifs have been shown to be clustered into ensemble repeats consisting of 20–40 residues where the glycine-rich region is terminated by poly-alanine stretches.[67–69] In the case of fully sequenced MaSp1 of the black widow spider, *L. hesperus*, it has been shown that these motifs maintain their repetitive pattern for the entire length between C- and N-terminal domains.[26] Interestingly, in *A. diadematus* two MaSp2-homologues, ADF-3 and -4 have been identified.[22] *Nephila* and *Latrodectus* MA spinning dopes comprise only 1–2% of proline since MaSp2 represents only its minor part. *Araneus* and *Argiope* MA glands are more elastin-like, containing proline residues in the range of 10–12%.[70,71]

A closer analysis of hydropathicity of these spidroins among the spider species reveals a pair of hydrophobic and hydrophilic counterparts.[72] The ratio of MaSp1 and MaSp2 varies between the species, with the tendency for MaSp1

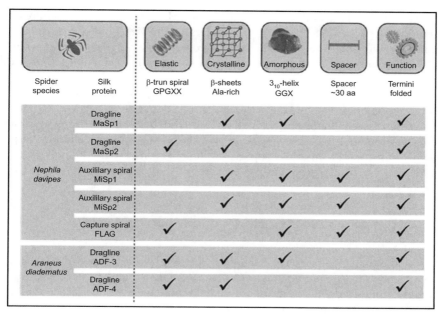

Spider species	Silk protein	β-trun spiral GPGXX (Elastic)	β-sheets Ala-rich (Crystalline)	3₁₀-helix GGX (Amorphous)	Spacer ~30 aa (Spacer)	Termini folded (Function)
Nephila davipes	Dragline MaSp1		✓	✓		✓
	Dragline MaSp2	✓	✓			✓
	Auxililary spiral MiSp1		✓	✓	✓	✓
	Auxililary spiral MiSp2		✓	✓	✓	✓
	Capture spiral FLAG	✓		✓	✓	✓
Araneus diadematus	Dragline ADF-3	✓	✓	✓		✓
	Dragline ADF-4	✓	✓			✓

FIG. 7. Structural motifs occurring within the primary structure of spider silk proteins determining their mechanical properties. X indicates a residue that may vary within or between proteins. The spacer represents nonrepetitive but conserved regions that disrupt the glycine-rich repeats. Reprinted from Ref. 65. Copyright 2008, with permission from Elsevier.

to be more abundant.[73] Therefore, one key difference between the spidroins of MA silk is their hydropathicity, as this is a conserved feature across all species, with the larger ratio component being more hydrophobic than the smaller ratio component. Moreover, immunostaining of radial dissections along the MA gland and thread of *N. clavipes* has revealed heterogeneous distribution of spidroin 2 clusters concentrated within the fiber core and embedded in a spidroin 1 matrix, indicating a phase separation.[74] This supports the concept that one key difference is the hydrophobicity of these proteins. Apart from the discontinuous distribution of two major components, the thread alone reveals a core–shell structure where the core of MaSps is covered by a skin composed of a MiSp-like protein followed by glycoprotein and lipid coats.[75]

The mentioned MaSp silk sequence features cannot be generalized among the different types of glands since, for example, tubuliform, aciniform, or egg case silk comprise substantially different repeats: that is, motifs based on serine or glutamine or very long ensemble repeats comprising 200–400 amino acids, which will be discussed later.[68] Considering MA silks only, the mentioned composition is typical for "modern" spiders of the Araneoidea superfamily; however, many non-orb weavers that are progressively further removed evolutionarily share fewer of the specific repeats seen in Araneoidea, but are however internally repetitive.[22]

The MI gland shares a similar morphology to the MA gland, and even the composition of MiSp 1 and 2 is very similar to the corresponding MaSps. They also comprise GGX and shorter A_n ($n = 2$–4) stretches (Fig. 7). MaSp's typical longer poly-A motifs are here replaced by longer $(GA)_n$ repeats. Additionally, in both MiSp proteins multiple copies of a unique conserved, 130-amino-acid, nonrepetitive, serine-rich spacer were found that alternate with the repetitive regions.[68,69]

D. Secondary Structures of MA and MI Spidroins in Glands and Fibers

Studies of native proteins in an intact MA gland were performed on several species such as *N. clavipes*, *N. edulis*, *A. diadematus*, and *L. hesperus*. Spidroins, while stored in the gland, are thought to be mainly intrinsically unfolded adopting loose and dynamic helical conformations. NMR, ultraviolet CD, and infrared spectroscopy experiments revealed that the structural content of the spinning dope is about 30% α-helices, 40% "random coils," and 30% β-turns, thus resembling the silk-I structure found in the *Bombyx mori* silkworm dope.[40,45,67,76] There is additional evidence that poly-alanine may adopt a polyproline II-like (PPII) conformation (3_1-helix).[77,78] The presence of PPII conformation could be important for the stability of the highly concentrated dope, since the PPII extended structure is stabilized by backbone hydrogen bonding with water molecules. Additionally, the dihedral angles of PPII are very close to those of a β-strand, and thus the energy barrier for the formation of β-sheets may be small.[79] Indeed, the MA spinning dope in the gland seems to be in a metastable state with a highly concentrated protein in a solution state, whereas it readily converts to the solid fiber if mild forces in the spinning duct are applied. Using polarized Raman spectroscopy, the conformational and orientational transformation was followed along the spinning apparatus of *N. clavipes*.[40] The major conformational change occurs in the third limb of the duct near to and beyond the drawdown taper (Fig. 5) as a result of shear forces applied against the cuticle. A significant portion of the β-sheet structure forms in this region with a continuous increase up to the spigot.[40]

Nowadays, it is quite firmly believed that the poly-alanine stretches in the solid thread adopt a β-sheet conformation and form the crystalline domains with sizes around $2 \times 5 \times 7$ nm and with antiparallel oriented strands packed into orthorhombic unit cells.[80–83] This conformation and orientation has been repeatedly confirmed using different NMR techniques[45,66,84–87] and X-ray diffraction (XRD)[80–82] in the late 1990s and early 2000s. According to these data, the β-sheet crystals represent the highly ordered fraction of the fiber oriented in line with the fiber axis. Nevertheless, it has been shown that in MA silk only 40% of alanine β-sheets are highly ordered and the other part (60%) exists as poorly aligned β-sheet regions.[84] The model of structural preorganization known as "string of beads" was proposed for antiparallel folding of poly-alanine stretches[84,88,89] assuming a reversal of the chain direction at the points in the sequence where GX occurs (where X is S or N), which results in a folded hairpin structure with about six peptide segments per fold and seven folds per MaSp. Molecular dynamic simulations on poly-alanine aggregation have also shown that the antiparallel orientation in the H-bonding direction and a parallel stacking in the side-chain direction lead to the most stable β-sheets.[90–92]

The conformation of $(GA)_n$ and GGX repeats in MA and MI silk is, in comparison to the alanine β-sheets, less understood. NMR studies provide evidence showing these motifs in a fraction of β-sheets as well as in a fraction of the less ordered helical structure.[66,87] XRD studies report on 10–15% crystalline volume fraction in MA silk,[80,93] but it is possible that only the poly-alanine runs are ordered enough to diffract, since NMR and Raman spectroscopy studies[66,84,94] have shown higher content of β-sheets ranging from 30% to 40%. According to these NMR data, a significant fraction of $(GA)_n$ and GGX motifs seems to be included in large, less ordered β-sheet regions.

The disordered glycine-containing fractions of spider MA silk were also addressed in several NMR studies with ^{13}C-enriched glycine.[83,86,87,95] A regular helical structure with threefold symmetry similar to the poly-glycine II 3_1-helical structure was proposed for these regions, which seem also to be oriented along the fiber axis.[87] The secondary structure of the GPGXX pentapeptide repeat found in MaSp2 and ADF-3/4 proteins is well understood. Main structural considerations come from studies on highly extensible flagelliform silk, since this motif represents its key structural feature.[96] The GPGXX unit likely forms β-turns and, in the case of successive repetition, a spiral is formed as suggested for elastin.[22] A further description of this motif is given in the section on flagelliform silk.

In summary, the $(GA)_n$ motifs are assigned usually to the ordered β-sheet structures of MI silk, or they are thought as flanking regions of MA poly-Ala stretches, whereas the helical structure is generally ascribed to GGX motifs, for example, GGA.

With the recent exponential increase in computer power, there has been a rush to develop bottom-up molecular dynamic approaches to enable modeling of nanostructures, their assembly, composition, and mechanical properties on an atomic level.[97] Thus, the experimental findings on the secondary structure content of MA silk were also modeled using replica exchange molecular dynamics (REMD) applied on simplified MaSp1 and MaSp2 sequences of *N. clavipes*. In agreement with the experimental model, poly-alanine segments have been shown to have an extremely high propensity for forming crystalline β-sheet structures, whereas glycine-rich chains are semi-extended and quite well oriented, and predominantly possess a 3_1 conformation. The density of hydrogen bonds in this region is significantly lower than in β-sheet nanocrystals. The higher proline content of MaSp2 leads to a more disordered structure in the amorphous region.[98]

E. Nanostructural Model of Silk

The observed mechanical properties of MA and MI fibers can be explained on the basis of silk-specific repetitive motifs and their macromolecular arrangement. It is generally accepted that there are three levels of silk structural order: first, the nanostructural model of distribution and interconnection between crystalline and less ordered/amorphous phases; second, the alignment of microscopic fibrillar structures within the fiber; third, the already mentioned macroscopic order of the fiber, consisting of a core–skin structure (Fig. 8).

The arrangement on the nanoscopic level is the most studied feature of the MA silk since it has probably the greatest influence on mechanical behavior of the silk. Tightly packed and aligned β-sheet crystals of poly-alanine or $(GA)_n$ repeats are thought to be responsible for the observed high-tensile strength. However, if comparing poly-alanine and $(GA)_n$ sheets, the glycine residues do not provide side chains to form the same hydrophobic interactions as alanine residues do, which results in a lower binding energy of strands included in $(GA)_n$ crystals.[89] This is in agreement with the lower tensile strength observed for MI silks compared to MA silks.[100] Nevertheless, both types of crystallites are embedded in an "amorphous" phase responsible for the extensibility of the fiber. The amorphous phase consists of GGX 3_1-helices and/or GPGXX β-turns and β-spirals.

The first such model was envisioned by Termonia in 1994, in which the crystallites act as multifunctional cross-linkers and create inside the amorphous regions a thin layer with a modulus higher than the bulk. The amorphous chains are connected to each other as well as to β-sheet crystals by hydrogen bonds, which gives the fiber its high initial modulus (Fig. 9A).[102] Predictions of this theoretical model were already in a good agreement with observations on the stress–strain behavior of the real dragline fiber. This model was refined as further experimental data on the dragline silk structure accumulated. On the basis of an NMR study,[84] less ordered β-sheets were added to the model, and later an alignment of the hydrogen-bonded "amorphous" phase (GGX helical structures) was recognized.[87] Moreover, the presence of smaller crystals rather than larger crystallites, predicted by Termonia to be crucial for the observed

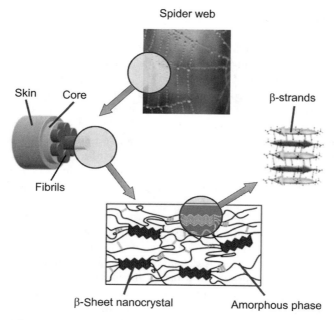

FIG. 8. Schematic representation of the hierarchical spider silk structure displaying key structural features of silk, including hydrogen-bonded β-strands, β-sheet nanocrystals, a hetero-nanocomposite of stiff nanocrystals interconnected with a softer semiamorphous phase, and silk fibrils, and the skin–core arrangement, which assembles into macroscopic silk fibers. Figure modified with permission from Ref. 98. Modified with permission from Ref. 99.

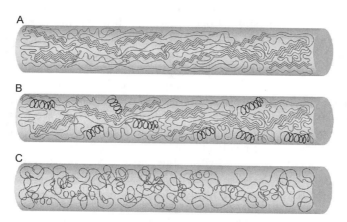

FIG. 9. Schematic models of silk structure. (A) MA, MI, and tubuliform silk. (B) Aciniform and piriform silk. (C) Flagelliform silk. Wavy lines represent β-sheets, helices represent α-helices, other lines in (A) and (B) represent the amorphous matrix and the unordered structure in (C). Reprinted with permission from Ref. 101. Copyright 2009 American Chemical Society.

mechanical properties, was confirmed by XRD and NMR experiments. Recent studies based on static and time-resolved FT-IR spectroscopy have brought further qualitative insight into this model. The crystals are tightly interconnected with worm-like chains (WLCs) of a glycine-rich matrix having a Gaussian distribution of prestrain. The resulting internal prestress is counterbalanced by the core periphery and the fiber skin.[103–105] The tight crystal–amorphous matrix interconnections provide immediate transduction of macroscopic stress to the microscopic level.[104] Apart from the reinforcement of the amorphous phase by β-sheets, nanocrystals additionally provide transfer of the lateral loading within WLCs, similarly to their function in other mechanical proteins. Such types of cohesion enable the amorphous phase to stretch significantly.[106,107] The model was further proven with MA and MI silk of different *Nephila* species, *A. diadematus*, *Cupiennius salei*, and *Nuctenea sclopetaria*. The lower prestrain observed for MI silk compared to MA silk seems to result in different initial elastic moduli and extensibilities.[105]

The above-mentioned high energetic cost for breaking hydrogen bonds within the crystalline regions is the main reason for the observed extraordinary toughness of the spider silk. It has been shown that reducing the crystal size by increasing the reeling speed or by metal infiltration has a significant influence on the toughness and ultimate strength of the fiber.[12,102] Recent molecular dynamics simulations allowed a closer look into the role of small nanocrystals in fiber reinforcement.[91,108,109] Generally, antiparallel β-sheet nanocrystals are held together by weak but numerous hydrogen bonds. It would be obvious to expect that arbitrarily increasing the size of the nanocrystals, for example, the number of amino acids per strand or the number of strands per nanocrystal, would lead to improved mechanical properties. However, an atomistic simulation reveals two different mechanisms for deformation of large and small nanocrystals, respectively. Larger nanocrystals are actually softer and fail at lower forces. The main source of the failure is crack-like flaw formation that causes rapid disintegration of the silk fiber. This may be further facilitated by water competing for hydrogen bonds.[110] Recent molecular modeling studies have suggested that the optimal length for β-sheets is eight residues, because above this distance the force applied to the sheets dissipates with the last residue not being affected by the applied force.[109] Additionally, it has been suggested that a process called "slip-stick motion" of the strand occurs within the nanocrystal, the strand slides, and then re-forms hydrogen bonds. This mechanism protects β-strands from exposure to water molecules and provides a sort of energy dissipation, resulting in enhanced mechanical properties.[91,98]

The microscopic fibrillar structure of the dragline silk was observed by SEM and AFM methods in a few cases such as *L. hesperus*,[111] *N. clavipes*,[43] and *N. pilipes*[82] and was included into a structural model of the dragline fiber.[87,104] Nanofibrils have a diameter of approximately 100 nm and are aligned with the fibril axis. The embedded fibrils are thought to provide a toughening mechanism similar to that in a rope.[112,113]

Apart from fibrils, which seem to be a topological feature of the dragline thread, coarse segmentation is apparent on the same microscopic level by methods such as transmission electron microscopy (TEM), transmission X-ray microscopy (TXM), and XRD. Larger crystals, ranging from 70 to 500 nm have been detected in TEM studies of *N. clavipes* and *L. hesperus* dragline silk. Their presence has also been deduced from analysis of XRD patterns obtained from bundles of *N. clavipes* dragline.[114–116] Recent studies on dragline silk of *L. hesperus*, exploiting scanning transmission X-ray microscopy (STXM), revealed a very fine microstructure of oriented and unoriented domains surrounded by a moderately oriented matrix, which would allow good connectivity and indicate the presence of the long-range order necessary to explain the mechanical properties of silk. Similar to previous observations, the most highly ordered regions display dimensions (average area of 1800 nm^2) that are too large to be solely formed by individual β-crystallites. The previously proposed skin[74,75] was here directly observed as an outer region with significantly enhanced orientation and a width of 120 nm.[101] All these observations of larger crystallites through the fiber support a theoretical model known as "nonperiodic lattice" (NPL) crystal model proposed for MA silk threads. The model considers alanine β-sheet regions of nearly (but not perfectly) repeated composition and order, embedded in imperfect glycine-rich areas, and in this way these regions contribute to certain material characteristics of the spider silk.[115]

F. Mechanical Properties

It is recognized that there is some variability in the physical properties of the same silks from the same spiders, and this can be related to diet, reeling speed, humidity, and temperature in some cases. As such, general figures tend to be given for the physical properties such as strength, Young's modulus, toughness, etc., and these vary between studies. Thus we mostly discuss the differences in general trends and try to compare numbers from the same studies.

The secondary structures that are assigned to different repetitive motifs have a direct influence on the macromolecular organization of the silk. If we summarize common patterns, MA and MI silk fibers could be characterized as a mixture of crystallites and amorphous flexible structures highly or less aligned with the fiber axis. Therefore, the thread could be assumed to be a composite material comprising hydrophilic/hydrophobic spidroins as major components of the core and, at least in the case of dragline fiber, also MI-like skin and outer layers of glycoproteins and lipids. All these features result in outstanding mechanical properties of spider silks, especially in the case of the dragline silk fiber (Table I).

Mechanical properties of a silk thread are probed by stretching a fiber at a specific rate, changing length per time unit (dl/dt) while measuring the force F required for a given extension Δl. The observed behavior can be expressed as stress–strain curve (Fig. 10A), where the engineering stress is defined as force per cross-sectional area ($\sigma = F/A$, the area A being averaged through the length of fiber) and strain is expressed as a normalized extensibility ($\varepsilon = \Delta l/l_0$, where Δl represents the change in length and l_0 the initial length). It has to be mentioned that the cross-sectional area of the silk fiber slightly decreases during the extension, but for calculations it can be considered to be constant. The slope of the curve provides the Young´s modulus E, representing the ratio of the stress to the strain ($E = \Delta\sigma/\Delta\varepsilon$) and expressing the stiffness of the fiber. The area under the curve expresses the toughness of the material and represents the energy taken by the fiber until it breaks. The stress–strain profile can often display sudden slope changes called "yield points," which are indicative of major structural transitions in the material.[28,38,68,117]

The typical stress–strain curve of dragline silk displays an early stiffening phase followed by an early yield point related to material softening and followed by a second stiffening phase (Fig. 10A). The short softening regime could also be observed immediately prior to failure. Moreover, if the fiber relaxes before it breaks, a viscoelastic behavior is observed. Therein, the loading and unloading paths are not the same and the relaxed thread does not recover to its original length, indicating the presence of a small plastic component.[118]

When comparing dragline silk with man-made fibers, the final strength of 1.1 GPa (*A. diadematus*) is in the range of that of high-tech materials (Table I). Nevertheless, high-tensile steel, Kevlar, or carbon fibers present several times higher strength and stiffness. However, the overall toughness, meaning how much energy is required to break the fiber, further depends strongly on the extensibility. In these terms, high-tech fibers appear to be more brittle compared to the more extensible (up to 30%) dragline threads. Assuming this and the relatively low density (e.g., 1.3 g cm^{-3} compared to 7.8 g cm^{-3} of the steel) dragline silk is by far the strongest material.[37] Mechanical properties of different *Araneus*, *Nephila*, *Lactrodectus*, *Argiope*, and other species have been extensively tested (Fig. 10B).[12,44,100,119,120] MI silk has been much less characterized since silk collection is more intricate. Data for *Nephila*, *Argiope*, and *Araneus* have been collected.[88,100,105,119] Although MA and MI silk are chemically very similar, they differ considerably in mechanical properties. MI silk has substantially lower strength (usually half that of MA silk) but, on the other hand, higher extensibility (doubled that of MA silk). In spite of the contradiction in stress–strain behavior, both silks possess similar toughness (Fig. 11).

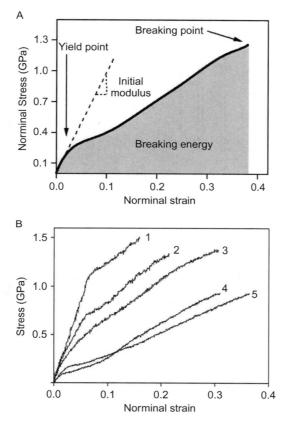

Fig. 10. (A) Typical stress–strain curve for spider dragline silk. Average data for silk collected from an adult *N. edulis* female are as follows: silk diameter, 3.35 ± 0.63 nm; tensile breaking strain, 0.39 ± 0.08; breaking stress, 1.15 ± 0.20 GPa; initial modulus, 7.9 ± 1.8 GPa; yield stress, 0.15 ± 0.06 GPa; and breaking energy, 165 ± 30 kJ kg^{-1}. (B) Stress–strain characteristics of dragline silk reeled from different web-building spiders: *Euprosthenops* sp. (Pisauridae) (1), *Cyrtophora citricola* (Araneidae) (2), *Latrodectus mactans* (Theridiidae) (3), *Araneus diadematus* (Araneidae) (4), and *Nephila edulis* (Tetragnathidae) (5). Reprinted by permission from Macmillan Publishers Ltd: Ref. 12, copyright 2001.

Generally, mechanical characteristics vary greatly between the same type of fiber from different spiders (Fig. 10B), but even the same spider can produce quite different quality of silk strongly dependent on various factors.[120] Environmental conditions may influence the amino-acid composition of the spinning dope as a result of differential regulation of the gene expression or of diet.[121–123] Additional influences such as humidity, temperature, and reeling

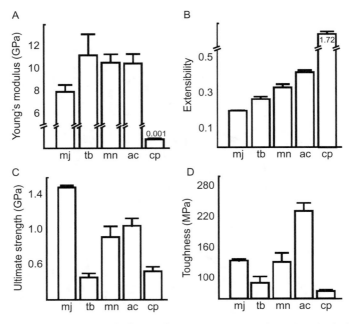

Fɪɢ. 11. Physical properties of silks spun by *Argiope argentata* shown (A–D) calculated from stress–strain curves. The silks are mj—MA silk; tb—tubuliform silk; mn—MI silk; ac—aciniform silk; and cp—capture spiral. Reproduced with permission from Ref. 100.

speed during the spinning process should be taken into account, as they can have large effects on the mechanical properties of the fiber as well as on the water uptake in the final steps of spinning.[82,124]

Another interesting feature of spider silk is its torsional dampening ability. When silk is twisted to a new equilibrium position, the thread does not oscillate around this position, like a Kevlar thread would. Additionally, the thread slowly returns to the original position, indicating that the thread has a "shape memory" of its original configuration. Most man-made shape-memory materials require some external stimuli to return to their original configuration; spider silk just requires time. All this indicates that spider silk contains a variety of torsional relaxations controlling this behavior.[125]

Additionally, dragline silk has the ability to undergo supercontraction. When a native dragline thread comes in contact with water or a relative humidity greater than 60%, the thread starts to swell radially, leading to an increase in diameter and a shrinkage in length by about 50%.[126,127] In nature, this characteristic property allows reorientation of hydrogen bonds between the spider silk protein molecules during the uptake of water, thereby plasticizing the thread and changing its mechanical properties (Fig. 12A).[124]

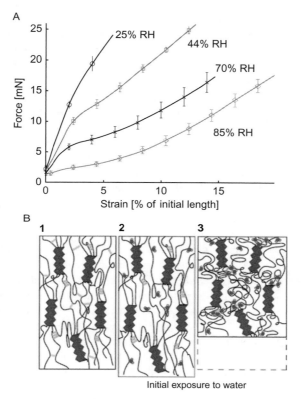

FIG. 12. (A) Effect of supercontraction on the tensile strength and elasticity of MA silk. With increasing relative humidity, MA silk responds as a more elastic material. (B) Hypothesized model of interaction of water molecules with spider silk proteins. (1) Virgin silk consists of multiple spidroins linked by crystals and the orientation of the crystals along the axis of the fiber is maintained by hydrogen bonding within the amorphous network. (2) When water molecules first penetrate the silk, it interacts primarily with hydrophilic amino acids in the random coils. The glycine-rich linker regions still maintain their overall orientation such that the random-coil network remains mostly oriented. (3) Once humidity exceeds a critical threshold of ~70%, water penetrates the glycine-rich linker regions and disrupts hydrogen bonding to allow the spidroins to reconfigure toward a higher entropic state and cause the entire silk fiber to suddenly contract in length while expanding in overall volume. The silk now behaves like a filled rubber with a relatively low modulus. (A) Reprinted from Ref. 118. Copyright 2007, with permission from Elsevier. (B) Modified with permission from Ref. 99.

The supercontraction is not a general property of spider silk and is not clearly observed in other silks. MI silk, for example, reveals a much lower degree of shrinkage upon wetting (upto 5%).[105,119,127]

A nanoscopic model sheds light on the silk supercontraction phenomenon,[99,124,126] where the change from viscoelastic to rubber-like mechanical properties is accompanied by a large decrease in elastic modulus, an increase in the extensibility, and a decrease in the stiffness (Fig. 12A). Different experimental techniques such as Raman spectroscopy,[128] birefringence,[70] XRD,[85,93] NMR,[95,129,130] and FT-IR[104] have been employed to follow molecular changes upon supercontraction. Modulating reeling speed during spinning was demonstrated to have impact on molecular alignment within the fiber[82,95] and consequently also on the fibers' capacity to shrink (Csh).[124] Formation of the supercontracted silk, comprising a poorly orientated amorphous phase, implies a ground state with Csh = 0. Soft native silk, formed at very low reeling speed, is already close to this state. Hard native silks comprise "amorphous" phases, where intermolecular hydrogen bonds fix chains in an aligned and partially elongated conformation. This process results in prestrain of the fiber. Water molecules penetrating the fiber are able to replace interchain hydrogen bonds in the loosely packed amorphous phase (Fig. 12B), leading to an increase in chain mobility and consequently to entropy-driven chain reorientation and release of the prestress reflected in higher Csh values upon wetting.[104,105,124,129] Since the crystalline poly-alanine regions are packed more tightly, they remain rigid when exposed to water.[95,130,131] Summarizing these insights, the chains of the amorphous phase collapse into so-called mobile entropic springs, whereas the crystalline phase remains intact while losing a small amount of its alignment (Fig. 12B). The apparent lack of supercontraction in the case of MI silk could be due to the observed significantly lower prestrain within the fiber.[105]

Interestingly, a certain correlation between the supercontraction ability of the dragline fiber and the proline content has an additional implication for the molecular model of spider silk. Comparisons between different spider species of three families *Araneidae, Tetragnathidae*, and *Theridiidae* revealed that the proline content is indeed not reflected in the mechanical properties of the dry as-spun fiber, but of the supercontracted fiber.[70,132] This again supports the notion that it is the difference in hydrophobicity of the two MaSps that is important for their roles in MA silk, but not their proline content. Experimental observations can be summarized as follows: the higher the proline content, the higher the Csh value and the "softer" the fiber. The role of proline residues in this behavior seems to be quite clear. Proline is known as a structure and β-sheet breaker[98] because of the torsion angles of the proline ring as well as the lack of the amide hydrogen, which is capable of hydrogen-bond formation. Thus, regions containing proline, that is, GPGXX motifs, are less ordered or deficient in β-sheets, and subsequently these regions are more accessible to water molecules, leading to facilitated formation of entropic springs. The entropic springs[95,103] are responsible for the high extensibility and rubber-like behavior of the supercontracted fiber (Fig. 12A). Applying stress on the supercontracted fiber causes the entropic springs to extend easily, which is revealed as a very low initial modulus in the stress–strain behavior.

The crystalline phase remains intact until the extension limit of the springs is reached. A similar stage is also reached upon extension of a native fiber; however, the initial situation in this case is quite different. Since this fiber contains very little water, hydrogen bonds interconnect crystallites with chains in the amorphous phase. Applied stress causes the interchain hydrogen bonds to break, which is reflected as a short, initial elastic phase in the stress–strain curve characterized by a high modulus value (Fig. 10A). The following short plastic phase has a modulus comparable to the elongation of supercontracted fibers, since it represents elongation of latent entropic springs after the hydrogen bonds have been broken. Thus both cases, supercontracted and native fibers, end up with fully extended chains and intact crystallites. Further extension leads to a stiffening phase due to chain rearrangement and breaking of the β-sheet hydrogen bonds.[95,105]

VI. Flagelliform Silk (aka Capture Silk and Viscid Silk)

Flagelliform silk is used in the capture spiral of orb weavers, specifically the ecribellate spiders. As such, it is often referred to as "capture silk," or also "viscid silk" because of its stickiness, which is related to the coating of aggregate silk, which is discussed later. Capture silk and viscid silk refer more to the composite of flagelliform and aggregate silk and thus are inaccurate terms to describe the silk thread alone.

Flagelliform silk is exclusively used by members of the Araneoidea super family, and it is referred to as a "wet silk" due to the fact that it is coated with droplets of aggregate silk. Because of its constant exposure to humidity in the natural web, it appears not to undergo supercontraction; however, a recent study of dried flagelliform silk indicates that it does undergo supercontraction.[133] As with MA silk, flagelliform silk consists of a core repetitive domain and N- and C-terminal nonrepetitive domains. The sequence is highly conserved within the repeats and also between species, indicating that concerted evolution maintains the sequence. The cDNA of the flagelliform silk from *N. clavipes* and *N. madagascariensis* indicates that the gene for flagelliform silk is split up over 13 exons covering 30 kb, rather than one huge exon as is the case with other spidroins. Between the two species studied, the nonrepetitive regions are the most highly conserved parts of the sequence, differing by less than 5%. The repetitive regions contain differences at the genetic level between the species, but there are clear indications of concerted evolution due to the high degree of similarity at the DNA level between the introns and exons within each species. As with MA sequences, there is an extreme level of bias in the use of codons toward those that end in C or G.[134] The main motif seen in the repetitive domain is Gly-Pro-Gly-Gly-(Xaa)$_n$ where Xaa is generally Ala, Ser, Tyr, or Val. This motif can appear over 40 times before it is interrupted with a second motif Gly-Gly-Xaa, which is only repeated a few times before a spacer, such as THEDLDITIDGADGPITISEELTI. The sequence then returns to the Gly-Pro-Gly-Gly-(Xaa)$_n$ motif. Each of these ensemble repeats contains approximately 440 amino acids and occurs 10 times in the entire protein.

The C-terminal domain of flagelliform silk is divergent and shorter than those of the other spidroins, while the N-terminal nonrepetitive sequence is very similar to that of other spidroins containing a signal peptide.[23] The sequence of the N-terminus is predicted to contain five α-helices, which, as previously mentioned, has been shown to be true for the MaSps.[35] As such, it is likely that the other silks have similar structures in their N-terminal domains. It can be assumed that flagelliform silk is the most recently evolved silk, as it is only expressed at high levels in the Araneoidea superfamily, but has been detected in spiders of the sister group Deinopoidea, even though it is not utilized. This would imply a recent development, approximately 136 million years ago,[135] in which these two groups split and then further evolved within the Araneoidea superfamily. The use of this silk in prey capture has resulted in its highly extensible property, being more clearly seen in comparison to the likely evolutionary predecessor pseudoflagelliform silk described later.

From the genetic sequence, flagelliform silk seems to lack the alanine and GA-rich repeats that are characteristic of MA and MI silk; so, as a consequence, this silk has no or very few crystalline regions as would be expected for a glycine-rich sequence (Fig. 9C). This has also been implied from solid-state NMR spectroscopy of capture threads from *A. diadematus*.[136]

FT-IR and CD spectra of flagelliform silk from *N. edulis* indicate a structure mainly comprising α-helices and β-turns, with some β-sheets and disordered structure. This agrees with the predicted structure from the amino-acid sequence of the repetitive region: GPGGXX is predicted to form β-turns which are concatenated into β-spirals.[137] Upon heating, the silk undergoes a two-step transition to an unfolded structure, most likely with the initial loss of the β-spiral structure before a further transition.[137,138] Polarized Raman spectroscopy of flagelliform silk from *N. clavipes* show amide I and amide III peaks which are broader than those of MA and MI silk as well as tubuliform silks. This indicates that flagelliform silk has a more disordered structure. Additionally, there is no detectable polarization in the spectra, indicating that there is no preferential molecular orientation in this silk in comparison to the others.[101] Raman spectroscopy of the spinning dope of flagelliform silk indicates that it has no specific structure, and maintains the same lack of structure in the final fiber structure.[139] Solid-state NMR studies of peptides based on the GPGGX repeat indicate that the structure is likely to be a mixture of type II (40%) and type I (20%) turns. This agrees with other studies, except for the proportion of each structure, which is likely due to the different ways the different techniques observe β-turns.

Flagelliform silk has been shown to have physical properties that are very different to that of the MA silks. The primary difference is the extensibility of this silk. It has been shown that it is possible for this silk to extend up to 475% of its original length. However, the silk has between a tenth and a quarter of the strength of MA silk with strengths in the order of 100–250 MPa depending on different studies and species.[17,75,140] Because of the extensibility of flagelliform silk, the toughness of this silk approaches that of MA silk at around 75 J cm^{-3} (Fig. 11D). On applying force, flagelliform silk initially offers a very low resilience until the thread has been stretched to at least 100% of its initial length, at which point it begins to stiffen.[17,100] Repeated extensions of flagelliform silk results in a slight shift to a higher resilience on subsequent extensions, but this can be reversed by allowing the silk to relax for a period of time, ~5 min,[17] highlighting the elastic properties of the material and that these properties are most likely related to noncovalent rearrangements rather than the rupturing of chemical bonds in the proteins. All these features are due to the requirement of the capture spiral of the orb web to decelerate a flying prey to a stationary state and dissipate most of that energy so that the web does not act as a trampoline launching the prey out of the web. How the molecular structure of flagelliform silk gives these physical properties will be discussed after the introduction of pseudoflagelliform silk.

VII. Pseudoflagelliform Silk and Cribellate Silk

Pseudoflagelliform silk first appeared as an extensively used material in the capture spiral of cribellate orb webs of the Deinopoidea family, and is the evolutionary predecessor of flagelliform silk used by the ecribellate spiders. It has been shown that these two silks are very similar to each other, and that pseudoflagelliform proteins are also related to MaSp proteins. MaSp, as previously mentioned, contain $GPG(X)_n$ sequences that act as turns and linkers between the crystallites. In pseudoflagelliform silk, up to 40 copies of $GPQ(X)_n$[141] are seen, while in flagelliform silk similar sequences of $GPG(X)_n$ are observed.[22] These sequences are proposed to act as molecular springs in the fibers formed, giving the material its elastic properties. As discussed with MaSp, these sequence motifs link the crystallites together; however, in flagelliform silk there are no poly-alanine and GA sequences to form β-sheet crystallites. As such, there are no crystallites in these silks, and it is not exactly clear where the strength of these materials originates. However, the extensive runs of sequences containing proline and glycine are easier to explain in their molecular action. By comparing the repetitive regions of MA silk, pseudoflagelliform, and flagelliform silk, it is clear that the $GP(G/Q)(X)_n$ repeats are involved in modulating the extensible properties of these silks. The least extensible of these three materials is MA silk with an extensibility of only 30%. Only one of the proteins in MA silk (MaSp2) contains repeats of this motif and then only in short sequence runs of at most nine copies. Interestingly, in the case of *A. diadematus* both MaSps contain these repeats; however, ADF-3 has very short repeats of only two or three copies. Furthermore, it is of interest that the MA silk from this species has the highest elasticity: whether this is significant or not is unclear.[140]

As mentioned, pseudoflagelliform silk is spun as the capture spiral of a web in combination with cribellate silk. As a consequence, the physical properties studied are that of the capture thread, consisting of two pseudoflagelliform threads (one from each spigot), with cribellate silk combed and packed between the two pseudoflagelliform core threads, creating a capture thread with a tufted appearance. Overall, cribellate capture threads exhibit an initial elastic behavior and defined yield similar to that of MA silk; however, the stiffness of these threads is one order of magnitude lower (approximately 1 GPa) than that of MA silk (approximately 10 GPa). After the yield point, the threads become rubber-like similar to that of flagelliform silk, but the region of high compliance is shorter and the axial fibers break after being stretched to 50–100% of their original length. However, the integrity of the thread is maintained by the cribellate silk, which allows the thread to be stretched to as much as 500% of its initial length. As a consequence, in the case of the capture silk of *Deinopis spinosa*, over 90% of the work is done by the cribellate silk.[142] This is likely due to the extension and failure of the tangled masses of cribellar fibrils giving rise to rapid increases and decreases in force.

Thus, the cribellate silk acts in two ways: first, it is the sticky coating to the pseudoflagelliform fibrils that captures prey through van der Waals interactions of this silk's large surface area; second, it acts as a method of continuously absorbing energy after the main axial fibers have failed. Thus, the 40 repeats of $GPQ(X)_n$ sequence provide an extensibility of the main threads of between 50% and 100%. In contrast, in flagelliform silk, where there are 40 repeats of $GPG(X)_n$, the extensibility of the threads goes up to 475%.[100] Clearly, the replacement of a glutamine residue in the pseudoflagelliform silk repeat grants a more elastic structure to the material. AFM pulling experiments of flagelliform silk have revealed that the fibers have a modular nature similar to that of extended proteins such as titin, with individual modules unfolding sequentially. The data from these experiments support the concept that the GPGXX sequences form β-spirals and that the spacers may act as cross-links to link the proteins together acting as one thread.[103]

Interestingly, the properties of pseudoflagelliform silk are quite variable. A study into four spiders from the two families of Deinopoidea, with three different web architectures, horizontal orb webs (*Uloborus diversus*), triangle webs held under tension by the spider (*Hyptiotes cavatus* and *Hyptiotes gertischi*), and, lastly, nets held by the spider and plunged over the prey (*D. spinosa*), has been carried out. The variation in the properties of the pseudoflagelliform silk of each species is clear, with axial fibers from *D. spinosa* being the weakest, but stretchier compared to the others. Thus the properties of the silk are tailored for their specific function. In the latter case, a net that can be stretched over the prey with high strength is not needed, as the prey is not moving so fast.[142]

From an evolutionary perspective, the high extensibility of flagelliform silk is likely related to the need to replace the composite material of pseudoflagelliform and cribellate silk with the single-material flagelliform silk with similar load bearing and extensibility, and a glue like substance to replace the cribellate silks stickiness.

VIII. Aciniform Silk

Aciniform silk is used in prey-wrapping and it has also been identified in the egg cases of some spiders acting as the internal soft layer.[143] Aciniform silk consists of a C-terminal, nonrepetitive domain and 14 repeats of a 200-amino-acid sequence,[119] the N-terminal region has not yet been identified. The C-terminal region is similar to the C-termini of the other spidroins; however, the repetitive regions have no substantial similarity to the other repetitive regions of spidroins. There are very few subrepeats, the most common being poly-serine. However, the full 200-amino-acid repeat is almost identical in all the copies of the repeat within the protein from one species, with over 99.9% similarity between the repeats at the DNA level.[119] This is again an indicator that the spidroins have undergone concerted evolution which maintains the sequence through recombination. Interestingly, unlike the other spidroins, there is no codon bias in this silk, potentially easing the issues of producing recombinant versions of this spidroin. Raman spectroscopy of aciniform spinning dope from *N. clavipes* indicates that the solution is nearly 50% α-helical.[139] The final fibers from *N. clavipes*[101,139] and also *L. hesperus*[144] indicate a β-sheet structure, with the former containing roughly 30% β-sheet. Additionally, unlike MA silk, between 24% and 30% α-helix content remains in the fiber. Polarized Raman spectroscopy indicates that the β-sheets and α-helices are parallel to the fiber axis. The remainder, as indicated by the amide I band in FT-IR spectra, seem to be less orientated, acting as turns and random chains (Fig. 9B).[101] Currently, it has not been possible to identify which parts of the sequence are responsible for the various secondary structures.

Aciniform silk has the highest toughness of all the silks at roughly 250 J cm^{-3} and a strength that is comparable to MI silk, which is about one-half that of MA silk (Fig. 11). Additionally, aciniform silk has an extensibility similar to that of pseudoflagelliform silk, being able to be stretched to between 50% and 80% of its original length.[100,119] The mixture of β-sheet structures parallel to the fiber axis and also α-helices parallel to the axis may indicate that these two components are providing much of the strength and elasticity, respectively. The presence of large amounts of β-sheets parallel to the fiber axis has been suggested to impart physical strength along the fiber axis. It is easy to see that the presence of α-helices introduces extensibility, due to their ability to unfold to an extended structure, a replacement for the β-spirals seen in pseudoflagelliform and flagelliform silk.

IX. Tubiliform Silk (aka Cylindriform Silk)

Unlike most silks, tubuliform silk is only produced by adult female spiders and matches the development of the ovaries.[145] The tubuliform spidroin (TuSp1) has been found in the tubuliform gland of several Araneoidea species[144,146–149] as well as putative orthologous transcripts in the total gland libraries of *U. diversus* and *D. spinosa*.[146] The C-terminal nonrepetitive region is similar in sequence to the C-terminus of most other C-termini of spidroins, which through sequence analysis clearly indicates that this protein is a spidroin. Additionally, TuSp1 in all the studied species is expressed as one huge exon as is common with most other spidroins. Again, as with other spidroins, there appears to be concerted evolution with consecutive repeats in the sequence being very similar to each other with similarities at the genetic level reaching over 98%.[146]

The repetitive region is different from that of the other spidroins with a serine-rich and glycine-poor composition relative to the other spidroins, with 26% alanine and 22% serine residues.[147] TuSp1 does not consist of the repeats that are commonly seen in the other spider silks; instead, the main repeats are S_n, $(SA)_n$, $(SQ)_n$, and GX (where X is Q, N, I, L, A, V, Y, F, D).[148–150] Also, even though alanine is one of the most abundant amino acids in TuSp1, there are no long alanine repeats, with A_3 being the longest repeat. Interestingly, the TuSp1 equivalents in *U. diversus* show expansion regions within these long repeats which have nearly perfect repeats of ASQAGSQA or GSQA, while *D. spinosa* has expansions of GAXAGA, where X is S, V, or I. These expansions may be due to the fact that the Deinopoid TuSp1 has a bias in its codon usage with a prevalence of adenine and thymine in the third codon position.[146] The repeats seen in TuSp1 are reminiscent of those seen in the mygalomorph *Euagrus chisoseus* which only produces one silk. This implies that TuSp1 might be an ancient silk predating MaSp and MiSp silks. However, a systematic study of silks indicates that TuSp1 evolved after MA and MI silks.[150] Within the tubuliform gland, the spidroin has an α-helical structure between a content of 50%[139] and 60%,[138] the remainder of the structure being random coil. The final fiber structure is very similar to MA silk, being dominated by β-sheet structure.[85,139]

Tubuliform fibers from *N. clavipes*, *A. gemmoides*, and *A. argantata* tend to be thicker than the MA and MI silk threads from the same spiders.[100,151] Tubuliform silk has a Young's modulus similar to that of MA silk at roughly 9 GPa with also a similar extensibility. However, the ultimate strength of this silk is much lower than that of MA silk (around 300 MPa), giving a toughness lower than that of the other silks, except capture silk, around 90 J cm^{-3} (Fig. 11).[100]

The lower toughness of tubuliform silk is postulated to be related to the lack of poly-alanine repeats and the reliance on large side chains at the interface of β-sheets, which may not perfectly align, resulting in a less optimal packing.[150,152] This is supported by XRD data, indicating that there is a larger b spacing in the crystallographic unit cell than seen in MA silk.[85,150] This is likely beneficial for this material, as it is intended to protect spider eggs in the egg case, and a brittle material that does not elastically respond would be more suited to protecting the eggs. An elastic material may deform well and then, when the force is removed, will flex back to the initial position and may, in turn, damage the eggs.

X. Egg Case Proteins

TuSp1 is expressed at all stages of the reproduction process in female spiders, but its mRNA levels reach their maximum at late vitellogenesis. At the same time, two other proteins are also expressed at higher levels in the tubuliform silk gland.[145] These two proteins are Egg Case Protein 1 and 2 (ECP-1 and ECP-2), both containing alanine stretches and GA couplets but lacking a homolog to the C-terminus of the silk proteins. Both proteins are also smaller than most silk proteins, at only 100 and 80 kDa, respectively. It is postulated that these two proteins may be involved in the cross-linking of TuSp1 in tubuliform silk threads, as both possess 16 cysteine residues in their N-terminal regions.[146,153]

XI. Aggregate Silk (aka Glue)

Aggregate silk is produced only by ecribellate spiders of the Orbiculariae clade, where it is placed as droplets on specific structures in their prey-capture devices. These droplets contain a mixture of proteins, the primary component being two glycoproteins, one of 38 kDa and the other of 65 kDa, named aggregate spider glue 1 and 2 (ASG-1 and -2), respectively. Interestingly, they are encoded by the same stretch of DNA but from strands opposite to each other. ASG-2 has a repeating sequence of GSSVSGLGV, while ASG-1 has a repeating sequence of EPETPSPETE. Both are predicted to have multiple O-glycosylation sites, and analysis of proteins recovered from the glue solution indicates that these two proteins are indeed glycosylated. These two are so far the only silk proteins that have been shown to be glycosylated, and it is theorized that the glycosylation is involved in the stickiness of the glue.[154] Interestingly, the stickiness of the glue droplets correlates well with the stretchiness of the capture spiral in orb weavers, with a release force of the glue between 20% and 70% of the breaking force of the thread. The detected variability is related to the surface the glue attaches to.[155] Thus a prey pulling away from the capture spiral does not destroy the orb web. Two additional small peptides have also been detected in the aggregate glue of *L. hesperus*, namely, spider coating peptide 1 and 2 (SCP1 and 2), with molecular weights of 3.8 and 2 kDa, respectively. SCP1 has been shown to possess metal binding properties, and the SCPs may act in a similar fashion to ECP-1 and -2 seen in tubuliform silk as cross-linkers. The exact function of these two peptides and their molecular structure has not been determined yet.

XII. Pyriform Silk

Pyriform silk is used to create disks for attachment of silk to various substrates, for example as anchor points for the dragline, the frame of a web, or a sheet, or to attach egg cases to a surface. Not much is known about this material; only recently has the gene encoding this spidroin been found in *L. hesperus*. The pyriform spidroin 1 (PySp1) shares a similar C-terminal domain with most other spidroins. The repetitive region of this spidroin contains novel repetitive blocks that are not seen in other spidroins. They consist of AAARAQAQAEARAKAE and AAARAQAQAE repeats. Additional short blocks of alanine consisting of up to three alanines are seen, as well as three larger 78-amino-acid iterations which contain an exceptionally high proportion of hydrophilic amino acids that interrupt the main repeats. Pyriform silk has the highest alanine content of all spidroins. However, again these are not in long alanine repeats as seen in the MaSp sequences. PySp1 also has the highest amount of charged residues of any spidroin, combined with an almost complete lack of glycine residues. As a consequence, pyriform silk repetitive sequences are the most different in sequence of all the spidroins.[156] The only available spectroscopy data on pyriform silk from *N. clavipes* indicate that, within the gland, it has a content of roughly 45% α-helices, and the final fiber has a mixture of α-helices and β-sheets, similar to those seen in aciniform silk.[139] It is unclear whether just this protein is involved in disks or whether it interacts with other small molecules and proteins. The amino-acid content of attachment disks does not match that of PySp1, indicating that there may well be more proteins present. The attachment disks have also been shown to contain SCP1, which was originally detected in the aggregate gland of *L. hesperus*.[156]

XIII. Silk Summary

In this section, we have described the spidroins that are predominantly found in the various spider silk glands and also mentioned some of the other proteins that have been found in silk materials. These additional proteins highlight the complexity of spider silks in general. Most of the identified proteins in spider silks are the spidroins, which are very high molecular weight repetitive proteins. However, there is abundant evidence that there are other constituents of spider silks that remain as yet unidentified. These include the fact that there are glycoproteins and lipids present on the surface of silk threads. Yet, only ASG-1 and -2 have been shown to actually be glycosylated, and, as such, there are likely still more proteins to be discovered. However, their apparently low abundance in the silk makes it hard for them to be identified. It must also be borne in mind that some of the spidroins have been detected in glands other than their identified gland, often only expressed at a basal level, but still present. Whether these are incorporated into fibers is unknown, but the fact that the majority of the spidroins contain a C-terminal nonrepetitive region that is homologous between all the silks implies that there is potential for interactions. Also, because of the size and codon bias observed in most spidroins, it has been difficult to obtain full-length gene transcripts for all the spidroins, and thus only some of the nonrepetitive N-termini have been identified, but those again appear to be homologous to one another. Therefore, attempts to produce mimics of spider silks in large quantities have had a number of challenges to surmount in the past. These challenges and how many of them have been overcome will now be described as well as the challenges still to be surpassed.

XIV. Producing Recombinant Silk Proteins

While spiders produce a wide range of high-performance silks, farming spiders is problematic, as most of them are not social creatures and have a tendency to be cannibalistic. Additionally, collecting silk from spiders can be difficult and time consuming, whereas silkworms (*B. mori*) can be easily farmed and produce a cocoon, making silk collection easy. As such, with the development of genetic technologies, scientists have attempted to transfer the silk genes from spiders to other organisms to produce high yields of recombinant proteins in an efficient manner. Spider silks are of great interest to material scientists because of their outstanding properties. There have been two general routes by which suitable protein expression systems have been developed; first, transfer of partial natural spidroin genes to host organisms; and second, the generation of new genes suited for high-yield biotechnological production that mimic the repeats in spidroins, the so-called engineered spidroins.[157,158]

A. Natural Partial Spidroins

Producing proteins using natural spidroin genes has presented a number of problems which need to be surmounted. First, the favorite bacterial host for industrial-scale production of proteins, namely, *Escherichia coli*, has a different codon usage in comparison to that of spiders. Additionally, bacteria undergo homologous recombination, which removes repetitive sequences from their genome. Due to the repetitive nature of partial MaSp1, the gene expression levels of these proteins in *E. coli* are lower than expected, and also the production of various truncated forms of the protein are observed.[159] By over-expressing the tRNAs in *E. coli* for GGU and GGC, which are the two main codons used for glycine in silk genes, it has been possible to overcome this problem. A partial MaSp1 spidroin from *N. clavipes* with up to 96 copies of the repeat motif has been expressed in these modified *E. coli*, giving a protein with a molecular weight of 285 kDa, which is similar in size to that of the natural spider silk protein.[160]

Eukaryotic cells have been used to express spidroins, as there is less of an issue with codon usage. Tobacco plants have been utilized to express fragments of the MaSp1 and MaSp2 from *N. clavipes*; however, the low yields of protein from these systems make this route uneconomical.[161]

Mammalian cell lines, bovine mammary epithelial alveolar cells, and baby hamster kidney cells have been used to express the same fragments of MaSp1 and 2. These cell lines have successfully managed to secrete 110- and 140-kDa silk proteins. However, as the size of the protein increased, the yield dramatically declined.[162] This may be related to a range of factors, including inefficient transcription, low copy numbers, or a limitation of the systems within the cell to cope with such large proteins.[159] As previously mentioned, it is known that there are gland-specific pools of tRNAs for glycine and alanine to deal with the extreme bias in amino acids used in spidroins.[163]

The baculovirus expression system in the insect cell line sf9 has been used by our research group to express the partial genes for ADF-3 and -4. It was found that ADF-3 was soluble because of its higher hydrophilicity, whereas ADF-4 was insoluble, forming small filaments within the cells.[72]

Silkworms produce silk naturally, extruding it from salivary glands rather than abdominal spigots; however, its physical properties are not as impressive as those of spider silk. The presence of a natural silk production system in silkworms means that they already have the capability to produce large repetitive proteins that have a similar amino acid and codon bias as spider silk. As a consequence, there have been attempts to transfer the *MaSp1* genes to silkworms. A baculovirus containing a *MaSp1* fragment has been shown to infect *B. mori*, which then produces reasonable yields of protein up to 3% of the total soluble protein in the larvae of silk worms.[164] An alternative has been to produce transgenic *B. mori* that have been transfected with a fragment of MaSp1. The MaSp1 protein is then expressed and incorporated into the cocoon of the silkworm.[165]

B. Engineered Spidroins

All the approaches so far described have attempted to utilize fragments of actual spider genes for spidroins. However, engineering new proteins is possible, as the amino-acid sequences of most spidroins are highly repetitive with short motifs (e.g., A_n, GPGXX) that are themselves repeated in a specific format to create modules which are then repeated a number of times to create the repetitive domain of the protein. These modules within the repetitive domain are highly conserved within the respective silks due to concerted evolution. As such, it is easy for researchers to identify the modules and create new artificial genes based on the modules. The basic strategy is to design short oligonucleotide sequences that encode the various motifs seen in a specific spidroin. These short sequences are then ligated to give a module that encompasses the various motifs seen in the spidroin. Engineered genes of various sizes can then be made by ligating multiple copies of the module to generate an engineered spidroin.[166] As a result, it is easy to test the effect of the protein size and components on its production yield and purification process. There are a large number of different systems in the literature: the vast majority use the sequences from MaSp1 and 2 of *N. clavipes*, or the sequences for ADF-3 and -4 from *A. diadematus*, but some have used the flagelliform repeats from *N. clavipes*.[167–170]

Most of these engineered proteins are produced in *E. coli,* but there are also examples that are produced in plants as well as eukaryotic cells and also transgenic animals. We will briefly describe here a couple of examples: first, an example that utilizes plants. In this system, 18 short oligonucleotides encoding different motifs from *N. clavipes* were assembled into six short gene constructs, which could then be fitted together to create synthetic spidroin genes of various sizes. The genes were transferred to tobacco and potato plants, and expression yields from these plants reached levels in excess of 2% of the total soluble protein. The recombinant silk protein could be easily purified by a combination of heat treatment, acidification, and ammonium sulfate precipitation.[171] It has been possible in a similar system using a different engineered spidroin that is targeted to the seeds of *Arabidopsis*[172] to raise the expression of the desired engineered spidroin to 18% of the total soluble protein, which makes this a viable means of protein production.[173] Most of the engineered systems utilize just the repetitive region of the spidroins to generate genes. However, in a few cases the C-terminal and in one case the N-terminal nonrepetitive natural domain have been appended to the engineered spidroin. The only example that uses both the N- and C-termini is a short, in comparison to others, protein NT4RepCT, which highlights the effect of these two domains on folding and solubility of the spidroins.[35]

Another system based on spidroin 1 of *N. clavipes* utilized *Pichia pastoris* as an expression system, as it is capable of producing large proteins of at least 3000 amino acids without truncation. In this case, a 101-amino-acid module was inserted as an 8mer or a 16mer gene under the control of the AOX1 promoter. This resulted in expression of the desired gene, but, due to a combination of high gene copy number and also module deletions at the gene level, a variety of multimers of the 101-amino-acid module was produced by the cells.[174]

Our research group has developed three module units that are based on the ADF-3 and ADF-4 sequences, A and Q for ADF-3 and C for ADF-4.[31,60] These modules are designed in a way that multiple modules can be combined. Additionally, it is possible to later add the coding sequences for N- and C-terminal domains (Fig. 13).[58] All these proteins are recombinantly produced in *E. coli,* and soluble proteins can be obtained which can then be rapidly purified using a heat step, followed by an ammonium sulfate precipitation step similar to the method used to purify the engineered spidroins from plants.[171]

C. Mechanical Properties of Fibers Made of Artificial Spidroins

We will only concentrate on the attempts to generate threads of recombinant spider silk proteins; however, material scientists have utilized silk proteins in general to create a wide range of materials that have been extensively

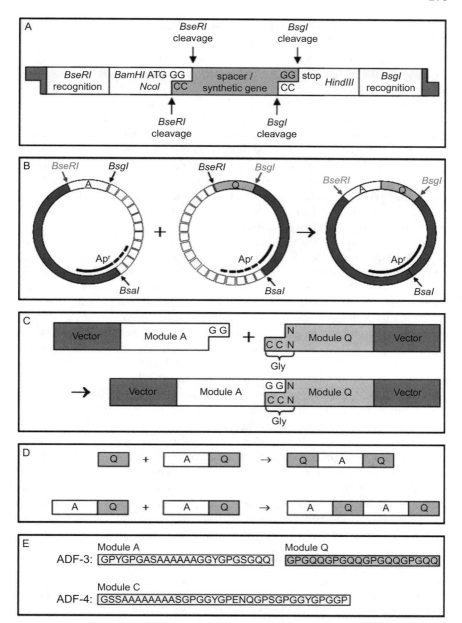

FIG. 13. (A) Cloning strategy for engineered spidroins. The cloning cassette contains restriction sites needed for module multimerization (BsgI and BseRI) and for excising assembled genes

described and reviewed elsewhere, including films, hydrogels, nonwovens, beads and capsules,[65] and composite materials.[175] There is now a range of systems for producing engineered spidroins based on MaSp or its homologs as well as a couple of systems that utilize truncated spidroin 1 genes as described above. As a result, it is now feasible to generate an artificial spinning dope of these recombinant proteins and draw fibers from these solutions and compare their physical properties to those of natural spider silk. Making individual threads from recombinant spidroins using organic solvents such as hexafluoro isopropanol (HFIP) is a routine method. This involves the extrusion of a high percentage (w/v) spidroin solution into a coagulation bath, usually methanol, and then drawing the fiber to several times its original length before allowing the thread to dry under tension.[65,158] However, the physical properties of such fibers, even using dissolved natural spider silk and extruding a new fiber, is lower than seen in the natural material.[176]

An alternative to using organic solvents is to use aqueous solvents. However, this is more challenging, as it is necessary to increase the protein concentration at several steps using buffers and salts to keep the spidroins in solution. Generally, the spidroins are dissolved into an aqueous denaturant such as urea,[162] guanidinium hydrochloride,[177] or guanidinium thiocyanate,[60] which is then replaced by a buffer, or the solution is extruded into a coagulation bath. Using this process, it is possible to manually draw fibers from an aqueous solution, and several groups have achieved this[58,162,167]; however, the properties of such fibers tend again to be poorer than the natural material.[178] Our research group has managed to utilize microfluidic devices to produce fibers of eADF3 and blends with eADF4 at a concentration of 20 mg mL^{-1}. The setup demonstrated the requirement for salt, pH, and shear stress to be applied to the spinning dope in specific orders to produce threads rather than aggregates.[60] Natural spinning dope can get up to 50% (w/v); however, reaching this percentage with man-made spidroins has proved problematic, and it has been necessary to utilize a variety of salts to reach values half of this. Processing the aqueous spinning dope to generate a fiber is currently difficult, but an understanding of how the spider undertakes this, combined with solubility studies of recombinant spidroins, has highlighted the likely processes required. However, the current techniques for spinning

---

(NcoI, BamHI, and HindIII). (B) Site-directed connection of two modules accomplished by ligating two appropriate plasmid fragments. The vector's ampicillin resistance gene *Ap^r* was reconstituted. (C) Nucleotides required for linking two modules confined within the first codon of each module. (D) Module multimers connected like single modules, resulting in controlled assembly of synthetic genes. (E) Amino acid sequences of designed silk modules derived from dragline silk proteins ADF-3 and ADF-4. Reprinted with permission from Ref. 31. Copyright 2004 American Chemical Society.

recombinant silk threads are not scalable, and there have been various attempts to create a process that is scalable to a level where industrial production of fibers is achievable.[37,179]

XV. Future Outlook

Our current understanding of spider silks concentrates on the MA and MI silks and flagelliform silk. These are probably the three silks to understand easily, as they consist of short repetitive units that are built up in a hierarchical manner to produce large proteins. As such, it has been possible to relate the protein sequences to molecular structures and then link these to the physical properties of the fibers. As a result, most of the materials research has concentrated on two of these three materials, MA silk and flagelliform silk, partly because of their outstanding physical properties, but also because more information is available on their structure. Thus we are now approaching the point where it is possible to generate recombinant proteins that closely mimic the properties of natural spidroins. Spider silks have developed under evolutionary pressure, and, as mentioned at the beginning of this chapter, the various silks have evolved to fulfill a specific task and not go beyond it, that is, dragline silk and flagelliform silk combine in orb webs to only capture a prey of a similar size and not larger than the spider.[17] It may well be possible to create a silk fiber with a higher tensile strength and toughness. There, the size of the engineered spidroin may be an important factor, as well as combinations of the various repeats seen in spider silks.

Most research has concentrated on orb-weaving spiders and also on a couple of species of cob-web weavers. It is possible that there are silks produced by other spiders that have developed higher strengths or toughness or other physical properties. Most likely, the best candidates for new materials are within the Orbicluraea clade, as only at this evolutionary point a good mixture of properties emerged in silks because of the need for high-performance materials for the construction of more efficient prey-capture devices. However, a large portion of the Orbicluraea clade no longer spin orb webs and have either abandoned webs for prey capture altogether or have developed new methods such as the cob web. Thus, it is likely that different spider species and web architectures will have different properties, some of which may exceed that of the currently known silks. An excellent example of this has recently come to light: the spider species *Caerostris darwini* has been found to spin orb webs across rivers with bridging threads covering up to 25 m. The MA silk from this species has an average toughness of 354 J cm^{-3} and a maximum of 520 J cm^{-3}.[180] This average toughness is over double that of most other spider species, which appears to be mostly due to the extraordinary elasticity of this MA silk. The molecular composition of this silk will therefore be of great interest and only goes to highlight the fact that, as more spiders are investigated, an even wider range of superb materials will be found to be provided by nature. Additionally, the study of this silk may help us to further understand how the spidroin sequences are related to the physical properties of the silk, and thus it will become easier to relate these to each other and develop even better materials.

Now that we have a better understanding of the sequences of all of the spidroins, it is clear that the strength, elasticity, and toughness of silks cannot in all cases easily be explained by β-sheet crystallites made from poly-alanine sequences and β-spirals from GPGXX sequences. The presence of other short repeats in tubuliform silk, such as S_n, SA_n, and the 200-amino-acid repeat of aciniform silk, which is the toughest silk of all, indicates that more complicated sequences may impart desirable physical properties. With these other silks, it should be possible in the future to apply the same techniques to produce recombinant proteins as being done nowadays with MaSp, and in the case of some, such as aciniform silk, it should be easier to produce recombinant proteins because of a lack of codon bias in their sequences. As such, it should be possible to soon mix and match different components to see how they affect the properties of artificially produced silk fibers.

With methods to produce recombinant spidroins on a large scale, it is necessary to produce more scalable methods of spinning the proteins into fibers in a reproducible manner. Currently, the methods used are not suitable for high-throughput spinning of fibers that have consistent properties. It is well known that spiders can tune their silks properties by changing the speed at which the thread is produced and the forces applied to it. While this is useful for a spider, it is a hindrance to industrial-scale production of threads. As such, stringent methods need to be developed to control the production of recombinant silk threads to avoid variability in the final produced threads.

It is likely that attempts to produce recombinant spider silks will eventually develop into viable materials, and the next challenge will, of course, then be to produce the materials in an economically viable manner for future applications.

ACKNOWLEDGMENTS

We would like to thank Lukas Eisoldt for constructive comments on this text. This work was financially supported by the DFG (SCHE 603/4-4) and the Bundesministerium für Bildung und Forschung (BMBF), Grant no. 13N9736.

References

1. Gerritsen VB. The tiptoe of an airbus. *Protein Spotlight Swiss Prot* 2002;**24**:1–2.
2. Sutherland TD, Young JH, Weisman S, Hayashi CY, Merritt DJ. Insect silk: one name, many materials. *Annu Rev Entomol* 2010;**55**:171–88.
3. Moon M-J. Microstructure of the silk spinning nozzles in the lynx spider, *Oxyopes licenti* (Araneae: Oxyopidae). *Integr Biosci* 2006;**10**:85–91.
4. Moon M-J. Microstructure of the silk apparatus in the coelotine spider *Paracoelotes spinivulva* (Araneae: Amaurobiidae). *Entomol Res* 2008;**38**:149–56.
5. Moon M-j, An J-S. Spinneret microstructure of the silk spinning apparatus in the crab spider, *Misumenops tricuspidatus* (Araneae: Thomisidae). *Entomol Res* 2005;**35**:67–74.
6. Moon M-J, An J-S. Microstructure of the silk apparatus of the comb-footed spider, *Achaearanea tepidariorum* (Araneae: Theridiidae). *Entomol Res* 2006;**36**:56–63.
7. Platnick NI. *The world spider catalog, version 12.0. American Museum of Natural History.* http://research.amnh.org/iz/spiders/catalog/INTRO1.html.
8. Meehan CJ, Olson EJ, Reudink MW, Kyser TK, Curry RL. Herbivory in a spider through exploitation of an ant-plant mutualism. *Curr Biol* 2009;**19**:R892–3.
9. Selden PA, Gall JC. A triassic mygalomorph spider from the northern vosges, France. *Palaeontology* 1992;**35**:211–35.
10. Coddington JA, Levi HW. Systematics and evolution of spiders (Araneae). *Annu Rev Ecol Syst* 1991;**22**:565–92.
11. Blackledge TA, Scharff N, Coddington JA, Szuts T, Wenzel JW, Hayashi CY, et al. Reconstructing web evolution and spider diversification in the molecular era. *Proc Natl Acad Sci USA* 2009;**106**:5229–34.
12. Vollrath F, Knight DP. Liquid crystalline spinning of spider silk. *Nature* 2001;**410**:541–8.
13. Peters HM. The spinning apparatus of Uloboridae in relation to the structure and construction of capture threads (Arachnida, Araneida). *Zoomorphology* 1984;**104**:96–104.
14. Hawthorn AC, Opell BD. Evolution of adhesive mechanisms in cribellar spider prey capture thread: evidence for van der Waals and hygroscopic forces. *Biol J Linn Soc* 2002;**77**:1–8.
15. Hawthorn AC, Opell BD. van der Waals and hygroscopic forces of adhesion generated by spider capture threads. *J Exp Biol* 2003;**206**:3905–11.
16. Griswold CE, Coddington JA, Platnick NI, Forster RR. Towards a phylogeny of entelegyne spiders (Araneae, Araneomorphae, Entelegynae). *J Arachnol* 1999;**27**:53–63.
17. Denny M. Physical-properties of spiders silk and their role in design of orb-webs. *J Exp Biol* 1976;**65**:483–506.
18. Witt P, Scarboro MB, Daniels R, Peakall DB, Gause RL. Spider web building in outer space evaluation of records from the skylab spider experiment. *J Arachnol* 1976;**4**:115–24.
19. Crews SC, Opell BD. The features of capture threads and orb-webs produced by unfed *Cyclosa turbinata* (Araneae: Araneidae). *J Arachnol* 2006;**34**:427–34.
20. Craig CL. The ecological and evolutionary interdependence between web architecture and web silk spun by orb web weaving spiders. *Biol J Linn Soc* 1987;**30**:135–62.
21. Guerette PA, Ginzinger DG, Weber BHF, Gosline JM. Silk properties determined by gland-specific expression of a spider fibroin gene family. *Science* 1996;**272**:112–5.
22. Gatesy J, Hayashi C, Motriuk D, Woods J, Lewis R. Extreme diversity, conservation, and convergence of spider silk fibroin sequences. *Science* 2001;**291**:2603–5.
23. Rising A, Hjalm G, Engstrom W, Johansson J. N-terminal nonrepetitive domain common to dragline, flagelliform, and cylindriform spider silk proteins. *Biomacromolecules* 2006;**7**:3120–4.
24. Motriuk-Smith D, Smith A, Hayashi CY, Lewis RV. Analysis of the conserved N-terminal domains in major ampullate spider silk proteins. *Biomacromolecules* 2005;**6**:3152–9.

25. Ittah S, Michaeli A, Goldblum A, Gat U. A model for the structure of the C-terminal domain of dragline spider silk and the role of its conserved cysteine. *Biomacromolecules* 2007;**8**:2768–73.

26. Ayoub NA, Garb JE, Tinghitella RM, Collin MA, Hayashi CY. Blueprint for a high-performance biomaterial: full-length spider dragline silk genes. *PLoS One* 2007;**2**:e514.

27. Challis RJ, Goodacre SL, Hewitt GM. Evolution of spider silks: conservation and diversification of the C-terminus. *Insect Mol Biol* 2006;**15**:45–56.

28. Rising A, Nimmervoll H, Grip S, Fernandez-Arias A, Storckenfeldt E, Knight DP, et al. Spider silk proteins–mechanical property and gene sequence. *Zoolog Sci* 2005;**22**:273–81.

29. Sponner A, Unger E, Grosse F, Weisshart K. Conserved C-termini of spidroins are secreted by the major ampullate glands and retained in the silk thread. *Biomacromolecules* 2004;**5**:840–5.

30. Sponner A, Vater W, Rommerskirch W, Vollrath F, Unger E, Grosse F, et al. The conserved C-termini contribute to the properties of spider silk fibroins. *Biochem Biophys Res Commun* 2005;**338**:897–902.

31. Huemmerich D, Helsen CW, Quedzuweit S, Oschmann J, Rudolph R, Scheibel T. Primary structure elements of spider dragline silks and their contribution to protein solubility. *Biochemistry* 2004;**43**:13604–12.

32. Hagn F, Eisoldt L, Hardy JG, Vendrely C, Coles M, Scheibel T, et al. A highly conserved spider silk domain acts as a molecular switch that controls fibre assembly. *Nature* 2010;**465**:239–42.

33. Bini E, Knight DP, Kaplan DL. Mapping domain structures in silks from insects and spiders related to protein assembly. *J Mol Biol* 2004;**335**:27–40.

34. Hedhammar M, Rising A, Grip S, Martinez AS, Nordling K, Casals C, et al. Structural properties of recombinant nonrepetitive and repetitive parts of major ampullate spidroin 1 from Euprosthenops australis: implications for fiber formation. *Biochemistry* 2008;**47**:3407–17.

35. Askarieh G, Hedhammar M, Nordling K, Saenz A, Casals C, Rising A, et al. Self-assembly of spider silk proteins is controlled by a pH-sensitive relay. *Nature* 2010;**465**:236–9.

36. Townley MA, Tillinghast EK. Developmental changes in spider spinning fields: a comparison between Mimetus and Araneus (Araneae: Mimetidae, Araneidae). *Biol J Linn Soc* 2009;**98**:343–83.

37. Heim M, Keerl D, Scheibel T. Spider silk: from soluble protein to extraordinary fiber. *Angew Chem Int Ed* 2009;**48**:3584–96.

38. Gosline JM, Guerette PA, Ortlepp CS, Savage KN. The mechanical design of spider silks: from fibroin sequence to mechanical function. *J Exp Biol* 1999;**202**:3295–303.

39. Ortlepp CS, Gosline JM. Consequences of forced silking. *Biomacromolecules* 2004;**5**:727–31.

40. Lefevre T, Boudreault S, Cloutier C, Pezolet M. Conformational and orientational transformation of silk proteins in the major ampullate gland of *Nephila clavipes* spiders. *Biomacromolecules* 2008;**9**:2399–407.

41. Hagn F, Thamm C, Scheibel T, Kessler H. pH-dependent dimerization and salt-dependent stabilization of the N-terminal domain of spider dragline silk—implications for fiber formation. *Angew Chem Int Ed* 2010;**50**:310–3.

42. Silvers R, Buhr F, Schwalbe H. The molecular mechanism of spider-silk formation. *Angew Chem Inter Ed* 2010;**49**:5410–2.

43. Augsten K, Muhlig P, Herrmann C. Glycoproteins and skin-core structure in *Nephila clavipes* spider silk observed by light and electron microscopy. *Scanning* 2000;**22**:12–5.

44. Vollrath F, Knight DP. Structure and function of the silk production pathway in the spider *Nephila edulis*. *Int J Biol Macromol* 1999;**24**:243–9.

45. Hijirida DH, Do KG, Michal C, Wong S, Zax D, Jelinski LW. 13 C NMR of *Nephila clavipes* major ampullate silk gland. *Biophys J* 1996;**71**:3442–7.
46. Knight DP, Vollrath F. Liquid crystals and flow elongation in a spider's silk production line. *Proc R Soc Lond B Biol Sci* 1999;**266**:519–23.
47. Willcox PJ, Gido SP, Muller W, Kaplan DL. Evidence of a cholesteric liquid crystalline phase in natural silk spinning processes. *Macromolecules* 1996;**29**:5106–10.
48. Knight DP, Vollrath F. Biological liquid crystal elastomers. *Philos Trans R Soc Lond B Biol Sci* 2002;**357**:155–63.
49. Hamley IW. Liquid crystal phase formation by biopolymers. *Soft Matter* 2010;**6**:1863–71.
50. Holland C, Terry AE, Porter D, Vollrath F. Comparing the rheology of native spider and silkworm spinning dope. *Nat Mater* 2006;**5**:870–4.
51. Kojic N, Bico J, Clasen C, McKinley GH. Ex vivo rheology of spider silk. *J Exp Biol* 2006;**209**:4355–62.
52. Bunning JD, Lydon JE. The cellular optical texture of the lyotropic nematic phase of the caesium pentadecafluoro-octanoate (CsPFO) water system in cylindrical tubes. *Liq. Cryst.* 1996;**20**:381–5.
53. O'Brien JP, Fahnestock SR, Termonia Y, Gardner KCH. Nylons from nature: synthetic analogs to spider silk. *Adv Mater* 1998;**10**:1185–95.
54. Knight DP, Knight MM, Vollrath F. Beta transition and stress-induced phase separation in the spinning of spider dragline silk. *Int J Biol Macromol* 2000;**27**:205–10.
55. Knight DP, Vollrath F. Changes in element composition along the spinning duct in a Nephila spider. *Naturwissenschaften* 2001;**88**:179–82.
56. Chen X, Knight DP, Vollrath F. Rheological characterization of nephila spidroin solution. *Biomacromolecules* 2002;**3**:644–8.
57. Vollrath F, Knight DP, Hu XW. Silk production in a spider involves acid bath treatment. *Proc R Soc Lond B Biol Sci* 1998;**265**:817–20.
58. Exler JH, Hummerich D, Scheibel T. The amphiphilic properties of spider silks are important for spinning. *Angew Chem Int Ed* 2007;**46**:3559–62.
59. Eisoldt L, Hardy JG, Heim M, Scheibel TR. The role of salt and shear on the storage and assembly of spider silk proteins. *J Struct Biol* 2010;**170**:413–9.
60. Rammensee S, Slotta U, Scheibel T, Bausch AR. Assembly mechanism of recombinant spider silk proteins. *Proc Natl Acad Sci USA* 2008;**105**:6590–5.
61. Slotta UK, Rammensee S, Gorb S, Scheibel T. An engineered spider silk protein forms microspheres. *Angew Chem Int Ed* 2008;**47**:4592–4.
62. Lin Z, Huang W, Zhang J, Fan JS, Yang D. Solution structure of eggcase silk protein and its implications for silk fiber formation. *Proc Natl Acad Sci USA* 2009;**106**:8906–11.
63. Xu M, Lewis RV. Structure of a protein superfiber—spider dragline silk. *Proc Natl Acad Sci USA* 1990;**87**:7120–4.
64. Sponner A, Schlott B, Vollrath F, Unger E, Grosse F, Weisshart K. Characterization of the protein components of *Nephila clavipes* dragline silk. *Biochemistry* 2005;**44**:4727–36.
65. Hardy JG, Romer LM, Scheibel TR. Polymeric materials based on silk proteins. *Polymer* 2008;**49**:4309–27.
66. Jenkins JE, Creager MS, Lewis RV, Holland GP, Yarger JL. Quantitative correlation between the protein primary sequences and secondary structures in spider dragline silks. *Biomacromolecules* 2010;**11**:192–200.
67. Lawrence BA, Vierra CA, Mooref AMF. Molecular and mechanical properties of major ampullate silk of the black widow spider, *Latrodectus hesperus*. *Biomacromolecules* 2004;**5**:689–95.
68. Hu X, Vasanthavada K, Kohler K, McNary S, Moore AMF, Vierra CA. Molecular mechanisms of spider silk. *Cell Mol Life Sci* 2006;**63**:1986–99.

69. Colgin MA, Lewis RV. Spider minor ampullate silk proteins contain new repetitive sequences and highly conserved non-silk-like "spacer regions". *Protein Sci* 1998;**7**:667–72.

70. Savage KN, Gosline JM. The effect of proline on the network structure of major ampullate silks as inferred from their mechanical and optical properties. *J Exp Biol* 2008;**211**:1937–47.

71. Creager MS, Jenkins JE, Thagard-Yeaman LA, Brooks AE, Jones JA, Lewis RV, et al. Solid-state NMR comparison of various spiders´ dragline silk fiber. *Biomacromolecules* 2010;**11**:2039–43.

72. Huemmerich D, Scheibel T, Vollrath F, Cohen S, Gat U, Ittah S. Novel assembly properties of recombinant spider dragline silk proteins. *Curr Biol* 2004;**14**:2070–4.

73. Brooks AE, Steinkraus HB, Nelson SR, Lewis RV. An investigation of the divergence of major ampullate silk fibers from *Nephila clavipes* and *Argiope aurantia*. *Biomacromolecules* 2005;**6**:3095–9.

74. Sponner A, Unger E, Grosse F, Klaus W. Differential polymerization of the two main protein components of dragline silk during fibre spinning. *Nat Mater* 2005;**4**:772–5.

75. Sponner A, Vater W, Monajembashi S, Unger E, Grosse F, Weisshart K. Composition and hierarchical organisation of a spider silk. *PLoS One* 2007;**2**:e998.

76. Hronska M, van Beek JD, Williamson PTF, Vollrath F, Meier BH. NMR characterization of native liquid spider dragline silk from *Nephila edulis*. *Biomacromolecules* 2004;**5**:834–9.

77. Lefevre T, Leclerc J, Rioux-Dube JF, Buffeteau T, Paquin MC, Rousseau ME, et al. Conformation of spider silk proteins in situ in the intact major ampullate gland and in solution. *Biomacromolecules* 2007;**8**:2342–4.

78. Pappu RV, Rose GD. A simple model for polyproline II structure in unfolded states of alanine-based peptides. *Protein Sci* 2002;**11**:2437–55.

79. Blanch EW, Gill AC, Rhie AGO, Hope J, Hecht L, Nielsen K, et al. Raman optical activity demonstrates poly(L-proline) II helix in the N-terminal region of the ovine prion protein: implications for function and misfunction. *J Mol Biol* 2004;**343**:467–76.

80. Grubb DT, Jelinski LW. Fiber morphology of spider silk: the effects of tensile deformation. *Macromolecules* 1997;**30**:2860–7.

81. Riekel C, Muller M, Vollrath F. In situ X-ray diffraction during forced silking of spider silk. *Macromolecules* 1999;**32**:4464–6.

82. Du N, Liu XY, Narayanan J, Li LA, Lim MLM, Li DQ. Design of superior spider silk: from nanostructure to mechanical properties. *Biophys J* 2006;**91**:4528–35.

83. Kummerlen J, vanBeek JD, Vollrath F, Meier BH. Local structure in spider dragline silk investigated by two-dimensional spin-diffusion nuclear magnetic resonance. *Macromolecules* 1996;**29**:2920–8.

84. Simmons AH, Michal CA, Jelinski LW. Molecular orientation and two-component nature of the crystalline fraction of spider dragline silk. *Science* 1996;**271**:84–7.

85. Parkhe AD, Seeley SK, Gardner K, Thompson L, Lewis RV. Structural studies of spider silk proteins in the fiber. *J Mol Recognit* 1997;**10**:1–6.

86. Holland GP, Creager MS, Jenkins JE, Lewis RV, Yarger JL. Determining secondary structure in spider dragline silk by carbon-carbon correlation solid-state NMR spectroscopy. *J Am Chem Soc* 2008;**130**:9871–7.

87. van Beek JD, Hess S, Vollrath F, Meier BH. The molecular structure of spider dragline silk: folding and orientation of the protein backbone. *Proc Natl Acad Sci USA* 2002;**99**:10266–71.

88. Vollrath F, Porter D. Spider silk as archetypal protein elastomer. *Soft Matter* 2006;**2**:377–85.

89. Hayashi CY, Shipley NH, Lewis RV. Hypotheses that correlate the sequence, structure, and mechanical properties of spider silk proteins. *Int J Biol Macromol* 1999;**24**:271–5.

90. Ma BY, Nussinov R. Molecular dynamics simulations of alanine rich beta-sheet oligomers: insight into amyloid formation. *Protein Sci* 2002;**11**:2335–50.

91. Keten S, Xu Z, Ihle B, Buehler MJ. Nanoconfinement controls stiffness, strength and mechanical toughness of beta-sheet crystals in silk. *Nat Mater* 2010;**9**:359–67.

92. Xiao SB, Stacklies W, Cetinkaya M, Markert B, Grater F. Mechanical response of silk crystalline units from force-distribution analysis. *Biophys J* 2009;**96**:3997–4005.

93. Yang Z, Grubb DT, Jelinski LW. Small-angle X-ray scattering of spider dragline silk. *Macromolecules* 1997;**30**:8254–61.

94. Lefevre T, Rousseau ME, Pezolet M. Protein secondary structure and orientation in silk as revealed by Raman spectromicroscopy. *Biophys J* 2007;**92**:2885–95.

95. Eles PT, Michal CA. A DECODER NMR study of backbone orientation in *Nephila clavipes* dragline silk under varying strain and draw rate. *Biomacromolecules* 2004;**5**:661–5.

96. Hayashi CY, Lewis RV. Spider flagelliform silk: lessons in protein design, gene structure, and molecular evolution. *Bioessays* 2001;**23**:750–6.

97. Buehler MJ, Keten S, Ackbarow T. Theoretical and computational hierarchical nanomechanics of protein materials: deformation and fracture. *Prog Mater Sci* 2008;**53**:1101–241.

98. Keten S, Buehler MJ. Nanostructure and molecular mechanics of spider dragline silk protein assemblies. *J R Soc Interface* 2010;**7**:1709–21.

99. Blackledge TA, Boutry C, Wong SC, Baji A, Dhinojwala A, Sahni V, et al. How super is supercontraction? Persistent versus cyclic responses to humidity in spider dragline silk. *J Exp Biol* 2009;**212**:1980–8.

100. Blackledge TA, Hayashi CY. Silken toolkits: biomechanics of silk fibers spun by the orb web spider *Argiope argentata* (Fabricius 1775). *J Exp Biol* 2006;**209**:2452–61.

101. Rousseau ME, Lefevre T, Pezolet M. Conformation and orientation of proteins in various types of silk fibers produced by *Nephila clavipes* spiders. *Biomacromolecules* 2009;**10**:2945–53.

102. Termonia Y. Molecular modeling of spider silk elasticity. *Macromolecules* 1994;**27**:7378–81.

103. Becker N, Oroudjev E, Mutz S, Cleveland JP, Hansma PK, Hayashi CY, et al. Molecular nanosprings in spider capture-silk threads. *Nat Mater* 2003;**2**:278–83.

104. Papadopoulos P, Solter J, Kremer F. Hierarchies in the structural organization of spider silk-a quantitative model. *Colloid Polym Sci* 2009;**287**:231–6.

105. Papadopoulos P, Ene R, Weidner I, Kremer F. Similarities in the structural organization of major and minor ampullate spider silk. *Macromol Rapid Commun* 2009;**30**:851–7.

106. Rief M, Gautel M, Oesterhelt F, Fernandez JM, Gaub HE. Reversible unfolding of individual titin immunoglobulin domains by AFM. *Science* 1997;**276**:1109–12.

107. Buehler MJ, Yung YC. Deformation and failure of protein materials in physiologically extreme conditions and disease. *Nat Mater* 2009;**8**:175–88.

108. Nova A, Keten S, Pugno NM, Redaelli A, Buehler MJ. Molecular and nanostructural mechanisms of deformation, strength and toughness of spider silk fibrils. *Nano Lett* 2010;**10**:2626–34.

109. Xiao S, Stacklies W, Debes C, Grater F. Force distribution determines optimal length of [small beta]-sheet crystals for mechanical robustness. *Soft Matter* 2011;**7**:1308–11.

110. Porter D, Vollrath F. The role of kinetics of water and amide bonding in protein stability. *Soft Matter* 2008;**4**:328–36.

111. Gould SAC, Tran KT, Spagna JC, Moore AMF, Shulman JB. Short and long range order of the morphology of silk from *Latrodectus hesperus* (Black Widow) as characterized by atomic force microscopy. *Int J Biol Macromol* 1999;**24**:151–7.

112. Putthanarat S, Stribeck N, Fossey SA, Eby RK, Adams WW. Investigation of the nanofibrils of silk fibers. *Polymer* 2000;**41**:7735–47.

113. Porter D, Vollrath F. Silk as a biomimetic ideal for structural polymers. *Adv Mater* 2009;**21**:487–92.

114. Thiel BL, Kunkel DD, Viney C. Physical and chemical microstructure of spider dragline—a study by analytical transmission electron-microscopy. *Biopolymers* 1994;**34**:1089–97.
115. Thiel BL, Guess KB, Viney C. Non-periodic lattice crystals in the hierarchical microstructure of spider (major ampullate) silk. *Biopolymers* 1997;**41**:703–19.
116. Trancik JE, Czernuszka JT, Bell FI, Viney C. Nanostructural features of a spider dragline silk as revealed by electron and X-ray diffraction studies. *Polymer* 2006;**47**:5633–42.
117. Ko FK, Jovicic J. Modeling of mechanical properties and structural design of spider web. *Biomacromolecules* 2004;**5**:780–5.
118. Vehoff T, Glisovic A, Schollmeyer H, Zippelius A, Salditt T. Mechanical properties of spider dragline silk: humidity, hysteresis, and relaxation. *Biophys J* 2007;**93**:4425–32.
119. Hayashi CY, Blackledge TA, Lewis RV. Molecular and mechanical characterization of aciniform silk: uniformity of iterated sequence modules in a novel member of the spider silk fibroin gene family. *Mol Biol Evol* 2004;**21**:1950–9.
120. Madsen B, Shao ZZ, Vollrath F. Variability in the mechanical properties of spider silks on three levels: interspecific, intraspecific and intraindividual. *Int J Biol Macromol* 1999;**24**:301–6.
121. Work RW, Young CT. The amino-acid compositions of major and minor ampullate silks of certain orb-web-building spiders (Araneae, Araneidae). *J Arachnol* 1987;**15**:65–80.
122. Vollrath F. Biology of spider silk. *Int J Biol Macromol* 1999;**24**:81–8.
123. Zax DB, Armanios DE, Horak S, Malowniak C, Yang ZT. Variation of mechanical properties with amino acid content in the silk of *Nephila clavipes. Biomacromolecules* 2004;**5**:732–8.
124. Liu Y, Shao ZZ, Vollrath F. Relationships between supercontraction and mechanical properties of spider silk. *Nat Mater* 2005;**4**:901–5.
125. Emile O, Le Floch A, Vollrath F. Biopolymers: shape memory in spider draglines. *Nature* 2006;**440**:621.
126. Perez-Rigueiro J, Elices M, Guinea GV. Controlled supercontraction tailors the tensile behaviour of spider silk. *Polymer* 2003;**44**:3733–6.
127. Jelinski LW, Blye A, Liivak O, Michal C, LaVerde G, Seidel A, et al. Orientation, structure, wet-spinning, and molecular basis for supercontraction of spider dragline silk. *Int J Biol Macromol* 1999;**24**:197–201.
128. Shao Z, Vollrath F, Sirichaisit J, Young RJ. Analysis of spider silk in native and supercontracted states using Raman spectroscopy. *Polymer* 1999;**40**:2493–500.
129. van Beek JD, Kummerlen J, Vollrath F, Meier BH. Supercontracted spider dragline silk: a solid-state NMR study of the local structure. *Int J Biol Macromol* 1999;**24**:173–8.
130. Holland GP, Jenkins JE, Creager MS, Lewis RV, Yarger JL. Solid-state NMR investigation of major and minor ampullate spider silk in the native and hydrated states. *Biomacromolecules* 2008;**9**:651–7.
131. Holland GP, Lewis RV, Yarger JL. WISE NMR characterization of nanoscale heterogeneity and mobility in supercontracted *Nephila clavipes* spider dragline silk. *J Am Chem Soc* 2004;**126**:5867–72.
132. Liu Y, Sponner A, Porter D, Vollrath F. Proline and processing of spider silks. *Biomacromolecules* 2008;**9**:116–21.
133. Guinea GV, Cerdeira M, Plaza GR, Elices M, Perez-Rigueiro J. Recovery in viscid line fibers. *Biomacromolecules* 2010;**11**:1174–9.
134. Hayashi CY, Lewis RV. Molecular architecture and evolution of a modular spider silk protein gene. *Science* 2000;**287**:1477–9.
135. Selden PA. Orb-web weaving spiders in the early cretaceous. *Nature* 1989;**340**:711–3.
136. Bonthrone KM, Vollrath F, Hunter BK, Sanders JKM. The elasticity of spiders webs is due to water-induced mobility at a molecular-level. *Proc R Soc Lond B Biol Sci* 1992;**248**:141–4.

137. Hayashi CY, Lewis RV. Evidence from flagelliform silk cDNA for the structural basis of elasticity and modular nature of spider silks. *J Mol Biol* 1998;**275**:773–84.
138. Dicko C, Knight D, Kenney JM, Vollrath F. Secondary structures and conformational changes in flagelliform, cylindrical, major, and minor ampullate silk proteins. Temperature and concentration effects. *Biomacromolecules* 2004;**5**:2105–15.
139. Lefèvre T, Boudreault S, Cloutier C, Pézolet M. Diversity of molecular transformations involved in the formation of spider silks. *J Mol Biol* 2011;**405**:238–53.
140. Kohler T, Vollrath F. Thread biomechanics in the 2 orb-weaving spiders Araneus-Diadematus (Araneae, Araneidae) and Uloborus-Walckenaerius (Araneae, Uloboridae). *J Exp Zool* 1995;**271**:1–17.
141. Garb JE, DiMauro T, Vo V, Hayashi CY. Silk genes support the single origin of orb webs. *Science* 2006;**312**:1762.
142. Blackledge TA, Hayashi CY. Unraveling the mechanical properties of composite silk threads spun by cribellate orb-weaving spiders. *J Exp Biol* 2006;**209**:3131–40.
143. Vasanthavada K, Hu X, Falick AM, La Mattina C, Moore AMF, Jones PR, et al. Aciniform spidroin, a constituent of egg case sacs and wrapping silk fibers from the black widow spider *Latrodectus hesperus*. *J Biol Chem* 2007;**282**:35088–97.
144. Hu XY, Lawrence B, Kohler K, Falick AM, Moore AMF, McMullen E, et al. Araneoid egg case silk: a fibroin with novel ensemble repeat units from the black widow spider, *Latrodectus hesperus*. *Biochemistry* 2005;**44**:10020–7.
145. Casem ML, Collin MA, Ayoub NA, Hayashi CY. Silk gene transcripts in the developing tubuliform glands of the Western black widow, *Latrodectus hesperus*. *J Arachnol* 2010;**38**:99–103.
146. Garb JE, Hayashi CY. Modular evolution of egg case silk genes across orb-weaving spider superfamilies. *Proc Natl Acad Sci USA* 2005;**102**:11379–84.
147. Huang W, Lin Z, Sin YM, Li D, Gong Z, Yang D. Characterization and expression of a cDNA encoding a tubuliform silk protein of the golden web spider *Nephila antipodiana*. *Biochimie* 2006;**88**:849–58.
148. Tian MZ, Lewis RV. Molecular characterization and evolutionary study of spider tubuliform (eggcase) silk protein. *Biochemistry* 2005;**44**:8006–12.
149. Zhao AC, Zhao TF, SiMa YH, Zhang YS, Nakagaki K, Miao YG, et al. Unique molecular architecture of egg case silk protein in a spider, *Nephila clavata*. *J Biochem* 2005;**138**:593–604.
150. Tian M, Lewis RV. Tubuliform silk protein: a protein with unique molecular characteristics and mechanical properties in the spider silk fibroin family. *Appl Phys A Mater Sci Process* 2006;**82**:265–73.
151. Stauffer SL, Coguill SL, Lewis RV. Comparison of physical-properties of 3 silks from *Nephila-Clavipes* and *Araneus-Gemmoides*. *J Arachnol* 1994;**22**:5–11.
152. Barghout JYJ, Thiel BL, Viney C. Spider (*Araneus diadematus*) cocoon silk: a case of non-periodic lattice crystals with a twist? *Int J Biol Macromol* 1999;**24**:211–7.
153. Hu XY, Kohler K, Falick AM, Moore AMF, Jones PR, Vierra C. Spider egg case core fibers: trimeric complexes assembled from TuSp1, ECP-1, and ECP-2. *Biochemistry* 2006;**45**:3506–16.
154. Choresh O, Bayarmagnai B, Lewis RV. Spider web glue: two proteins expressed from opposite strands of the same DNA sequence. *Biomacromolecules* 2009;**10**:2852–6.
155. Agnarsson I, Blackledge TA. Can a spider web be too sticky? Tensile mechanics constrains the evolution of capture spiral stickiness in orb-weaving spiders. *J Zool* 2009;**278**:134–40.
156. Blasingame E, Tuton-Blasingame T, Larkin L, Falick AM, Zhao L, Fong J, et al. Pyriform Spidroin 1, a novel member of the silk gene family that anchors dragline silk fibers in attachment discs of the black widow spider, *Latrodectus hesperus*. *J Biol Chem* 2009;**284**:29097–108.

157. Vendrely C, Scheibel T. Biotechnological production of spider-silk proteins enables new applications. *Macromol Biosci* 2007;7:401–9.
158. Teule F, Cooper AR, Furin WA, Bittencourt D, Rech EL, Brooks A, et al. A protocol for the production of recombinant spider silk-like proteins for artificial fiber spinning. *Nat Protoc* 2009;4:341–55.
159. Arcidiacono S, Mello C, Kaplan D, Cheley S, Bayley H. Purification and characterization of recombinant spider silk expressed in Escherichia coli. *Appl Microbiol Biotechnol* 1998;49:31–8.
160. Xia XX, Qian ZG, Ki CS, Park YH, Kaplan DL, Lee SY. Native-sized recombinant spider silk protein produced in metabolically engineered *Escherichia coli* results in a strong fiber. *Proc Natl Acad Sci USA* 2010;107:14059–63.
161. Menassa R, Hong Z, Karatzas CN, Lazaris A, Richman A, Brandle J. Spider dragline silk proteins in transgenic tobacco leaves: accumulation and field production. *Plant Biotechnol J* 2004;2:431–8.
162. Lazaris A, Arcidiacono S, Huang Y, Zhou JF, Duguay F, Chretien N, et al. Spider silk fibers spun from soluble recombinant silk produced in mammalian cells. *Science* 2002;295:472–6.
163. Candelas GC, Arroyo G, Carrasco C, Dompenciel R. Spider silk glands contain a tissue-specific alanine transfer-RNA that accumulates in vitro in response to the stimulus for slk protein-synthesis. *Dev Biol* 1990;140:215–20.
164. Zhang YS, Hu JH, Miao YG, Zhao AC, Zhao TF, Wu DY, et al. Expression of EGFP-spider dragline silk fusion protein in BmN cells and larvae of silkworm showed the solubility is primary limit for dragline proteins yield. *Mol Biol Rep* 2008;35:329–35.
165. Wen HX, Lan XQ, Zhang YS, Zhao TF, Wang YJ, Kajiura Z, et al. Transgenic silkworms (Bombyx mori) produce recombinant spider dragline silk in cocoons. *Mol Biol Rep* 2010;37:1815–21.
166. Scheibel T. Spider silks: recombinant synthesis, assembly, spinning, and engineering of synthetic proteins. *Microb Cell Fact* 2004;3:14.
167. Teule F, Furin WA, Cooper AR, Duncan JR, Lewis RV. Modifications of spider silk sequences in an attempt to control the mechanical properties of the synthetic fibers. *J Mater Sci* 2007;42:8974–85.
168. Miao YG, Zhang YS, Nakagaki K, Zhao TF, Zhao AC, Meng Y, et al. Expression of spider flagelliform silk protein in *Bombyx mori* cell line by a novel Bac-to-Bac/BmNPV baculovirus expression system. *Appl Microbiol Biotechnol* 2006;71:192–9.
169. Zhou YT, Wu SX, Conticello VP. Genetically directed synthesis and spectroscopic analysis of a protein polymer derived from a flagelliform silk sequence. *Biomacromolecules* 2001;2:111–25.
170. Heim M, Ackerschott CB, Scheibel T. Characterization of recombinantly produced spider flagelliform silk domains. *J Struct Biol* 2010;170:420–5.
171. Scheller J, Guhrs KH, Grosse F, Conrad U. Production of spider silk proteins in tobacco and potato. *Nat Biotechnol* 2001;19:573–7.
172. Barr LA, Fahnestock SR, Yang JJ. Production and purification of recombinant DP1B silk-like protein in plants. *Mol Breed* 2004;13:345–56.
173. Yang JJ, Barr LA, Fahnestock SR, Liu ZB. High yield recombinant silk-like protein production in transgenic plants through protein targeting. *Transgenic Res* 2005;14:313–24.
174. Fahnestock SR, Bedzyk LA. Production of synthetic spider dragline silk protein in *Pichia pastoris*. *Appl Microbiol Biotechnol* 1997;47:33–9.
175. Hardy JG, Scheibel TR. Composite materials based on silk proteins. *Prog Polym Sci* 2010;35:1093–115.
176. Seidel A, Liivak O, Calve S, Adaska J, Ji GD, Yang ZT, et al. Regenerated spider silk: processing, properties, and structure. *Macromolecules* 2000;33:775–80.

177. Shao ZZ, Vollrath F, Yang Y, Thogersen HC. Structure and behavior of regenerated spider silk. *Macromolecules* 2003;**36**:1157–61.
178. Arcidiacono S, Mello CM, Butler M, Welsh E, Soares JW, Allen A, et al. Aqueous processing and fiber spinning of recombinant spider silks. *Macromolecules* 2002;**35**:1262–6.
179. Keerl D, Hardy JG, Scheibel TR. Biomimetic spinning of recombinant silk proteins. *Mater Res Soc Symp Proc* 2010;**1239**:VV07–20.
180. Agnarsson I, Kuntner M, Blackledge TA. Bioprospecting finds the toughest biological material: extraordinary silk from a giant riverine orb spider. *PLoS One* 2010;**5**:e11234.

Protein Modifications Giving Rise to Homo-oligomers

Georg E. Schulz

Institut für Organische Chemie und
Biochemie, Albert-Ludwigs-Universität,
Freiburg im Breisgau, Germany

After the structures of numerous proteins have been established at the atomic level and after a multitude of proteins can be produced with almost no restrictions, the time seems ripe to apply this knowledge for engineering purposes. An apparently simple task is the designed association of protein molecules to form homo-oligomers. A number of worked examples are presented. The associations split into flexible *versus* rigid designs and also into fixed *versus* switchable ones. It is shown that the practical work is tightly governed by the multiplicity concept, which in turn is interwoven with symmetry. The available symmetries and multiplicities are explained. Unfortunately, the most desirable contacts with a multiplicity of one, which lead to asymmetric assemblies with 5–50 nm spacings, are most difficult to achieve. Emerging rules for the required surface properties are put forward. Suitable mutations for changing such surfaces are discussed.

I. Proteins as Building Blocks

Living organisms consist of four types of molecules, namely, nucleic acids, proteins, lipids, and saccharides. These compounds interact in various ways with each other, forming covalent and noncovalent complexes. Here we are only interested in noncovalent aggregates. Well-known examples are the ribosome (RNA–protein), the transcription complex (DNA–protein), the pyruvate

Progress in Molecular Biology
and Translational Science, Vol. 103
DOI: 10.1016/B978-0-12-415906-8.00006-6

dehydrogenase complex (protein–protein), exocytose complexes (protein–lipid), membranes (lipid–lipid), bacterial cell wall (protein–saccharide), and cellulose (saccharide–saccharide).

Besides the natural complexes, artificial aggregates have also been designed. This concerns, in particular, DNA–DNA complexes based on the association of complementary DNA strands. Such complexes can be programmed by using partially complementary nucleotide sequences.[1] Nowadays, DNA strands of any sequence can be synthesized at ease, so that intricate association patterns can be designed and produced. This has resulted in many double-stranded DNA constructs with specific ramifications giving rise to multiple two- and three-dimensional topologies such as, for instance, bottles.[2] These constructs are of limited conformational and chemical stability. They can be studded with different enzymes such that, for example, the enzymes of a reaction chain come spatially close together.

In a design with lipids, vesicles were coupled to each other via integral membrane proteins that form protein–protein links (Fig. 1). This particular association could be switched on and off by disulfide bridge formation and splitting under oxidizing and reducing conditions, respectively.[4] It allows attaching tightly a vesicle loaded with a membrane channel protein of interest onto a semiconductor device in order to analyze the channel properties.[5]

Among the four types of biological materials, proteins are universal because they can form a virtually unlimited number of structures and fulfill a multitude of functions, in contrast to the other materials. Most proteins have well-defined, sturdy structures that are suitable as building blocks in larger associations.[6] This property has given rise to a large number of protein crystals, which were analyzed by X-ray diffraction yielding the structures at atomic resolution.

If a crystal structure is known, the rigidity of all parts of a protein can be estimated from the so-called temperature factor assigned to each atom. This factor describes the atomic mobility. It results from the X-ray diffraction analysis in addition to the atomic coordinates. If the asymmetric unit of a crystal contains more than one protein, the conformational differences between these proteins outline mobile regions. This is of particular importance for large-scale motions such as domain displacements within proteins and may not manifest in the temperature factors. The mobility of local parts of a protein can also be determined by nuclear magnetic resonance measurements,[7] by hydrogen–deuterium exchange experiments,[8,9] or by molecular dynamics calculations.[10] The rigidity of a region is a most welcome property for the design of protein–protein contacts because rigid regions do not require a high entropy reduction on contact formation.

More intricate are proteins that lose their structure partially, for instance, to form complexes that are highly specific but have only a low binding energy (Fig. 2).[11] Such complexes can dissociate again easily as suggested early and

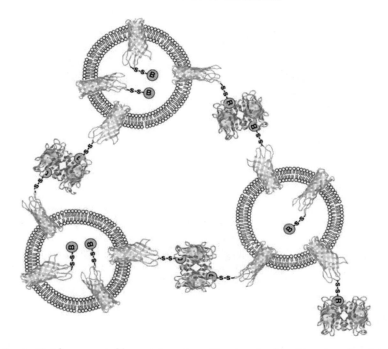

FIG. 1. Vesicles connected by protein anchors. OmpA molecules with an introduced cysteine at the most exposed position were connected via an S–S bridge to a tethered biotin (B, see Fig. 13) and incorporated into vesicles.[4] The addition of streptavidin, which binds biotin tightly, aggregates the vesicles. Aggregation and dissolution can be controlled by the redox potential closing and opening the disulfide bridges. The construct is intended for fixing membranes containing channel proteins of interest tightly onto electronic sensors.[5] (For color version of this figure, the reader is referred to the web version of this chapter.)

shown later to be important for substrate recognition in enzymes.[12] Proteins with functionally important mobile parts are now well established.[13,14] It seems clear that such proteins should be avoided when designing aggregates because their behavior on association is virtually unpredictable.

The task of a protein determines its structure. Proteins forming a scaffold, for instance the bacterial cell wall, have predetermined contacts.[15] Such contacts can be used to design artificial associations.[16] Proteins that transport solutes have usually at least two conformations, one with the cargo and the other without it. Designing an aggregate exclusively for the conformation with cargo then may allow switching the aggregation on and off by adding or removing the cargo, respectively. A well-known example is hemoglobin showing distinct oxy and a deoxy conformations.[17]

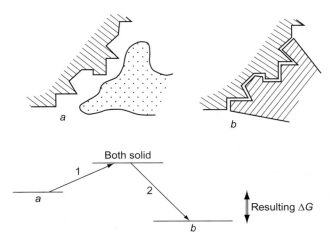

FIG. 2. Highly specific binding with a low binding energy.[11] If a binding partner is mobile but can freeze to assume the complementary surface of the other partner, the complex is highly specific but can still be easily opened again as the resulting binding energy ΔG is small because of the required entropy reduction. This principle is useful for enzymes catalyzing reactions in which the products are very similar to the incoming reactants.[12]

Proteins that catalyze metabolic reactions are likely to assume several conformations during a catalytic cycle. If such enzymes are not rigid, it may become very difficult to engineer them into complexes. An extreme example is a kinase that gives rise to a complete movie showing the drastic conformational changes during a reaction cycle (Fig. 3).[18] On the other hand, the motions during catalysis may be locally confined or they may be small and remain within a restricted conformational boundary (Fig. 4).[19,20] Any engineering attempt that intends to keep the aggregate catalytically active therefore has to leave enough space for the required motions.

As a further possibility, a presumed rigid protein may break up into two or more domains, which can adopt different relative geometries. Such a case is known for viral coats where the icosahedral limit of 60 asymmetric units was tripled by using different relative domain orientations in a so-called quasi symmetry.[21,22] Other cases are proteins with multiple binding domains that are mobile against each other for functional reasons.[23] As a general rule, all mobile parts of a protein, such as domains, surface patches, main chain terminals, or large side chains, should be avoided when designing protein associations because an appreciable entropy reduction is needed to fix them in a complex. Even more cumbersome, they may solidify in more than one conformation giving rise to heterogeneity, which is probably the reason why fusion proteins usually resist crystallization.[24] In summary, designs should be

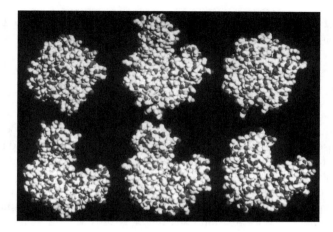

FIG. 3. Domain motions in an enzyme during catalysis. The six pictures of closely related homologous adenylate kinases with different ligands and in different crystal packings illustrate the extensive motions during catalysis of this enzyme group. An ordering of these and further structures resulted in the first movie showing the conformational changes over a complete catalytic cycle.[18] (For color version of this figure, the reader is referred to the web version of this chapter.)

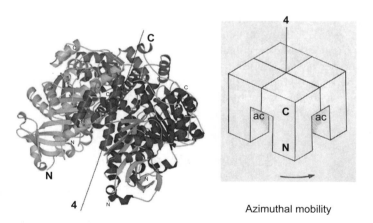

Azimuthal mobility

FIG. 4. Restrained mobility of an enzyme domain.[19] The crystal structure of a C_4-symmetric aldolase revealed a restrained mobility of the N-terminal domains (here azimuthal around the fourfold axis). It was suggested that these domains harness random solvent fluctuations and direct them to domain motions squeezing and opening the active center (ac), which supports the catalysis mechanically. The proposal was confirmed by a mutational analysis.[20] (For color version of this figure, the reader is referred to the web version of this chapter.)

restricted to well-structured units. Given the difficulties in producing suitable contacts, it is most advisable to stay with protein building blocks that are internally rigid and also have a rigid surface.

II. Hetero- Versus Homo-oligomers

Proteins may form heterocomplexes with other proteins or homocomplexes with themselves. Hetero-oligomers are mostly transient and involved in various cellular functions, whereas homo-oligomers are predominantly permanent.[25] Hetero-oligomers are always asymmetric and therefore difficult to produce via random mutations during evolution. Model calculations of the evolutionary development of oligomers resulted in only about 1% hetero- as compared to homo-oligomers.[26] For the same reason, it is also very difficult to design such contacts. In nature, quite a number of heterocomplexes, such as that between the α- and β-subunits of hemoglobin,[17] have most likely evolved from a homo-oligomer, here from the tetramer of a myoglobin-like molecule. Such a detour circumvents the major obstacle of forming asymmetric contacts from scratch.[26] Therefore, it may indicate a practicable pathway for engineering an asymmetric contact.

In the case of homo-oligomers, a given contact or a given contact ensemble has the lowest energy. At the energy minimum, this contact is adopted throughout by all protein units. The contact is symmetric if it connects the same surface patch of two identical protein molecules, which generates a twofold axis along the interface (Fig. 5). Any other contact is asymmetric. After self-assembly, both symmetric and asymmetric contacts always give rise to symmetric arrangements except in cases of steric hindrance (Fig. 6). Note that the selection of the best contacts depends on the environmental conditions such as, for example, the pH or the ionic strength of the solvent. Therefore, many assemblies can be switched on and off just by changing the environmental conditions.

III. Symmetric Associations

Humans tend to equate symmetry with beauty. The most popular symmetry element is a vertical mirror, which also fits an even human body. It is used in numerous representative buildings. More playful is a cyclic symmetry, which is found, for instance, in multiple forms in the Baptistry of Pisa, a Renaissance building shown in Fig. 7. Multicyclic symmetry is also possible for self-assembly, as demonstrated by a bacterial needle complex with combined 15- and 24-fold circles of proteins [28] or by a bacteriophage portal vertex.[29] We are here interested in symmetries formed by chiral particles in a self-assembly process.

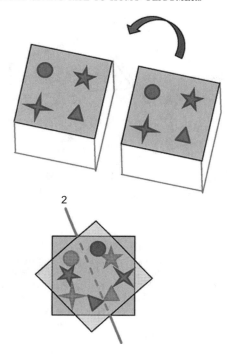

FIG. 5. Symmetric association of two identical molecules at identical surface patches. A centered contact between two identical surface patches is C_2-symmetric in all relative orientations. The twofold axis lies along the interface. The two molecules are sketched as blocks. In the superposition the upper molecule is half transparent. (For color version of this figure, the reader is referred to the web version of this chapter.)

Such a process has to be directed by an ensemble of suitable contacts that repeat endlessly: that is, either in a cyclic or in an infinite mode. The resulting aggregates can be zero, one, two, or three dimensional.

Zero-dimensional symmetry is called point group because the units are arranged around a central point. There are only five types of point groups (without mirror operations) that are available for the assembly of chiral units (Fig. 8). These are C_n (cyclic n-fold rotation comprising n units), D_n (dihedral n-fold rotation with n twofold axes perpendicular to the central axis comprising $2n$ units), T (twofold and threefold axes arranged to fit a regular tetrahedron built from 12 units), O (two, three, and fourfold axes arranged to fit a regular octahedron formed by 24 units), and Y (two, three, and fivefold axes arranged to fit a regular dodecahedron or icosahedron consisting of 60 units). Natural examples for all these point groups are deposited in the Protein Data Bank (PDB).[30] A multitude of homo-oligomeric enzymes are in point group C_2.

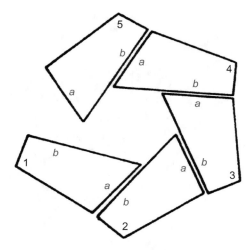

Fig. 6. Linear groups. With one asymmetric contact *a–b*, a protein assembles into a linear group. If unit 5 is above (below) unit 1, this develops to an endless left (right) handed helix. If a sixth unit cannot associate because it would collide with unit 1, the pitch is too small for a helix and the association ends in an asymmetric aggregate. At certain angles, the circle closes, for instance, for 72° to a pentamer that is C_5-symmetric. In case of a collision, a random evolution is likely to develop a cyclic symmetry because this stabilizes the aggregate.[27] (For color version of this figure, the reader is referred to the web version of this chapter.)

A much smaller number of them are in C_n with $n = 3$, 4, etc., because this requires an asymmetric contact (Fig. 6). Many homotetrameric enzymes are in D_2 because this group involves only symmetric contacts (Fig. 5). Fewer homo-oligomers are in D_n with $n = 3$, 4, etc. for the same reasons as for the corresponding cyclic associations. More seldom are associations in point groups *T*, *O*, and *Y*. Most of them have special functions such as the container ferritin or a viral coat, etc. (Fig. 8).

One-dimensional symmetry is called linear group. Linear groups are always endless helices that are defined by radius, pitch, and handedness (or sign of the pitch). They are based on a single asymmetric contact between the units (Fig. 6). All contacts give rise to helices except for cases where the pitch is so small that steric hindrance prevents finishing the first turn.[27] On the other hand, an exact closure of the first turn is point group C_n in zero instead of one dimension. A natural example for an endless helix is F-actin (PDB code 2ZWH). There are many examples for length-limited helices which, however, require a second molecule as a ruler as for instance a length-determining double-stranded DNA with a repeating binding sequence for a tandem repressor (PDB code 3H0D).

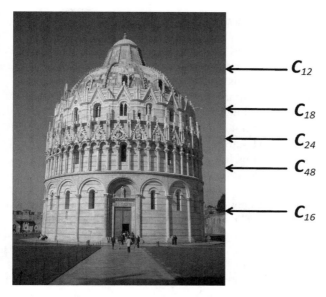

C_{12}

C_{18}

C_{24}

C_{48}

C_{16}

FIG. 7. Multiple cyclic symmetry in the Baptistry of Pisa, a Renaissance building (courtesy of Wikipedia). The cyclic symmetries change from the bottom to the top. Not all of them are in register. Multiple cyclic symmetries are well known in nature.[28,29] (For color version of this figure, the reader is referred to the web version of this chapter.)

Bio-point groups

Cn -	**c**yclic	Mycobact. porin (n = 8)	1UUN
Dn -	**d**ihedral	Clp protease (n = 7)	3P2L
T -	**t**etrahedral	Superoxide dismutase	3BFR
O -	**o**ctahedral	Ferritin	3NOZ
Y -	**i**cosahedral	Tomato bushy stunt virus	2TBV

FIG. 8. Point groups are arrangements around a central point without any translational element. They are zero dimensional. For chiral objects (no mirror image), there exist only five types of such groups, the so-called bio-point groups. Natural examples are given together with their PDB codes.

It should be noted that point groups can be combined with linear groups if a contact pair giving rise to a linear group is added to those forming the point group. For instance, C_n and D_n assemblies can associate along their central axes and form a rod, which is an endless linear group as it has been observed as a transient assembly of a kinase (Fig. 9).

Two-dimensional symmetry is named planar group. There are a total of 17 planar groups for achiral units whereas chiral units that do not obey inversion or mirror symmetry elements are restricted to the five planar groups with pure rotations $p1$, $p2$, $p3$, $p4$, and $p6$. These are the only cyclic symmetries that can combine with translations filling a plane completely like a parquet. The available planar groups require at least two contacts, one for each dimension. In nature, planar aggregates can be found in membranes. Most famous is bacteriorhodopsin in $p3$ (PDB code 1FBB) the structure of which was established by electron diffraction.[32] The bacterial S-layers (PDB code 3CVZ) are further examples.[15,16,33] The production of planar crystals for the structure determination of membrane proteins by electron diffraction is a popular, though not always successful, method.[34]

Three-dimensional symmetry is called space group. It results in crystals. Achiral units can crystallize in 230 space groups, whereas chiral units in the absence of their mirror image are restricted to 65 among them. There are a

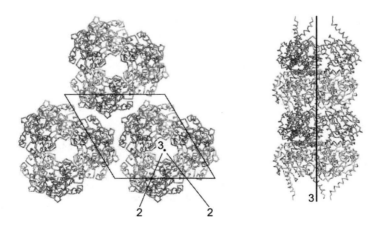

FIG. 9. Diurnal assembly of a kinase in a plant cell. The crystal structure showed rods of the enzyme adenylate kinase (PDB code 1ZAK), which were suggested to assemble in the absence of the substrate AMP at night.[31] The rods are head-to-head assemblies of almost exactly D_3-symmetric hexamers stacked along their threefold axes. The rod structure is incompatible with the known motions during a reaction cycle (Fig. 3). Therefore, it must dissolve in the presence of AMP during photosynthesis at daytime. (For color version of this figure, the reader is referred to the web version of this chapter.)

couple of examples of racemic protein mixtures, all of which crystallize in the primitive space group $P1^-$(PDB codes 3ODV, etc.). Since crystallization with subsequent X-ray diffraction analysis is the most efficient method for protein structure determination at the atomic level, a very large number of proteins have been crystallized and analyzed. Today, all 65 space groups have been realized for at least one protein deposited in the PDB.[30] The symmetric association is so relevant for crystallization that crystals with a large number of molecules in the asymmetric unit show usually a local symmetry between these molecules.[35-37] For entropic reasons, crystals with more than one protein in their asymmetric unit should grow from preformed asymmetric units which means from protein complexes. Like in solution, here also the assembly of an asymmetric complex is much less likely than a symmetric one,[26] which explains the abundance of local symmetry in crystals.

Only a few natural protein crystals are known. One example is a protein storage form in plants.[38] Another is the cellular storage form of insulin in micrometer-sized crystals.[39] A crystal is also very likely formed by the NADH peroxidase as an efficient organelle for destroying molecular oxygen in an anaerobic bacterium (Fig. 10).[40]

IV. Lessons from Protein Crystallization

With the aim to elucidate protein structures by X-ray diffraction, a large amount of work has been and is still being done in the production of well-ordered crystals which, after all, are protein–protein associations. Crystallization is usually a trial-and-error method using a multitude of environmental conditions. However, numerous ways have been found that facilitate such a process. The gathered experience should be used as a guide for engineering protein–protein contacts.

The most important requirement for crystallization is a protein ensemble with a uniform conformation. Homogeneous samples have been obtained after purification of original biological material of some very abundant proteins such as myoglobin, hemoglobin, lysozyme, ribonuclease, chymotrypsin, and others. Around 1980, however, it became clear that lengthy isolation procedures of less abundant proteins result in heterogeneous conformations. A simple example is the partial loss of a prosthetic group. Heterogeneity is detrimental for crystallization because molecules that are different but resemble each other may associate to a growing crystal face, where they then present an unsuitable surface for the next layer. Thus, very small amounts of heterogens on a crystal surface can stop crystal growth effectively. The wealth of protein crystals available since around 1990 is clearly due to the widespread production of recombinant proteins using either the original underlying gene, or a mutated

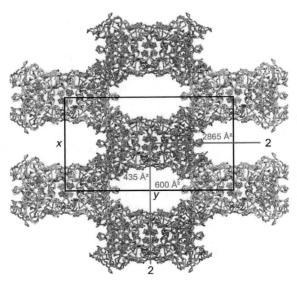

Fig. 10. Putative crystalline organelle removing dioxygen in an anaerobic bacterium (PDB code 2NPX).[40] The large channels allow the diffusion of the main reactant NADH. The x-axis of 77 Å defines the scale. The (invisible) channels along the x- and y-axes are about the same size. NADH can reach the active center through the channels. As the reaction is a mere electron transfer, the motions of the enzyme during catalysis are negligible. Apart from a very tight C_2-symmetric dimer contact (2865 Å^2 interface), the second best contact is asymmetric (600 Å^2) and connects single subunits to linear filaments as indicated by colors yellow and green. The third crystal contact (435 Å^2) is large enough to induce crystallization. (For interpretation of the references to color in this figure legend, the reader is referred to the Web version of this chapter.)

gene, or a fully synthetic gene.[41] Such proteins can often be produced in large quantities and thus have a good chance of keeping their conformational homogeneity during a short purification process.

Furthermore, a protein of interest should be monodisperse in solution so that it can associate in any required orientation. Fortunately, cytosolic proteins are already selected for solubility because otherwise they would be deleterious for a cell, as demonstrated by the example of sickle cell hemoglobin.[42,43] Proteins can be checked by dynamic light scattering for their tendency to associate.[44] Since associations depend on the environmental conditions, however, the conditions in such a test have to meet the crystallization conditions reasonably well.

Another difficulty with protein crystallization was experienced with the elongation factor Tu that had been incubated as an intact molecule but crystallized only after being attacked by a microbial impurity in the crystallization drop.[45] It turned out that this attack removed mobile parts of the protein that had prevented the formation of suitable crystal contacts.[46] The removal of mobile parts can of course be repeated in a defined manner, for example, by a time-limited digestion with trypsin of a given concentration.[45] Alternatively, the mobile part can be deleted by changing the underlying gene.

A further omnipresent source of problems is solvent-exposed cysteines that can form unique or multiple (for one or more thiols per unit) disulfide bridges producing a disulfide-bridged dimer with internal mobility (for one thiol) or heterogeneity (for several thiols). Both cases are detrimental for crystal growth.[47] Solvent-exposed cysteines have to be recognized either by a comparison with homologous structures or by a solvent-accessibility prediction based on the amino acid sequence.[48] They should be mutated, for example to serines, because it is difficult to proceed with crystallization and an X-ray analysis under strictly anaerobic conditions.

The crystallization of membrane proteins is particularly difficult because suitable crystal contacts have to be found within the rather limited polar surface area of these proteins. By structure prediction from the amino acid sequence or by biochemical experiments, it is usually possible to locate the polar surface regions and introduce mutations there. These mutations follow a couple of general rules. One rule is the introduction of tyrosines because they are amphiphilic and not very mobile. Tyrosines are like glue, as demonstrated by their abundance in the antigen binding sites of antibodies. Another rule, for example, is the elimination of flexible side chains such as lysines, etc. The first attempt with membrane proteins was most successful.[49] The method was later refined and is now applied to all types of proteins.[50,51]

Structure prediction from the amino acid sequence was established early on for α-helices, β-sheets, and turns.[52] It is also available for flexible regions [53] and for solvent exposure.[48] If structure prediction fails to give reasonable results although several related protein sequences are known, exposed parts of a protein can also be located by their commonly increased rate of mutations during evolution.[54] This method was applied successfully for the crystallization of a mycobacterial porin.[55]

In another case, a channel-forming protein crystallized in many different crystal forms as a monomer but never as the oligomeric channel of interest.[37] In all crystals, the same 12-residue portions of the N- and C-terminal tails remained unstructured. It seems likely that the addition of three N-terminal residues, which had been necessary for the required production as a fusion protein, prevented channel formation. This problem may be overcome by a newly designed synthetic gene that allows establishing the exact N-terminus after splitting off the fused protein. This case contradicts the commonly adopted view that the N- and C-terminal ends of proteins can be modified without any problems, for instance, with His tags to ease the purification process.

An X-ray analysis is more valuable if the ordering of the crystal is high. Thus, there were attempts to improve the ordering by introducing further contacts based on the known packing geometry of less well-ordered crystals. In one case, this increased the crystallization speed by a factor of 40 but, unfortunately, did not improve the ordering.[56] This experiment, like others, demonstrates that the dynamics of associations can also be engineered,[57] which may become an important aspect in future designs.

In summary, the vast experience collected with protein crystallizations offers the following messages for the engineering of defined associations: (i) the protein should be produced in large amounts in order to keep it homogeneous; (ii) the protein should be monodisperse in solution so that no contact is occluded; (iii) the protein should not contain mobile parts in the designed contacts and it must not show domain mobility; (iv) the protein should not contain solvent-exposed cysteines; (v) the knowledge of the protein surface of interest is most helpful for random and designed mutations; (vi) inconspicuous details such as incorrect chain ends may destroy a design; and (vii) a design need not be restricted to slowly forming permanent contacts because quickly associating constructs are quite possible.

V. The Analysis of Known Protein–Protein Contacts

The availability of numerous protein crystal structures in the PDB[30] gave rise to various attempts to characterize the protein–protein contacts observed in crystals.[58–63] Unfortunately, all these analyses could not yet be developed into a useful set of tools for contact design. The identification of such contacts can be done visually.[64] A more practical search is provided by the program PISA, [65] which takes a PDB file and lists all crystal contacts, describing them in terms of buried solvent-accessible surface area, involved amino acid residues, and calculated strength of attachment. As a general rule, the crystal contacts contain for the most part polar residues and they are rather weak. Consequently, they form only at high protein concentrations such as, for example, those applied in crystallization drops. Moreover, they depend very much on the used solvent. High salt concentrations reduce the strength of the polar contacts, whereas low salt together with organic water binders such as poly(ethylene glycol) increases their strength. The pH value is crucial, as it determines the charges at the protein surface repelling or attracting each other, which can cause substantial crystal packing changes.[66] Even the temperature is important as shown, for instance, by a reversible space group change between 8 and 20 °C.[67]

In many cases, the crystals contain the protein assembled in point groups that associate via crystal contacts. In space groups with genuine rotation axes (possibly besides screw axes as in the abundant space group $P2_12_12$), protein point group complexes may take part in the overall crystal symmetry. Alternatively, they may obey merely noncrystallographic (local) symmetry. It is often not possible to distinguish between contacts that are kept in solution and contacts that are only assumed at the high concentrations in the crystallization drop. There have been several proposals to distinguish between them, but no clear-cut solution of the problem has been found because the calculation of the contact strength from its geometry is still stricken by large errors.[68] In the case of noncrystallographic symmetry, one would expect that the complex has a greater chance to be stable in solution because it crystallizes most likely from point group complexes preformed in solution. Interestingly, noncrystallographic symmetries need not be local point groups but can also be linear groups extending infinitely in one dimension as observed with a glycosyl hydrolase.[69]

In summary, the crystal structures show contacts that are so weak that they are formed only at high concentrations (multiples of three for the three dimensions), or as contacts of intermediate stability in noncrystallographic symmetry, or as oligomeric contacts that are stable in solution. The proposal that NADH peroxidase forms crystals in a bacterial cell is based on the fact that the crystal contains large channels for diffusion and, more importantly, that it contains an unusually strong crystal contact with an area of 600 Å^2 that seems to be natural rather than adopted by chance (Fig. 10).[40]

VI. Self-assembly in Nature

Besides the numerous hetero-oligomeric assemblies in nature, which probably took a long time to be formed during evolution, we are here interested only in the much simpler homo-oligomers. Homo-oligomeric enzymes are very abundant. Most of them form point groups, which are based on repeating a given number of contacts in a given relative geometry. Interestingly, not all symmetries come with the same frequency. As a simple example, tetramers in D_2 symmetry outnumber by far those in C_4 symmetry.[70] The reason lies in the evolutionary development. While D_2 tetramers can evolve from C_2 dimers allowing an intermediate stable step in the process, the C_4 tetramers have to be produced in a single step essentially from a linear group that closes a ring (Fig. 6).[27]

The assembly into point groups is often a shield against a rough environment, as demonstrated by the observation that most monomeric enzymes of organisms living at normal temperatures are oligomeric in thermophiles.[71] Some enzymes have been driven on purpose to high stability by random mutations and selection as, for instance, the pyruvate oxidase developed for a medical test set.[72] In this case, the prosthetic group thiamine pyrophosphate (TPP) had to be fixed because it dissociated too easily in the test-set material.

As a big surprise, the obtained mutants stabilizing TPP were not at the binding site of TPP but were at the interface of the tetramer.[73] This shows that the interfaces of oligomers can be rather sensitive to environmental changes. A further reason for oligomerization is the ease of introducing cooperativity by slight changes of the contacts without much change within the polypeptide subunits.[17,74] Such cooperativity usually changes the chemical properties of a catalyst, increasing regulatory options and thus efficiency.

Some of the assemblies are transient as, for instance, the crystalline storage of plant proteins [38] or the storage of insulin.[39] The crystalline assembly reduces appreciably the required space and also the osmotic pressure. In both cases, only one cycle of assembly and disassembly is needed. A famous transient oligomerization is the sickle cell hemoglobin that assembles in the deoxy form and disassembles on dioxygenation.[42,43] This system was developed to defeat the malaria parasite during its life time in erythrocytes. In another case, the assembly is transient and recycles daily. It removes a kinase of a plant that is needed for CO_2 uptake only during sunlight but superfluous for the Calvin cycle running at night. This kinase is suggested to assemble each night into rods in order to release the water bound to the surface of the monomeric enzyme, as water is a precious commodity for plants.[31] A two-dimensional equivalent of the water release is the crystalline bacteriorhodopsin patches assembled in the membrane in order to free lipids for diffusion into the membrane where they are needed to keep the flexibility.[32]

It should be noted that the polypeptide folding process resulting in structured proteins is also a self-assembly, although it is supported by folding helpers such as the heat-shock proteins.[75,76] Here, loosely associated oligomers tend to start by folding the subunits separately and finish with their combination into an oligomer. In contrast, tightly bound oligomers are likely to start with an oligomeric nucleus. As an example, trimeric porins with a very solid hub and a surrounding fully meandering (all-next-neighbor-antiparallel) β-barrel indicate clearly that the trimeric hub with its strong apolar interior starts the folding process and the three β-barrels follow later on.[77-79] This is a rather direct self-assembly process that starts from three structureless polypeptide chains fastened at both ends and results in the C_3-symmetric oligomer (Fig. 11).

VII. Self-assembling Protein Constructs

Designed self-assemblies are pursued with numerous building blocks.[80] We are here interested in self-assembling proteins that need no further partner for channeling the assembly process into a given direction such as, for instance, a ruler for limiting the length of linear groups. Such simple

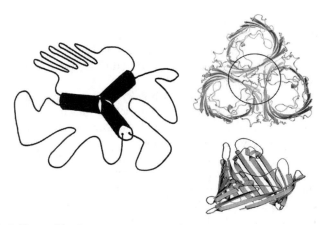

FIG. 11. Self-assembly of C_3-symmetric porins from a bacterial outer membrane (PDB code 1PRN). The folding process is suggested to start with the association to a trimeric central hub, which contains three lipid-binding sites and fixes all six ends of the three polypeptides. Subsequently, the three all-next-neighbor-antiparallel β-barrels can be formed.[78] The hub (encircled) is a solid construction reminiscent of a globular protein.[79] The connectivity of a single β-barrel as viewed from the threefold axis (vertical) is shown at the bottom right. (For color version of this figure, the reader is referred to the web version of this chapter.)

assemblies always result in group symmetry, which can be zero, one, two, or three dimensional. As described above, the generation of three-dimensional groups in the form of X-ray grade crystals has been pursued intensely for protein structure analysis, yielding a number of general rules that apply to all types of assemblies. In essence, the building block should be monodisperse in solution and have a rigid structure in its interior and a rigid and chemically inert surface.

Monodisperse means, for instance, that membrane proteins have to carry a detergent micelle which covers part of their apolar surface so that the geometric options for contacts are largely reduced.[81] A rigid interior must not consist of several domains that are mobile against each other. Useful contacts require rigid surfaces, which excludes mobile parts, as for instance those participating in a local induced-fit movement of an enzyme. A mobile region is likely to solidify in multiple conformations introducing heterogeneity and it also needs an entropy reduction which lowers the binding strength. In addition, the side chains on the surface should be relatively rigid (no lysines) so that the entropy reduction needed for fixing them in contacts is limited.

An early demonstration that proteins can be assembled to form a predetermined crystal was reported for a D_2-symmetric lectin and disaccharides connecting two saccharide binding sites of this lectin. The experiment was successful. The resulting crystals could be visualized under an electron microscope and the unit cell could be established by X-ray diffraction.[82]

The assembly was a one-step procedure. However, the size of the resulting crystals could be controlled by the pH value during assembly. Most likely, the pH value had a strong influence on the saccharide-binding strength.

In a more elaborate approach, the modified C_4-symmetric enzyme rhamnulose-1-phosphate aldolase (RhuA, Fig. 12) was used to produce a two-dimensional crystal belonging to the planar group $p4$. Here, each of the four enzyme subunits was labeled with two biotin molecules tethered to a cysteine by a disulfide bridge (Fig. 13). For labeling, the enzyme RhuA was mutated in the following manner. Cys126 was replaced by a serine because it showed a free thiol at the wrong position, which would disturb the design. Residues Asn133 and Lys261 were converted to cysteines, as they were at the locations that could bind a streptavidin via tethered biotin labels connected to these cysteines. The C-terminus ends in a His tag at the top of the molecule (Ct) such that binding to an NTA column does not affect the four sides where the streptavidins bind. When the tethered biotin precursor binds to free thiols, it releases a colored pyridine thiolate so that the reaction can be followed photometrically.

The construction principle of the planar network is sketched in Fig. 14. The C_4-symmetric labeled enzymes were connected via the D_2-symmetric streptavidin, which binds tightly to biotin. The selected residues for the labels at RhuA were such that the twofold axes were parallel and perpendicular to the fourfold axis, which kept all proteins in one plane. Appropriately mutated RhuA was fastened via one of its four His tags to an NTA column and decorated by

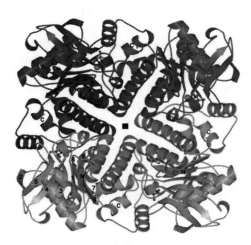

FIG. 12. The enzyme rhamnulose-1-phosphate aldolase (RhuA) is a C_4-symmetric tetramer (PDB code 1OJR). It is shown as a ribbon plot with colored subunits viewed along the symmetry axis. A regular repeating of RhuA in the paper plane may give rise to the planar group $p4$.(For color version of this figure, the reader is referred to the web version of this chapter.)

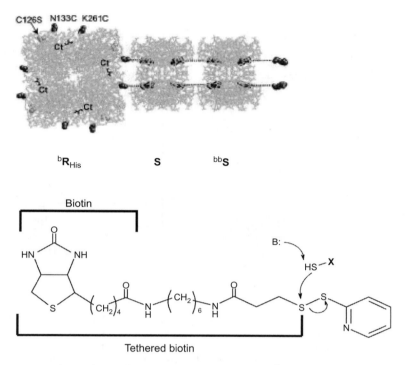

Fig. 13. A designed network belonging to the planar group *p4*.[3] The enzyme RhuA was mutated and labeled with eight tethered biotins (bR$_{His}$). Streptavidin (S) was added. Further streptavidin loaded with double-headed tethered biotins (bbS) prolonged the extension. The depicted precursor of the tethered biotin label allows photometric monitoring of the labeling reaction.

applying the tethered biotin precursor (Fig. 13) and washing the column afterwards. At this first stage, the labeled protein was detached from the column using imidazole and stored for further usage. The procedure was then repeated up to this first stage (no. 1) and continued by applying streptavidin to the column followed by washing. Using imidazole, the stage no. 2 product streptavidin-decorated RhuA was then released and mixed with the product of stage no. 1 to yield a network belonging to the planar group *p4* with one streptavidin as a spacer. The mesh size of this network can be increased by extending the spacer to two, three, or more streptavidin molecules connected by double-headed tethered biotin, which is produced in a reaction between (oxidized) precursor and disulfide-reduced precursor (Fig. 13). The construction described in Fig. 14 can be pursued in a rather defined manner because reactions can be driven to completion by monitoring photometrically and surplus chemicals can be removed by column washing.

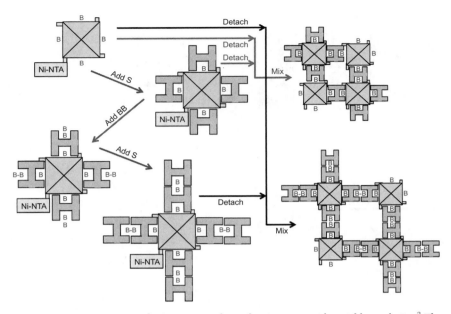

FIG. 14. A construction of *p4*-symmetric sheets forming a net with variable mesh size.[3] The sheet consists of the C_4-symmetric modified enzyme RhuA (green), D_2-symmetric streptavidin (orange), and double-headed tethered biotins B–B. RhuA carries two introduced thiols per subunit, which were decorated with tethered biotins. For clarity, only one of the two biotin sites is sketched. Each subunit also contains a His tag for the attachment to a Ni–NTA column. Streptavidin (add S) and double-headed tethered biotin (add BB) are added on a Ni–NTA column, so that every surplus of the chemicals can be washed out. For the formation of a net, the appropriate building blocks are detached from the column by imidazole and mixed. Depending on the decoration with double-headed tethered biotin and streptavidin, two-dimensional networks with different spacings can be built up. (See Color Insert.)

The results of this work are shown with electron micrographs after negative staining in Fig. 15. Several small aggregates could be produced, and averaging confirmed clearly the design.[3] However, only small rather irregular nets could be produced. This was first considered as a problem arising from the three-dimensional mixing process for building a two-dimensional construct. Therefore, the two-dimensional buildup was performed at the planar surface of a drop of water covered by a lipid monolayer containing NTA lipids. The NTA lipids bound RhuA at its His tag as sketched in Fig. 15E. The resulting nets shown in Fig. 15F were larger than after three-dimensional mixing but still not satisfactory. It was then concluded that a regular buildup was hindered by the strong attachment of biotin to streptavidin, which made each contact essentially irreversible. In this way, association errors could not be recovered. Weaker bonds would have allowed opening incorrect bonds, for instance, by changing

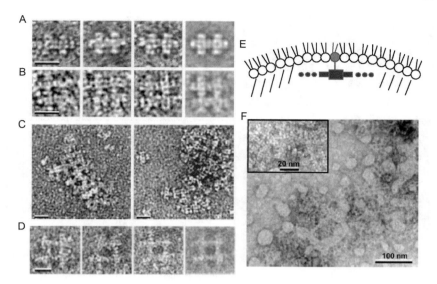

FIG. 15. The results of the RhuA–biotin–streptavidin system (Fig. 14) were negatively stained and viewed using an electron microscope.[3] (A) Single RhuA doublets, the average structure at the right side has the correct dimensions. (B) The same as panel A but with RhuA quartets with one streptavidin as spacer. (C) Larger aggregates of one streptavidin spacer association after three-dimensional mixing. (D) The same as panel B but with two streptavidin spacers. (E) A two-dimensional construction under the surface of a water drop covered with a lipid monolayer that contained some NTA lipids to bind RhuA at its His tag. (F) The results of the two-dimensional association were larger and cleaner than those of the two-dimensional version in panel C. The applied lipid disturbed the negative staining. (For color version of this figure, the reader is referred to the web version of this chapter.)

the environmental conditions. As with crystallization, a slow transition from weak to strong binding contacts is most likely much more appropriate for net formation. A redesign of this system would certainly use a mutated streptavidin with a much weaker binding affinity to biotin.[3]

In addition to the network with different but distinct spacer lengths, a further network with switchable mesh size was designed. The construction used the enzyme 6-phospho-β-galactosidase (PGAL), which is a monomeric globular protein with a mass of 51 kDa and thus clearly visible in negatively stained electron micrographs. As sketched in Fig. 16, two PGAL molecules were fused via a piece of a β-helix taken from a bacterial protease (PDB code 1SAT). The β-helix conformation requires two Ca^{2+} ions per turn and unfolds on Ca^{2+} removal through the application of EDTA. The respective electron micrographs demonstrate that this design was successful (Fig. 16). The construction is to be extended by adding a tetramer-forming α-helix at both ends as

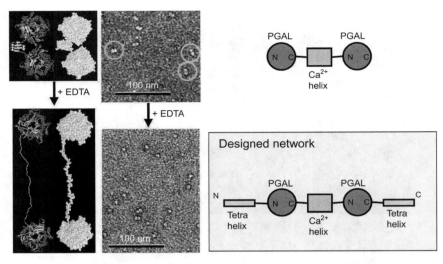

Fig. 16. Switchable constructs.[3] A fusion protein connecting two 6-phospho-β-galactosidase (PGAL) molecules via a Ca^{2+}-dependent β-helix (PDB code 1SAT) was imaged with and without Ca^{2+} ions (removed by EDTA). The electron micrographs confirm switching. The design is to be continued by adding to both ends α-helices that form tetramers (PDB code 1AIE). A preliminary test with PGAL fused to a C-terminal α-helix showed tetramer formation under the electron microscope. The full construction is not yet finished. It should provide a net with switchable mesh size. (See Color Insert.)

sketched (PDB code 1AIE). The association of these α-helices to tetramers should result in a net of fourfold hubs connected by spacers the length of which can be switched by applying Ca^{2+} or EDTA, respectively.

Besides natural protein–ligand contacts such as lectin–saccharide [82] or streptavidin–biotin,[3] it is also possible to use natural protein–protein contacts. Domain swaps, for example, use the internal interface of a two-domain protein (Fig. 17). Such swaps may occur if the binding strength of a domain interface is relatively low so that it can be opened in an appropriate environment and then recombined with other open molecules. This may happen in crystals if the crystallization conditions open the protein. Examples are dimers of a monomeric bacterial toxin,[36] trimers of the monomeric enzyme barnase,[83] and large oligomers of ribonuclease A.[84] Such domain swaps are disliked because they cause deviations from the natural protein structure. However, domains swaps have also been designed.[85] If the peptide chain between the two domains is long enough, the two domains can adopt almost unrestricted relative orientations so that not only dimers but also cyclic oligomers or polymers can be produced (Fig. 17).

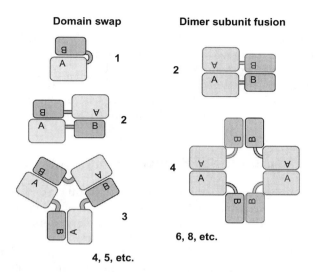

Domain swap   **Dimer subunit fusion**

FIG. 17. Assembly after domain swapping and assembly of fusion proteins. Opening up a two-domain protein (1) may lead to dimers (2), or trimers (3), up to filamentous polymers. All constructs use the established internal domain–domain contact A–B. The fusion of two dimer subunits may give rise to dimers (2, red means rear side of the molecule), or tetramers (4), or to filamentous polymers. For highly flexible linkers, both types tend to associate to irregular aggregates. (For interpretation of the references to color in this figure legend, the reader is referred to the Web version of this chapter.)

A construction similar to domain swapping can be designed with fusion proteins consisting of the subunits of two different dimeric proteins connected via a polypeptide spacer of predetermined length (Fig. 17). For long spacers allowing a broad range of relative orientations, this gives rise to irregular aggregation. Short spacers that define a fixed relative orientation between the two different units result in an endless helix if the relative angles are larger than 90° (Fig. 17). It is conceivable that a fixed relative orientation can be constructed by fusing a further rigid domain in between the two dimer subunits. It is, of course, possible to combine subunits of oligomers of any kind of point groups to form aggregates in two or three dimensions. This has been done with subunits of dimers coupled to subunits of trimers trying to achieve a relative geometry of the twofold and threefold axes that fit the point group T.[86] The attempt was probably successful because it yielded single globules of approximately the correct size as visualized under an electron microscope.

It should be noted, however, that the binding strengths of natural oligomers is usually very high so that fusion proteins consisting of subunits of different oligomers tend to aggregate immediately after production. Any attempt to separate the conglomerate again using urea or similar chemicals is likely to cause irreversible partial denaturation of the subunits. Therefore, in a practicable approach the binding strengths of the interfaces should first be diminished by mutations such that the subunits can be associated and dissociated at two environmental conditions A and B, respectively. The self-assembly of a regular network can then be achieved by a slow change from condition B to condition A, as it is common with crystallization, where a low protein concentration corresponds to condition B and a high one to condition A.

VIII. Design of Protein–Protein Contacts

Since long tethers are generally needed for permitting a desired interaction (Fig. 17), the use of natural protein–ligand or protein–protein complexes often results in aggregates that are flexible and thus not very useful. It is therefore reasonable to go a step further and try to develop novel direct protein–protein contacts. This has been performed successfully by random production procedures with a subsequent selection process, for instance, using mice for monoclonal antibody production,[87] a bacterial phage display,[88] or a ribosomal display system [89] for the detection of enzyme inhibitors. These methods are efficient in generating a binding contact *per se* but cannot be used for producing a particularly desired contact between two selected surface patches.

In a more specific experiment, the chains of two different DNA-*endo*-nucleases were fused and mutated to generate a new DNA-*endo*-nuclease of higher specificity as a versatile tool for gene manipulations.[90] Moreover, a group of homologous, structurally known enzyme–inhibitor complexes was analyzed for mutations giving rise to new binding partnerships and also for inhibitors that bind to a large group of these enzymes.[91] A similar experiment improved the specificity of calmodulin for a certain myosin, with a concomitant reduction of the binding strength to other natural partners such as spectrin.[92] Completely novel proteins and binding partners have been designed and produced by using one of the rare known peptide–peptide recognition complexes, namely, the knob-into-holes α-helix association.[93–97] One of the designs resulted in filaments.[85]

A further step forward was the design of self-associations of structurally known proteins by mutating a surface patch considered suitable for oligomerization.[98] These experiments showed that the success of a contact design depends critically on the contact multiplicity, which is explained in Fig. 18. In this engineering campaign, a monomer was converted into a designed C_2-symmetric dimer, two different dimeric proteins were converted to designed D_2-symmetric tetramers, and a C_4-symmetric tetramer was converted to several octamers. Moreover, an engineered C_8-symmetric octamer was found surprisingly to be a D_8-symmetric hexadecamer in solution. The results are

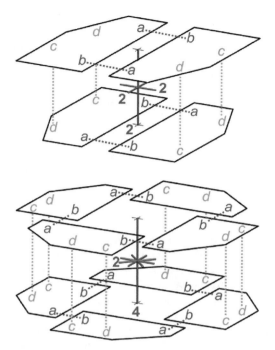

FIG. 18. The multiplicity of contacts.[98] The top figure illustrates a multiplicity of 4 for a D_2-symmetric tetramer which requires two contact types, here a–b and c–d. One plane of this sketch shows a multiplicity of 2 for a C_2-symmetric dimer with contact a–b. The bottom figure shows a multiplicity of 8 for a D_4-symmetric octamer that needs two types of contacts a–b and c–d. One plane of the drawing represents a multiplicity of 4 for a C_4-symmetric tetramer using contacts a–b. C_4-symmetry is difficult to evolve via random mutations because the a–b contact has to produce an angle of exactly 90° between the units. (See Color Insert.)

summarized in Fig. 19. All complexes were produced under relatively mild conditions and confirmed by dynamic light scattering and by size-exclusion chromatography. However, not all complexes could be crystallized. Unfortunately, some of the complexes yielded crystals that contained the original (mutated) molecule and not the designed oligomer, indicating that the newly formed complex was stable in the usual mild environment, but was disrupted under the harsher crystallization conditions.

The symmetric association of C_1-, C_2-, C_4-, and C_8-symmetric molecules required correspondingly fewer mutations because the multiplicity of the contacts were 2, 4, 8, and 16, respectively. Obviously, it is comparatively easy to produce a complex with high multiplicity by a few mutations, whereas an association at low multiplicity is difficult to achieve. A multiplicity of 1 was not

		Multiplicity	Mutations
Galactosidase	C_1 monomer $\longrightarrow$ C_2 dimer	2	FF, WF, FWFF, FWMF
Urocanase, sulfhydrylase	C_2 dimer $\longrightarrow$ D_2 tetramer	4	IIA, NTIV, NVLV, NILLN
Aldolase	C_4 tetramer $\longrightarrow$ D_4 octamer	8	F, Y, MF, WS
Rim of porin MspA	C_8 octamer $\longrightarrow$ D_8 hexadecamer	16	None

FIG. 19. Results of a protein–protein association campaign stating the used proteins and their point groups as well as the multiplicity of the designed contacts and the residues introduced through mutations.[98]

even tried in this series of experiments.[98] In the reverse, C_4- or C_8-symmetric molecules in a cell are hazardous because a single mutation may cause a harmful aggregation. This may be one of the reasons why there are so few C_4-symmetric tetramers as compared to D_2-symmetric ones.[70]

The engineering campaign started with the construction of C_2-symmetric dimers of the monomeric enzyme PGAL (PDB code 1PBG). The design was based on two large crystal contacts with interfaces of 770 and 640 Å^2 containing noncrystallographic twofold axes (Fig. 20). Noncrystallographic symmetry contacts are considered more stable than common crystal contacts because they are likely to be formed before crystallization. These contacts were fortified by introducing large apolar residues such as phenylalanine or tryptophan. The positions of the mutations are shown in Fig. 21. The changes in the surface of contact no. 2, where three charged residues were replaced by apolar ones, are illustrated in Fig. 22.

The resulting oligomerization was analyzed qualitatively by dynamic light scattering and also quantified by size-exclusion chromatography. Figure 23 demonstrates that the dimerization was incomplete. The dimer fraction of contact no. 1 with two mutations was much smaller than that of contact no. 2 with four mutations, presumably due to the larger apolar contact surface. In both cases, the dimer fraction could be rechromatographed without the reappearance of monomers. This indicates that there is no dynamic equilibrium: once formed, the contacts are stable under the storage and chromatography conditions. However, despite very extensive trials, these dimers could not be crystallized although they had been designed across noncrystallographic twofold axes in the original crystal form and should therefore fit the established packing scheme. It has to be concluded that the dimerization introduced structural heterogeneity, for instance, a nonuniform conformation at the contact, which tends to poison any crystal growth.

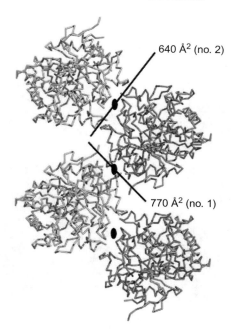

FIG. 20. The two C_2-symmetric noncrystallographic contacts in a crystal of 6-phospho-β-galactosidase (PGAL).[69] Both crystal contacts were developed to dimers in solution by introducing large apolar side chains.[98] (For color version of this figure, the reader is referred to the web version of this chapter.)

Introducing large apolar side chains as done with PGAL seems to enforce contact formation. However, it also creates two major problems. First, the enforced contact is likely to adopt several different conformations, which exclude crystallization and also more elaborate constructions. As a further experience, it was found that the large apolar residues decreased the production rate of PGAL appreciably. Most likely, they interfered with the folding process. For both reasons, such mutations are rather hazardous. They should be avoided whenever possible.

The design of D_2-symmetric tetramers was performed for two C_2-symmetric dimeric proteins, O-acetylserine sulfhydrylase (OASS, PDB code 2V03, Fig. 24) and urocanase (PDB code 2V7G, Fig. 25). Both enzymes were aligned "tail to tail" along their molecular twofold axes. The dimers were rotated against each other in search of suitably fitting patches. These patches were then mutated. In the light of the presumed conformational heterogeneity caused by large apolar residues in PGAL, the newly introduced residues were predominantly small apolar ones. Both enzymes gave rise to tetramers as

Contact no. 1   Contact no. 2

2 mutations   4 mutations

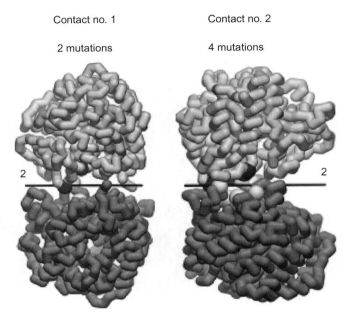

FIG. 21. The designed contacts of 6-phospho-β-galactosidase (PGAL).[98] The molecules are depicted as C_α chains with an unnaturally thick diameter. The designed twofold axes coincide with the noncrystallographic twofold axes shown in Fig. 20. (For color version of this figure, the reader is referred to the web version of this chapter.)

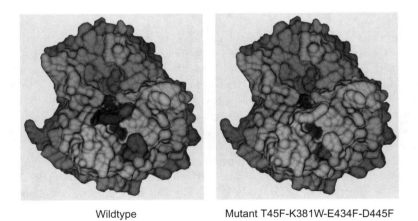

Wildtype   Mutant T45F-K381W-E434F-D445F

FIG. 22. The surface of the region of contact no. 2 of PGAL before and after the mutations.[98] The representation shows all nonhydrogen atoms. The colors denote residues that are negative (red), positive (blue), neutral polar (green) and apolar (yellow). (See Color Insert.)

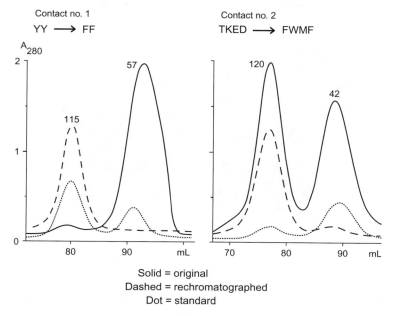

Solid = original
Dashed = rechromatographed
Dot = standard

FIG. 23. The size-exclusion chromatography results of the engineered mutants of contacts no. 1 and no. 2 of PGAL.[98] The dimer fraction of the two-mutation construct is much smaller than that of the four-mutation construct. In both cases, the rechromatography of the dimer peaks showed the pure dimer, indicating that there exists no dynamic monomer–dimer equilibrium. The applied dimer standard was a tandem fusion protein of PGAL.

confirmed by size-exclusion chromatography (Fig. 26). Urocanase was completely tetrameric, whereas the tetramer fraction of OASS was only 6% but could be rechromatographed without any reappearance of the original dimer. Consequently, there exists no dynamic oligomerization equilibrium under the storage and chromatography conditions as was observed with PGAL. For OASS and urocanase, the tetramer fractions yielded X-ray grade crystals. Unfortunately, OASS dissociated under the crystallization conditions showing the original (mutated) molecule in the crystal and thus failed to reveal the established contact at atomic resolution. On the other hand, the urocanase tetramer contact could be analyzed at high resolution (Fig. 25). The contact itself came out as designed. As a surprise, however, the tetramer did not belong to the designed point group D_2 because the contacting domains of each urocanase subunit had been displaced against the bulk of the protein. Obviously, the newly constructed contact was, on one hand, stronger than the internal interdomain contact but, on the other, it deviated slightly from the symmetric design. This resulted in an unexpected disruption of the protein.

Contact no. 1   **Contact no. 2**

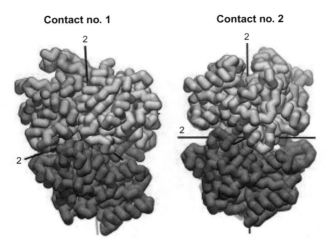

FIG. 24. Tetramerization of the dimeric *O*-acetylserine sulfhydrylase (OASS) by a tail-to-tail association along the molecular twofold axis.[98] The molecule is represented by unnaturally thick Cα chains with color-coded subunits. The mutations are indicated (purple). The two depicted relative orientations were found suitable for an engineering attempt. They correspond to contacts no. 1 and no. 2. (For interpretation of the references to color in this figure legend, the reader is referred to the Web version of this chapter.)

Design   **Reality**

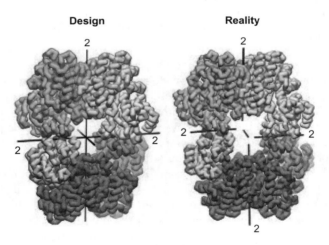

FIG. 25. Design and realization of the tetrameric association of the dimeric urocanase.[98] The molecule is represented by unnaturally thick C_α chains with color-coded subunits. The mutations are indicated (purple). The design of a tail-to-tail association aiming to produce a D_2-symmetric tetramer worked out in principle but caused an unexpected internal displacement between the core domain and the NAD$^+$ domain that broke the expected point symmetry D_2 down to C_2. (See Color Insert.)

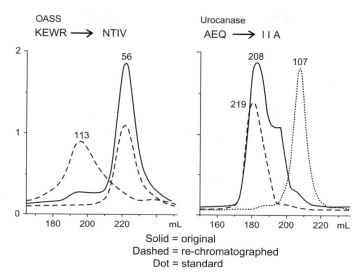

FIG. 26. Size-exclusion chromatography of OASS and urocanase mutants.[98] The OASS mutant shows only a small tetramer fraction which, however, rechromatographed as tetramer. Moreover, the dimer rechromatographed as a dimer. This excludes a dynamic dimer–tetramer equilibrium under the storage and chromatography conditions. The urocanase mutant is completely tetrameric and stays tetrameric in the rechromatography. As derived from later experiments, the shoulder in the original curve is a problem that arose only in this particular chromatography run.

The next step upward was the design of D_4 octamers from the C_4-symmetric tetrameric enzyme rhamnulose-1-phosphate aldolase (RhuA, Figs. 4 and 12). All designed contacts had a multiplicity of eight. One of the tail-to-tail designs required merely the single mutation Ala88 → Phe for forming the octamer at normal conditions (pH 7.0). A size-exclusion chromatography run showed that all mutated RhuA molecules were associated to octamers. The resulting octamer mass was confirmed by the addition of wild-type RhuA in the size-exclusion chromatography (Fig. 27). The octamer was crystallized and its structure was established by an X-ray analysis. It turned out that the contact followed the design closely. A second crystal form of this mutant was obtained at a pH of 4.5. It was analyzed using X-rays and revealed a slightly different arrangement which, however, kept the designed D_4-symmetry. A further size-exclusion chromatography at pH 4.5 demonstrated that the RhuA mutant remained fully associated at pH 4.5 (Fig. 27). The displacement is a rotation around the fourfold axis, as sketched in Fig. 28.

A further designed tail-to-tail association used the mutation Gln6 → Tyr. Size-exclusion chromatography showed again an association to 100% at normal conditions. Moreover, this RhuA mutant yielded also crystals. An X-ray analysis

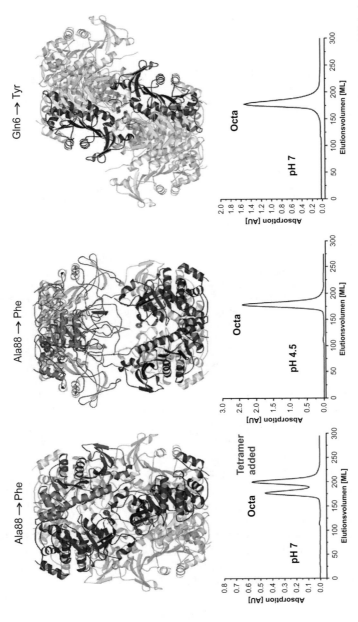

FIG. 27. Tail-to-tail assembly of the C_4-symmetric rhamnulose-1-phosphate aldolase (RhuA) for two mutants and two pH values as established with crystal structures.[98] The size-exclusion chromatography shows single octamer peaks, which in one case were confirmed by adding the wild-type tetramer to the sample. (For color version of this figure, the reader is referred to the web version of this chapter.)

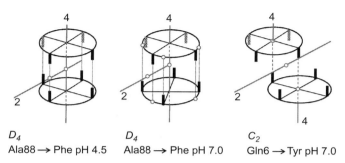

D_4 D_4 C_2
Ala88 → Phe pH 4.5   Ala88 → Phe pH 7.0   Gln6 → Tyr pH 7.0

Fig. 28. Sketch illustrating the tail-to-tail assemblies of RhuA. The pH difference causes only a slight change of the designed D_4-symmetric association. Mutant Gln6 → Tyr showed only C_2 symmetry instead of the designed D_4-symmetry (PDB code 2UYV). In this case, the preferred contact interface contains generally more rigid side chains than the designed interface in D_4 symmetry. It is therefore favorable, although it is a lower symmetry contact. Note that this contact is not a crystal packing effect because the size-exclusion chromatography shows a clear octamer (Fig. 27). (For color version of this figure, the reader is referred to the web version of this chapter.)

then showed a very unexpected association across a twofold axis that differed strongly from the design, as shown in Fig. 27. The expected point group D_4 had been broken down to point group C_2 with offset local C_4-symmetries, as sketched in Fig. 28. Such a symmetry breakage should occur only if the adopted contact is much stronger than the designed one.

One possibility for a stronger contact would be a lower entropy reduction on association. In order to check this point, a mobility analysis of all side chains was performed. This was possible because the numerous engineering trials with RhuA gave rise to 10 crystal structures with different crystal contacts.[98] The side-chain angle χ_1 and χ_2 of these structures were calculated and their spread was determined. In Fig. 29, the spread is plotted along the polypeptide chain, and in Fig. 30 it is shown in color code on the molecular surface. It turned out that there existed some hot spots. A comparison between the contacting residues in the tail-to-tail mutants Ala88 → Phe (D_4-symmetry, pH 7) and Gln6 → Tyr showed that the angular spread was 43° and 32°, respectively. Consequently, the latter required a lesser entropy reduction and is therefore more favorable than the D_4-symmetric contact found with Ala88 → Phe. This difference seems to explain the low-symmetry association of RhuA mutant Gln6 → Tyr.

Apart from the tail-to-tail associations, head-to-head associations using mutations Lys248 → Trp and Ala273 → Ser were also performed. This mutant showed a 60% fraction of octamer in size-exclusion chromatography. The octamer was purified by rechromatography, crystallized, and analyzed by

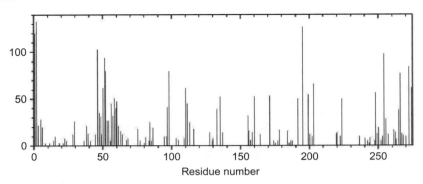

FIG. 29. The conformational mobility of the side chains of RhuA derived from 10 crystal structures with different packings.[98] The mobility is represented by the average angular distance of the side chain angles χ_1 and χ_2 from their averages in the 10 structures.

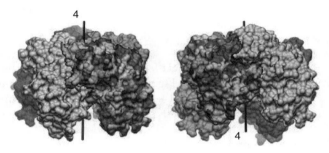

FIG. 30. The conformational mobility of the side chains of RhuA color-coded from blue to red on the surface of one subunit.[98] The mobility corresponds to the angular spread of χ_1 and χ_2 (Fig. 29). The subunit is viewed from two angles. The distribution is highly inhomogeneous with a few hot spots (red). (For color version of this figure, the reader is referred to the web version of this chapter.)

X-ray diffraction. Unfortunately, the resulting crystal structure showed the original (mutated) tetramer and not the crystallized octamer, indicating that the achieved contact strength was not high enough to withstand the crystallization conditions. The addition of the tail-to-tail mutation Ala88 → Phe to the head-to-head mutations then yielded filaments that could be viewed under the electron microscope (Fig. 31). They showed the expected diameter and the expected grating, which thus confirmed that the head-to-head contact had formed essentially as designed. The size-exclusion chromatography run for the filamentous head and tail aggregates then showed a broad distribution of masses beginning at the void volume containing the long filaments and ending in the octamer region. Since the tail-to-tail contact was known to be tight, the head and tail mutant was at least octameric (Fig. 31), as expected.

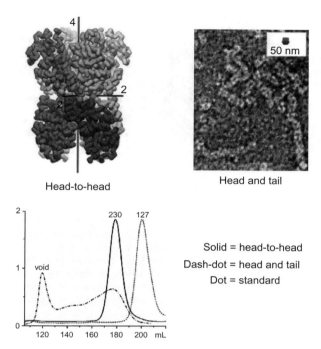

FIG. 31. The head-to-head assembly of RhuA with mutations Lys248 → Trp and Ala273 → Ser worked out under normal conditions as shown by size-exclusion chromatography.[98] The diagram shows the rechromatography of the original octamer fraction of 60% which, however, broke up on crystallization. The head and tail association containing the head-to-head mutations and the tail-to-tail mutation Ala88 → Phe yielded filaments of the expected size and grating, which confirms that the geometry of the head-to-head design (upper left side) was adopted. (See Color Insert.)

Using a C_8-symmetric octameric mycobacterial porin (Fig. 32), the engineering campaign was continued to a multiplicity as high as 16 in a designed D_8 hexadecamer. The structural analysis of this porin had already indicated that suitable crystals required an association, which then turned out as a dominant crystal contact around the eightfold axis.[55] In a separate project, it was planned to convert the Rim domain of this porin to a soluble carrier for large apolar molecules or complexes such as, for instance, an organic metal complex with catalyzing capability (Fig. 33). This construction resembled the cyclodextrins that enclose small apolar molecules with around a dozen carbon atoms. The Rim domain octamer showed a special type of β-barrel that was tightly connected (Fig. 32). The Rim domain was obtained by bridging the polypeptide chain between residues 69 and 122, removing the membrane-immersed residues 70 through 121 (Fig. 33).

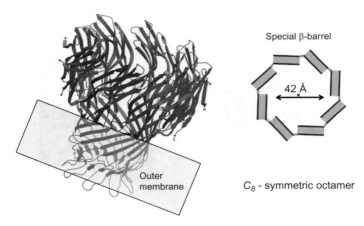

FIG. 32. The structure and the location of the C_8-symmetric mycobacterial porin MspA (PDB code 1UUN).[98] The crystal packing of MspA contains a tail-to-tail association as a dominant contact.[55] The contacting loops are here depicted below the membrane. The extra-membrane part consists of eight Rim domains that form a special strong β-barrel. It is a suitable building block for nanotubes. Moreover, it could be made apolar in its interior and used as an oversized cyclodextrin-type compound that may solubilize large apolar complexes such as organic catalysts. (See Color Insert.)

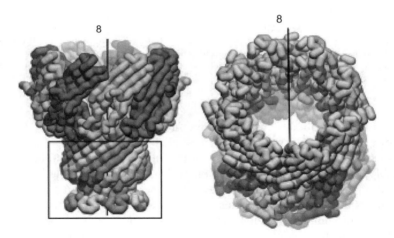

FIG. 33. Assembly of a C_8-symmetric ring of Rim domains (PDB code 2V9U)[98] that were dissected from a mycobacterial porin.[55] The Rim domains were produced by removing the membrane-immersed part (in rectangle). During protein production and/or purification, the C_8-symmetric octamer associated spontaneously to form a stable D_8-symmetric hexadecamer as established in crystals and by size-exclusion chromatography (Fig. 34). (See Color Insert.)

The construct could be produced in large quantities and crystallized. X-ray analysis showed a D_8-symmetric hexadecamer in the crystal. This was most welcome because the association doubled the internal space and also stabilized the carrier, both of which were desirable properties. In the future, the construct has to be further developed converting the polar internal surface into an apolar one. It is hoped that this change can be done by mutations which, however, must not disturb the folding process. Conceivable is the introduction of a couple of cysteines that could be modified using apolar labels.

Given the experience that complexes in crystals are not always also complexes in solution, and being aware of the fact that a protein complex mass of about 200 kDa cannot be derived safely from a size-exclusion chromatography run, it was necessary to compare the presumably hexadecameric construct with the underlying octamer. For this purpose, mutation Ala35 → Arg was introduced that split the spontaneously formed hexadecamer into octamers as demonstrated in size-exclusion chromatography runs with hexadecamers and octamers (Fig. 34). This confirmed the hexadecamer and also demonstrated that it was stable in solution.

The lesson learned from the experiments creating direct protein–protein contacts is rather clear: the higher the multiplicity, the higher is the chance to obtain a strong binding interface with a small number of mutations (Fig. 18). This observation was rationalized at the level of multiplicities 1 and 2. During a random evolutionary process, asymmetric contacts (multiplicity = 1) occur much more rarely than C_2-symmetric contacts (multiplicity = 2).[26,99,100] With random mutations, C_2 symmetry gives rise to a much broader range of interaction energies than asymmetry because each mutation counts doubly so that the

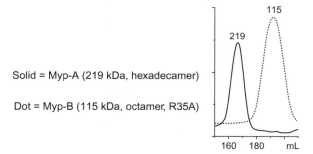

Solid = Myp-A (219 kDa, hexadecamer)

Dot = Myp-B (115 kDa, octamer, R35A)

FIG. 34. The size-exclusion chromatography of the C_8-symmetric protein ring of Rim domains defined in Figs. 32 and 33.[98] The product appeared at 219 kDa, corresponding to the expected mass of the hexadecamer observed in the crystals. However, since mass determinations in this range can be rather inaccurate, the underlying octamer was produced by destroying the contact with the mutation Arg35 → Ala. The resulting octamer ran at 115 kDa and thus confirmed the hexadecamer and also demonstrated its stability in solution.

resulting binding energy behaves like a distribution of the sum of n "2s" (multiplicity = 2) with random signs. This distribution is much broader than that of the sum of $2n$ "1s" (multiplicity = 1) with random signs. A broader range means a higher chance to find a good binder in the low-energy tail of the distribution. This purely mathematical effect dominates the formation of homodimers in nature.[26] It is clear that this effect is equally valid for a comparison between higher multiplicities (e.g., 8) with lower ones (e.g., 4), as was experienced in the engineering campaign (Fig. 19).[98]

IX. Future Applications

The peptide chain masses of proteins range between 20 and 100 kDa because shorter chains have difficulties adopting a defined structure and larger chains are endangered by translation errors. For globular proteins, this relates to diameters of around 5 nm. Designed protein complexes produce therefore patterns with features in the 5–50 nm range, which is 10–100 times smaller than the wavelength of light. Owing to the resolution limit of light, which is around its wavelength, such patterns cannot be produced with the usual optical devices. An alternative would be vacuum UV light with wavelengths in the 5 nm range to be produced in a synchrotron which, however, cannot be handled appropriately because there are no efficient lenses available. It is therefore conceivable that protein layers will be used in original (or chemically fortified) constructs directly (or in form of replicas) as covers on a silicon surface for etching a designed pattern. As an alternative, such a pattern could be engraved by an electron beam, which, however, is too expensive for mass production.

Up to now, the design is generally in form of self-assembling patterns that are necessarily symmetric and/or repetitive and therefore of small information- al content. A big step forward would be the design of asymmetric features. However, such design would require asymmetric contacts that have been demonstrated to be most difficult to produce.

A further extension would be the design of dynamic assemblies. An exam- ple is the diurnal assembly/disassembly of a kinase in a plant cell,[31] which is switched by the presence and absence of AMP (Fig. 9). It is also conceivable to introduce breathing motions as has been demonstrated for a protein dimer switched open and closed by the calcium ion concentration of the environment (Fig. 16).[3] This permits constructing nets of switchable mesh size, which may become useful for filtering solutes.

The most direct possibility is the use of one-, two-, or three-dimensional protein networks for arranging enzymes in a suitable order so that reaction chains can work much more efficiently.[101] For the simple case of an enzyme with small diffusible substrates, a crystal with free active sites and large diffusion channels would generate a very high enzyme concentration, which is most desirable in a reactor vessel. The NADH peroxidase, which appears to be a very compact organelle for dioxygen removal, is an example for such an application (Fig. 10).[40] Many further possibilities can be expected. The protein engineering field is wide open for all kinds of new ideas on building scaffolds in the 5–50 nm range, which is not easily reachable with other methods.

ACKNOWLEDGMENT

I thank C. Schleberger, P. Ringler, D. Grueninger, J. Koetter, S.-M. Schulze, N. Treiber, and M. Ziegler for technical help and S. Gerhardt and O. Einsle for discussions. This project was supported by the Wacker-Chemie, München.

REFERENCES

1. Chen J, Seeman NC. Synthesis from DNA of a molecule with the connectivity of a cube. *Nature* 1991;**350**:631–3.
2. Han D, Pal S, Nangreave J, Deng Z, Liu Y, Yan H. DNA origami with complex curvatures in three-dimensional space. *Science* 2011;**332**:342–6.
3. Ringler P, Schulz GE. Self-assembly of proteins into designed networks. *Science* 2003;**302**:106–9.
4. Ringler P, Schulz GE. OmpA membrane domain as a tight-binding anchor for lipid bilayers. *Chembiochem* 2002;**3**:463–6.
5. Sackmann E, Tanaka M. Supported membranes on soft polymer cushions: fabrication, characterization and applications. *Trends Biotechnol* 2000;**18**:58–64.
6. Papapostolou D, Howorka S. Engineering and exploiting protein assemblies in synthetic biology. *Mol Biosyst* 2009;**5**:723–32.
7. Rehm T, Huber R, Holak TA. Application of NMR in structural proteomics: screening for proteins amenable to structural analysis. *Structure* 2002;**10**:1613–8.
8. Spraggon G, Pantazatos D, Klock HE, Wilson IA, Woods Jr. VL, Lesley SA. On the use of DXMS to produce more crystallizable proteins: structure of the *T. maritima* proteins TM0160 and TM1171. *Protein Sci* 2004;**13**:3187–99.
9. Horn JR, Kraybill B, Petro EJ, Coales SJ, Morrow JA, Hamuro Y, et al. The role of protein dynamics in increasing binding affinity for an engineered protein–protein interaction established by H/D exchange mass spectroscopy. *Biochemistry* 2006;**45**:8488–98.
10. Smith GR, Sternberg MJE, Bates PA. The relationship between flexibility of proteins and their conformational states on forming protein–protein complexes with an application to protein–protein docking. *J Mol Biol* 2005;**347**:1077–101.
11. Schulz GE. Nucleotide binding proteins. In: Balaban M, editor. *Molecular mechanisms of biological recognition*. Amsterdam: Elsevier/North Holland Biomed. Press; 1979, p. 79–94.
12. Müller CW, Schlauderer GJ, Reinstein J, Schulz GE. Adenylate kinase motions during catalysis, an energetic counterweight balancing substrate binding. *Structure* 1996;**4**:147–56.
13. Dunker K, Obradovic Z. The protein trinity—linking function and disorder. *Nat Biotechnol* 2001;**19**:805–6.
14. Chouard T. Breaking the protein rules. *Nature* 2011;**471**:151–3.

15. Åvall-Jääskeläinen S, Hynönen U, Ilk N, Pum D, Sleytr UB, Palva A. Identification and characterization of domains responsible for self-assembly and cell wall binding of the surface layer protein of *Lactobacillus brevis* ATCC 8287. *BMC Microbiol* 2008;**8**:165–79.
16. Sleytr UB, Egelseer EM, Ilk N, Pum D, Schuster B. S-layers as a basic building block in a molecular construction kit. *FEBS J* 2007;**274**:323–34.
17. Ackers GK, Doyle ML, Myers D, Daugherty MA. Molecular code for cooperativity in hemoglobin. *Science* 1992;**255**:54–63.
18. Vonrhein C, Schlauderer GJ, Schulz GE. Movie of the structural changes during a catalytic cycle of nucleoside monophosphate kinases. *Structure* 1995;**3**:483–90.
19. Kroemer M, Merkel I, Schulz GE. Structure and catalytic mechanism of L-rhamnulose-1-phosphate aldolase. *Biochemistry* 2003;**42**:10560–8.
20. Grueninger D, Schulz GE. Domain motions supporting the reaction catalyzed by L-rhamnulose-1-phosphate aldolase. *Biochemistry* 2008;**47**:607–14.
21. Harrison SC, Olson AJ, Schutt CE, Winkler FK. Tomato bushy stunt virus at 2.9 Å resolution. *Nature* 1978;**276**:368–73.
22. Johnson JE, Speir JA. Quasi-equivalent viruses: a paradigm for protein assemblies. *J Mol Biol* 1997;**269**:665–75.
23. Treiber N, Reinert DJ, Carpusca I, Aktories K, Schulz GE. Structure and mode of action of a mosquitocidal holotoxin with its four ricin B type lectin domains. *J Mol Biol* 2008;**381**:150–9.
24. Smyth DR, Mrozkiewicz MK, McGrath WJ, Listwan P, Kobe B. Crystal structures of fusion proteins with large-affinity tags. *Protein Sci* 2003;**12**:1313–22.
25. Kim PM, Lu LJ, Gerstein MB. Relating three-dimensional structures to protein networks provides evolutionary insights. *Science* 2006;**314**:1938–41.
26. Schulz GE. The dominance of symmetry in the evolution of homo-oligomeric proteins. *J Mol Biol* 2010;**395**:834–43.
27. Schulz GE, Schirmer RH. *Principles of protein structure. Advanced text in chemistry.* New York: Springer Verlag; 1979, pp. 94.
28. Schraidt O, Marlovits TC. Three-dimensional model of *Salmonella*'s needle complex at subnanometer resolution. *Science* 2011;**331**:1192–5.
29. Cerritelli ME, Trus BL, Smith CS, Cheng N, Conway JF, Steven AC. A second symmetry mismatch at the portal vertex of bacteriophage T7: 8-fold symmetry in the procapsid core. *J Mol Biol* 2003;**327**:1–6.
30. Berman HM, Battistuz T, Bhat TN, Bluhm WF, Bourne PE, Burkhardt K, et al. The Protein Data Bank. *Acta Crystallogr D* 2002;**58**:899–907.
31. Wild K, Grafmüller R, Wagner E, Schulz GE. Structure, catalysis and supramolecular assembly of adenylate kinase from maize. *Eur J Biochem* 1997;**250**:326–31.
32. Henderson R, Unwin PNT. Three-dimensional model of purple membrane obtained by electron microscopy. *Nature* 1975;**257**:28–32.
33. Fagan RP, Albesa-Jove D, Qazi O, Svergun DI, Brown KA, Fairweather NF. Structural insights into the molecular organization of the S-layer from *Clostridium difficile*. *Mol Microbiol* 2009;**71**:1308–22.
34. Engel A, Hoenger A, Hefti A, Henn C, Ford RC, Kistler J, et al. Assembly of 2-D membrane protein crystals: dynamics, crystal order, and fidelity of structure analysis by electron microscopy. *J Struct Biol* 1992;**109**:219–34.
35. Kroemer M, Schulz GE. The structure of L-rhamnulose-1-phosphate aldolase (class II) solved by low-resolution SIR phasing and twentyfold NCS-averaging. *Acta Crystallogr D* 2002;**58**:824–32.
36. Reinert DJ, Carpusca I, Aktories K, Schulz GE. Structure of the mosquitocidal toxin from *Bacillus sphaericus*. *J Mol Biol* 2006;**357**:1226–36.

37. Ziegler K, Benz R, Schulz GE. An α-helical porin from *Corynebacterium glutamicum*. *J Mol Biol* 2008;**379**:482–91.

38. Shepardson S. Ultrastructure of protein crystals in potato and tomato trichomes. *Ann Bot* 1982;**49**:503–8.

39. Georgiou DK, Vekilov PG. A fast response mechanism for insulin storage in crystals may involve kink generation by association of 2D clusters. *Proc Natl Acad Sci USA* 2006;**103**:1681–6.

40. Stehle T, Ahmed SA, Claiborne A, Schulz GE. The structure of NADH peroxidase from *Streptococcus faecalis* 10C1 refined at 2.16 Å resolution. *J Mol Biol* 1991;**221**:1325–44.

41. Baedeker M, Schulz GE. Overexpression of a designed 2.2 kb gene of eukaryotic phenylala-nine ammonia-lyase in *Escherichia coli*. *FEBS Lett* 1999;**457**:57–60.

42. Wishner BC, Ward KB, Lattman EE, Love WE. Crystal structure of sickle-cell deoxyhemo-globin at 5 Å resolution. *J Mol Biol* 1975;**98**:179–94.

43. Makowski L, Magdoff-Fairchild B. Polymorphism of sickle cell hemoglobin aggregates: structural basis for limited radial growth. *Science* 1986;**234**:1228–31.

44. D'Arcy A. Crystallizing proteins—a rational approach? *Acta Crystallogr D* 1994;**50**:469–71.

45. Gast WH, Leberman R, Schulz GE, Wittinghofer A. Crystals of partially trypsin digested elongation factor Tu. *J Mol Biol* 1976;**106**:943–50.

46. Kabsch W, Gast WH, Schulz GE, Leberman R. Low resolution structure of partially trypsin-degraded polypeptide elongation factor, EF-Tu, from *Escherichia coli*. *J Mol Biol* 1977;**117**:999–1012.

47. Schwede T, Bädeker M, Langer M, Rétey J, Schulz GE. Engineering high quality crystals of histidine ammonia-lyase. *Protein Eng* 1999;**12**:151–3.

48. Rost B, Sander C. Conservation and prediction of solvent accessibility in protein families. *Proteins Struct. Funct. Genet.* 1994;**20**:216–26.

49. Pautsch A, Vogt J, Model K, Siebold C, Schulz GE. Strategy for membrane protein crystalli-zation exemplified with OmpA and OmpX. *Proteins Struct. Funct. Genet.* 1999;**34**:167–72.

50. Derewenda ZS. Rational protein crystallization by mutational surface engineering. *Structure* 2004;**12**:529–35.

51. Derewenda ZS, Vekilov PG. Entropy and surface engineering in protein crystallization. *Acta Crystallogr D* 2006;**62**:116–24.

52. Schulz GE, Barry CD, Friedmann J, Chou PY, Fasman GD, Finkelstein AV, et al. Comparison of predicted and experimentally determined secondary structure of adenylate kinase. *Nature* 1974;**250**:140–2.

53. Karplus PA, Schulz GE. Prediction of chain flexibility in proteins: a tool for the selection of peptide antigens. *Naturwissenschaften* 1985;**72**:212–3.

54. Ma B, Elkayam T, Wolfson H, Nussinov R. Protein–protein interactions: structurally con-served residues distinguish between binding sites and exposed protein surfaces. *Proc Natl Acad Sci USA* 2003;**100**:5772–7.

55. Faller M, Niederweis M, Schulz GE. The structure of a mycobacterial outer membrane channel. *Science* 2004;**303**:1189–92.

56. Mittl PRE, Berry A, Scrutton NS, Perham RN, Schulz GE. A designed mutant of the enzyme glutathione reductase shortens the crystallization time by a factor of forty. *Acta Crystallogr D* 1994;**50**:228–31.

57. Selzer T, Albeck S, Schreiber G. Rational design of faster associating and tighter binding protein complexes. *Nat Struct Biol* 2000;**7**:537–41.

58. Lo Conte L, Chothia C, Janin J. The atomic structure of protein–protein recognition sites. *J Mol Biol* 1999;**285**:2177–98.

59. Nooren IMA, Thornton JM. Diversity of protein–protein interactions. *EMBO J* 2003;**22**:3486–92.

60. Ofran Y, Rost B. Analysing six types of protein–protein interfaces. *J Mol Biol* 2003;**325**:377–87.
61. Halperin I, Wolfson H, Nussinov R. Protein–protein interactions: coupling of structurally conserved residues and of hot spots across interfaces. Implications for docking. *Structure* 2004;**12**:1027–38.
62. Schueler-Furman O, Wang C, Bradley P, Misura K, Baker D. Progress in modeling of protein structures and interactions. *Science* 2005;**310**:638–42.
63. Frederick KK, Marlow MS, Valentine KG, Wand JA. Conformational entropy in molecular recognition by proteins. *Nature* 2007;**448**:325–9.
64. Emsley P, Cowtan K. COOT: model-building tools for molecular graphics. *Acta Crystallogr D* 2004;**60**:2126–32.
65. Krissinel E, Henrick K. Inference of macromolecular assemblies from crystalline state. *J Mol Biol* 2007;**372**:774–97.
66. Dreusicke D, Schulz GE. The switch between two conformations of adenylate kinase. *J Mol Biol* 1988;**203**:1021–8.
67. Stehle T, Schulz GE. Temperature-dependent space group transitions in crystals of guanylate kinase from yeast. *Acta Crystallogr B* 1992;**48**:546–8.
68. Janin J, Rodier F, Chakrabarti P, Bahadur RP. Macromolecular recognition in the Protein Data Bank. *Acta Crystallogr D* 2007;**63**:1–8.
69. Wiesmann C, Schulz GE. Infinite non-crystallographic symmetries in crystals of a globular protein. *Acta Crystallogr D* 1997;**53**:274–8.
70. Levy ED, Erba EB, Robinson CV, Teichmann SA. Assembly reflects evolution of protein complexes. *Nature* 2008;**453**:1262–5.
71. Vonrhein C, Bönisch H, Schäfer G, Schulz GE. The structure of a trimeric archaeal adenylate kinase. *J Mol Biol* 1998;**282**:167–79.
72. Muller YA, Schulz GE. Structure and catalysis of the thiamine- and flavin-dependent enzyme pyruvate oxidase. *Science* 1993;**259**:965–7.
73. Muller YA, Schumacher G, Rudolph R, Schulz GE. The refined structures of a stabilized mutant and of wild-type pyruvate oxidase from *Lactobacillus plantarum*. *J Mol Biol* 1994;**237**:315–35.
74. Monod J, Wyman J, Changeaux J-P. On the nature of allosteric transitions: a plausible model. *J Mol Biol* 1965;**12**:88–118.
75. Hartl FU. Protein folding. Secrets of a double-doughnut. *Nature* 1994;**371**:557–9.
76. Daugaard M, Rohde M, Jäättelä M. The heat shock protein 70 family: highly homologous proteins with overlapping and distinct functions. *FEBS Lett* 2007;**581**:3702–10.
77. Weiss MS, Wacker T, Weckesser J, Welte W, Schulz GE. The three-dimensional structure of porin from *Rhodobacter capsulatus* at 3 Å resolution. *FEBS Lett* 1990;**267**:268–72.
78. Schulz GE. Porins: general to specific, native to engineered passive pores. *Curr Opin Struct Biol* 1996;**6**:485–90.
79. Kreusch A, Neubüser A, Schiltz E, Weckesser J, Schulz GE. The structure of the membrane channel porin from *Rhodopseudomonas blastica* at 2.0 Å resolution. *Protein Sci* 1994;**3**:58–63.
80. Whitesides GM, Grzybowski B. Self-assembly at all scales. *Science* 2002;**295**:2418–21.
81. Schulz GE. A new classification of membrane protein crystals. *J Mol Biol* 2011;**407**:640–6.
82. Dotan N, Arad D, Frolow F, Freeman A. Self-assembly of a tetrahedral lectin into predesigned diamond-like protein crystals. *Angew Chem Int Ed* 1999;**38**:2363–6.
83. Zegers I, Deswarte J, Wyns L. Trimeric domain swapped barnase. *Proc Natl Acad Sci USA* 1999;**96**:818–22.
84. Libonati M, Gotte G. Oligomerization of bovine ribonuclease A: structural and functional features of its multimers. *Biochem J* 2004;**380**:311–27.

85. Ogihara NL, Ghirlanda G, Bryson JW, Gingery M, DeGrado WF, Eisenberg D. Design of three-dimensional domain-swapped dimers and fibrous oligomers. *Proc Natl Acad Sci USA* 2001;**98**:1404–9.

86. Padilla JE, Colovos C, Yeates TO. Nanohedra: using symmetry to design self assembling protein cages, layers, crystals, and filaments. *Proc Natl Acad Sci USA* 2001;**98**:2217–21.

87. Köhler G, Milstein C. Continuous cultures of fused cells secreting antibody of predefined specificity. *Nature* 1975;**256**:495–7.

88. Sidhu SS, Fairbrother WJ, Deshayes K. Exploring protein–protein interactions with phage display. *Chembiochem* 2003;**4**:14–25.

89. Schweizer A, Roschitzki-Voser H, Amstutz P, Briand C, Gulotti-Georgieva MN, Prenosil E, et al. Inhibition of Caspase-2 by a designed ankyrin repeat protein: specificity, structure and inhibition mechanism. *Structure* 2007;**15**:625–36.

90. Chevalier BS, Kortemme T, Chadsey MS, Baker D, Monnat Jr. RJ, Stoddard BL. Design activity and structure of a highly specific artificial endonuclease. *Mol Cell* 2002;**10**:895–905.

91. Kortemme T, Joachimiak LA, Bullock AN, Schuler AD, Stoddard BL, Baker D. Computational redesign of protein–protein interaction specificity. *Nat Struct Mol Biol* 2004;**11**:371–9.

92. Shifman JM, Mayo SL. Modulating calmodulin binding specificity through computational protein design. *J Mol Biol* 2002;**323**:417–23.

93. Crick FHC. The packing of α-helices: simple coiled coils. *Acta Crystallogr* 1953;**6**:689–97.

94. Fairman R, Chao H-G, Lavoie TB, Villafranca JJ, Matsueda GR, Novotny J. Design of heterotetrameric coiled coils: evidence for increased stabilization by Glu–Lys+ ion pair interactions. *Biochemistry* 1996;**35**:2824–9.

95. Harbury PB, Plecs JJ, Tidor B, Alber T, Kim PS. High-resolution protein design with backbone freedom. *Science* 1998;**282**:1462–7.

96. Havranek JJ, Harbury PB. Automated design of specificity in molecular recognition. *Nat Struct Biol* 2003;**10**:45–52.

97. Papapostolou D, Smith AM, Atkins EDT, Oliver SJ, Ryadnov MG, Serpell LC, et al. Engineering nanoscale order into a designed protein fiber. *Proc Natl Acad Sci USA* 2007;**104**:10853–8.

98. Grueninger D, Treiber N, Ziegler MOP, Koetter JWA, Schulze M-S, Schulz GE. Designed protein–protein association. *Science* 2008;**319**:206–9.

99. Lukatsky DB, Zeldovich KB, Shaknovich EI. Statistically enhanced self-attraction of random patterns. *Phys Rev Lett* 2006;**97**:178101.

100. Andre I, Strauss CEM, Kaplan DB, Bradley P, Baker D. Emergence of symmetry in homo-oligomeric biological assemblies. *Proc Natl Acad Sci USA* 2008;**105**:16148–52.

101. Dueber JE, Wu GC, Malmirchegini GR, Moon TS, Petzold CJ, Ullal AV, et al. Synthetic protein scaffolds provide modular control over metabolic flux. *Nat Biotechnol* 2009;**27**:753–9.

Alpha-Helical Peptide Assemblies: Giving New Function to Designed Structures

ELIZABETH H.C. BROMLEY* AND
KEVIN J. CHANNON[†]

*Department of Physics, Durham
University, Durham, United Kingdom
[†]Department of Physics, Cavendish
Laboratory, University of Cambridge,
Cambridge, United Kingdom

The design of alpha-helical tectons for self-assembly is maturing as a science. We have now reached the point where many different coiled-coil topologies can be reliably produced and validated in synthetic systems and the field is now moving on towards more complex, discrete structures and applications. Similarly the design of infinite or fiber assemblies has also matured, with the creation fibers that have been modified or functionalized in a variety of ways. This chapter discusses the progress made in both of these areas as well as outlining the challenges still to come.

Progress in Molecular Biology
and Translational Science, Vol. 103
DOI: 10.1016/B978-0-12-415906-8.00001-7

231

I. Introduction

The design of self-assembling complex systems offers an excellent opportunity to reflect and improve on our ability to understand, mimic, and redesign natural systems. Knowledge of nature can be advanced through a cycle of observing, deducing the rules by which complex structure is formed, attempting to create synthetic analogs, and comparing the results to the original template. Where this approach has been successful in duplicating natural systems, a platform is created from which new versions can be made, with improved or varied functionality, and with technological applications across society. This area of research falls under the broad umbrella of synthetic biology.

A. Synthetic Biology

The research field of synthetic biology was proposed by Wacław Szybalski in 1974[1] as the next phase of what had previously been a more descriptive study of molecular biology. The rise of recombinant DNA technology at this time opened the door to the possibility of creating synthetic genomes and organisms, and as such the field has its origins in metabolic and genetic engineering. However, the strength of synthetic biology has been the incorporation of a wide range of ideas from various disciplines including biological, chemical, and conventional engineering, systems biology, and protein design.

The field is united by an interdisciplinary, goal-driven approach that aims to both design and fabricate biological components and systems that do not already exist in the natural world, and to redesign and fabricate existing biological systems.

There are many diverse approaches being used under the umbrella of synthetic biology but they all have a few key points in common. First, these approaches are attempts to produce a system that is nonnatural but exhibits some aspect of natural behavior. Second, they derive from the engineering perspective that biology is modular. Modularity is reflected at many of the length scales at which natural systems are present, including individuals in a population, cells in a tissue, and proteins assembled into transcription machinery all the way down to the individual amino acids in a protein chain. It therefore makes sense to map out attempts to create synthetic biology with reference to the point in the natural hierarchy of structure at which alterations cause divergence from nature.[2]

The y-axis of the map in Fig. 1 represents the biomolecular and systems hierarchies in *natural* biology. Starting with basic building blocks—such as the nucleotides, amino acids, carbohydrates, and lipids; moving through oligonucleotides and polypeptides, which we term *tectons*; onto folded, assembled, and functional biomolecules—including nucleic acids, proteins and assemblies thereof—and lipid vesicles; and up to cells, in which these various components are brought together, encapsulated, organized, and orchestrated. The x-axis

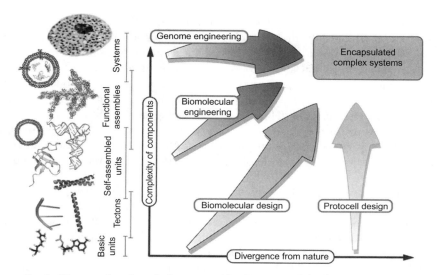

FIG. 1. Diagram of synthetic biology space. The divergence of the design components from their natural equivalent is represented on the x-axis and the complexity of constituent components is represented on the y-axis. Several pathways are shown with labels that show the approximate areas of study that they encompass. Reproduced with permission from Channon et al.[3] copyright Elsevier (2008). (For color version of this figure, the reader is referred to the web version of this chapter.)

represents increasing diversity from nature: the idea being that if nonnatural amino acids are incorporated into a protein, the divergence from the natural system would increase as the mutation was passed up the hierarchy.

Several pathways are indicated in the diagram, which correspond to broad areas of synthetic biology. At the highest level of the hierarchy is the *genome engineering* approach, in which constructed genomes are inserted into host cells.[4] This approach was successfully developed by Venter and colleagues at the J Craig Venter Institute using chemically synthesized fragments of DNA followed by *in vivo* recombination to produce full-length synthetic chromosomes. The synthetic chromosomes were injected into a host cell which had previously had all genetic material removed. The new DNA took control of the host cell's molecular machinery, causing them to produce progeny which contain only the synthetic chromosomes.[5]

The next pathway down, *biomolecular engineering*, uses the same concept of inserting new function into a host cell at the level of individual biomolecular components or pathways.[6] A key example of work in this area is that of Keasling and colleagues at Berkley on the antimalarial drug artemisinin.[7] In this study, a pathway that creates the drug precursor in plants has been transferred and adapted to function in yeast. The fast growing properties of yeast can then be exploited to produce the drug in far higher quantities at a much lower cost. The critical feature of this approach is that biological pathways are *modular* and can be cut, transposed, and added to new organisms with a minimal amount of interference with existing pathways. The BioBricks project aims to put this on a sound engineering footing by producing a catalog of genetic building blocks whose properties are well defined and can be plugged into new contexts.[8]

The next pathway is that of *Biomolecular design*, which will be discussed in more detail in the rest of this chapter. Essentially, it applies the same principle as above but at the length scale of individual biomolecules rather than whole pathways.

Finally, at the right of the diagram is the *protocell engineering* approach.[9] The idea behind this work is to capture the defining features of natural cells in biomimetic systems: that is, it is (1) an encapsulated system, (2) blueprinted by some molecular-based store of information, and (3) harnesses energy from its environment and performs some form of metabolism. Ultimately, protocells might also have the ability to pass on their blueprint for the construction of successive generations.[10,11] Ideally, though bioinspired, none of these aspects would use natural biomolecules: that is, no DNA/RNA-based information stores or transfers; no carbohydrate-based or similar metabolism; no protein structures, binders, or catalysts; and though this appears to be a less stringent stipulation, no natural lipids as membrane components.

B. Biomolecular Design

Biomolecular designers take the view that stripped-down or *de novo* biomolecules provide useful modular units for building novel structure and/or function. There are two main types of biomolecule to work with, namely, DNA or proteins, and hence two building blocks that can be considered, namely, DNA bases or amino acids. In this chapter, only the use of proteins and amino acids is discussed; however, it should be noted at this point that there is a great deal of exciting work being done creating new structures and machines out of RNA, DNA, and DNA analogs.[12–17]

The process of protein design involves a cycle of observing natural proteins, deducing the rules that cause them to fold and function, and then using these rules to develop analogs of natural proteins that are amenable to manipulation.

It is possible to look for patterns of amino-acid sequences that give rise to structural motifs at all levels of the protein structural hierarchy; however, this chapter focuses on the α-helical secondary structural element and its onward association to quaternary structures known as "coiled coils."

A   B

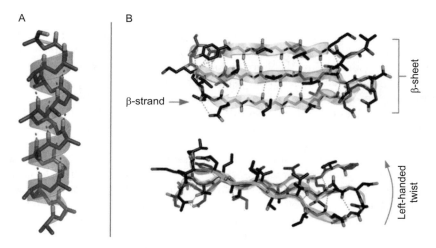

β-strand →

β-sheet

Left-handed twist

FIG. 2. Cartoons of (A) an α-helix and (B) a β-sheet formed from three β-strands. Hydrogen bonds are shown in each structure by dashed lines. The α-helix contains only *local* backbone hydrogen bonds, between *i* and *i* + 4 positions. The β-sheet contains many *nonlocal* hydrogen bonds. (For color version of this figure, the reader is referred to the web version of this chapter.)

C. The α-Helix as a Naturally Occurring Tecton for Self-Assembly

Focusing on α-helices might seem restrictive but, for several reasons, α-helices make excellent tectons. First, they are easier to work with than β-strands as they are stabilized by *local* interactions. An isolated β-strand is generally not a stable chain configuration, as there are a great many "dangling" hydrogen bonds to be satisfied (see Fig. 2). To stabilize the strand, these bonds must be satisfied by another strand and/or other nearby secondary structural element. This property makes the β-sheet a nonlocal secondary structure as many nonadjacent areas of the peptide chain may be involved. These considerations mean that *de novo* design of all but the simplest β-sheet structures can require a predictive knowledge of the behavior of the entire peptide, something that is not currently possible. However, a single helix can be stable individually, with the hydrogen bond network formed between amino acids that are near to each other in the primary sequence. Specifically, the hydrogen bonds are formed between the carbonyl oxygen of residue *i* and the amide proton of residue *i* + 4.

The α-helix is also relatively rigid, meaning that interactions at one end of the helix are fairly independent of those at the other end or, indeed, anywhere along the length.

Finally, the width of the helix and the length of each amino acid conspire to make a repeating pattern of side chains running up the outside of the helix. Specifically, the α-helix has 3.6 amino acids per turn, which leads to amino acid "$i + 7$" being almost directly above amino acid "i." This is a very useful property, as it means that repeated sequences are arranged regularly in space. It provides the opportunity to use a repeated pattern of hydrophobic (h) and polar (p) residues to produce a helix that is amphipathic (that has one polar face and one hydrophobic face). To achieve this in the α-helix, a pattern of $(hpphppp)_n$ is used along the chain (see Fig. 3A).

There are many examples of such amphipathic helices being used as tectons for self-assembly in nature. However, perhaps the most directly applicable use is in the *coiled coil*. In this structure, the hydrophobic stripes of two (or more) helices come together and wrap around each other in order to bury the hydrophobic side chains. Specifically, the side chains in the hydrophobic core pack tightly in a regime known as "knobs into holes" (KIH) packing.[18] The details of this packing determine the structure of the resulting coiled coil in

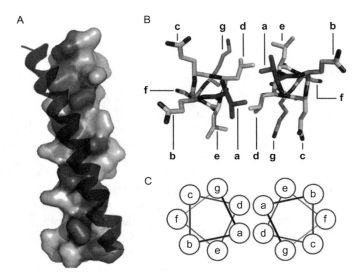

FIG. 3. Hydrophobic core interactions in coiled coils. (A) Solvent-accessible surface of an α-helix in a coiled coil illustrating how hydrophobic a and d core residues (highlighted) are aligned into a "hydrophobic stripe" on one side of the helix. The second helix of the coiled coil is shown docked along this stripe, through contact with its own hydrophobic stripe. (B) A view of a single helical repeat (seven residues) of the coiled coil from above, showing the organization of the side chains. The a and d residues point into the core, interacting with their opposite number. (C) The helical wheel diagram used to diagrammatically represent the structure of a coiled coil. (For color version of this figure, the reader is referred to the web version of this chapter.)

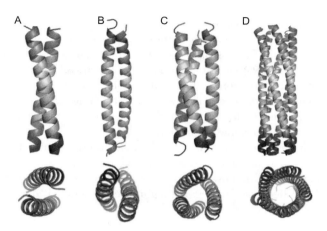

FIG. 4. Some examples of coiled-coil structures shown from orthogonal views. (A) A parallel dimer from 2ZTA. (B) An antiparallel dimer from 1HF9. (C) A parallel trimer from 1BB1. (D) A parallel pentamer from 2GUV. Chains are colored from blue at the N-terminus to red at the C-terminus. (For interpretation of the references to color in this figure legend, the reader is referred to the Web version of this chapter.)

terms of the orientation of the helices and the number of helices involved. Many different topologies of coiled coil are found in nature and these have been categorized in two searchable databases.[19,20] Some of the simpler coiled-coil structures are shown in Fig. 4, although this shows only a small subset of the possible geometries.

As one would expect for such an obvious natural tecton, coiled coils are ubiquitous as they are found in every compartment of plant cells and in all prokaryotic and eukaryotic cells.[21] Within these various contexts, coiled coils provide a wide range of structures and functions. Shorter coiled coils offer molecular recognition, bringing together other proteins and hence functions in specifically defined combinations. Examples of biological function within the cell include their use as DNA transcription factors, binding to DNA to either repress or promote gene transcription, and the self-assembly of signaling complexes, including ion channels.[22,23] Molecular recognition by coiled coils is also used to fuse transported vesicles to their target membranes using, for example, proteins of the SNARE family.[24]

The rigidity of the coiled-coil structure allows the use of longer coiled coils as structural components. Often, these coiled coils have defined lengths set by the length of the sequences, as is the case with bacterial cell wall spacers.[25] However, coiled coils also form components of fibrillar assemblies such as the intermediate filaments[26] and spectrin which form two- and three-dimensional scaffolds that support the cell.[27]

Finally, coiled coils within the cell are dynamic, both in terms of their ability to responsively mediate other protein–protein interactions but also more directly in the form of motor proteins. The three main classes of cytoskeletal motor proteins, namely, kinesins, myosins, and dyneins, all contain long coiled-coil domains.[23] The function of these coiled coils can be both structural—for example, controlling aspects of the stalks attaching cargo to the motor domains[28]—or functional—for example, in using rearrangements in the coiled-coil structure to achieve motive force.[29]

It is clear from the ways in which coiled coils are used in nature that they could provide an extremely useful tool for engineering biology. The following sections explore in more detail the knowledge that has been extracted from natural systems and how this has been used to begin designing new functional components.

II. Designing Discrete Helical Assemblies

In this section, the rules that link sequence to structure in coiled coils, which are the most ubiquitous helical self-assemblies in nature, are explored in more detail.

A. Lessons from Nature

The KIH packing found in coiled coils is a motif that can be searched for computationally in protein structures, using software such as SOCKET.[30] The basis of the packing is that a hydrophobic "knob" residue on one helix slots into a hole formed in the center of a diamond of four "hole" residues displayed on an opposing helix (see Fig. 5). By locating structures matching specific criteria, one is able to extract detailed sequence to structure information.[31] The repeating pattern of hydrophobic and polar amino acids $(hpphppp)_n$ can be examined in more detail and assigned the positional nomenclature *abcdefg*, with the hydrophobic residues occupying positions *a* and *d*. Further, it is often useful to visualize the heptad repeat by displaying it on a helical wheel as shown in Fig. 3C.

By examining the frequency with which various hydrophobic residues are used in these positions, it is found that, as well as being the main driving force for helix association, they control the oligomer state of the coiled coil. For example, a combination of isoleucine and asparagine at *a* and leucine at *d* favors dimer formation; using isoleucine at both *a* and *d* positions preferentially creates trimers; and using leucine at *a* positions and isoleucine at *d* positions favors the tetramer.[32] Higher order oligomers are also promoted by the inclusion of extra hydrophobic amino acids flanking the core region. The two residues (positions *e* and *g*) on either side of the hydrophobic interface tend to be occupied by charged amino acids, allowing a range of ionic interactions both between the helices of the coiled coil and within each helix.[33]

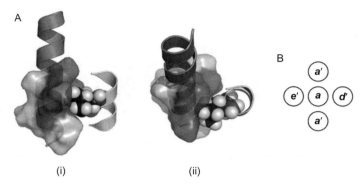

FIG. 5. Knobs-into-holes (KIH) interactions. (A) A KIH interaction between two helices, seen from (i) the side and (ii) the top. The hole is shown as a solvent-accessible surface; blue indicates the surface of hydrophobic core residues, and red is a charged lysine in an e' position. The knob is donated by the right-hand helix. A Van der Waals representation of the knob is shown, and it can be seen that there is a very close steric interaction between the knob residue and the hole. (B) The hole comprises residues a', d', e', and a' (from the next helical repeat), corresponding to residues in positions i, $i+3$, $i+4$, and $i+7$, and the knob is formed by residue a. (For interpretation of the references to color in this figure legend, the reader is referred to the Web version of this chapter.)

Recently,[34] it has been shown that these rules may not be generally applicable outside the context of the specific sequence originally examined (GCN4). One significant variation on these rules is caused by the existence of trigger sequences that are capable of specifying the oligomer state of the coiled coil.[35,36] These are small pieces of sequence that can fold independently into helical fragments before the oligomerization of the coiled coil occurs. It has recently been shown that, in the case of GCN4, the insertion of trigger sequences specifying oligomeric state may be more important than the use of individual specifying residues in determining the final structure.[37] Overall, the current situation is that caution must still be applied when designed peptides are required to exhibit a specific oligomerization state, and that experimental verification of structure is prudent.

As well as influencing oligomer state, the details of the hydrophobic core packing can determine the helix orientation creating both parallel and antiparallel coiled coils. Recently, progress has been made in analyzing how sequences are related to the helix orientation in coiled coils, and prediction algorithms are improving.[38] The thermodynamic preferences for various combinations of side chains packing into the core have been explored[39] and the design of antiparallel coiled coils is an expanding field.[40]

The most frequently occurring coiled coils are dimeric, and many of these find function in the cell as transcription factors. In particular, the basic leucine zipper domain proteins (bZIP) are a large collection of parallel dimeric coiled coils. These proteins are of interest to the design process because they exhibit various levels of specificity: that is to say, many of the sequences bind preferentially to only a few partner sequences in the collection. This specificity is a key property necessary in making designed self-assembly that is modular and not promiscuous. An interactome for bZIP proteins has been mapped out by Keating and colleagues using a microscale protein array technique, in which the interactions of 49 human bZIP proteins and 10 from *Saccharomyces cerevisiae* were measured.[41] This technique involves printing plates with each of the proteins under conditions in which they are expected to be monomeric, and then exposing the plates to fluorescently labeled analogs of each of the proteins. The resulting level of fluorescence retained on the plate is used to calculate the interaction strength (Fig. 6).

This study provides a wealth of information on how specificity of interaction is achieved, and the Keating group has gone on to produce a number of computer algorithms aimed at using this data to predict binding between bZIP proteins. These algorithms variously use combinations of electrostatic information, empirically determined weighting by sequence, and calculations of structural stability made from atomic resolution models.[42,43]

From these studies and others, it has further been deduced that specificity is achieved through three main methods. First, the use of asparagine at *a* positions (which also specifies for dimers) is used. The introduction of this polar side chain into the hydrophobic core of the coiled coil is destabilizing, however, this effect can be mitigated by partnering with another asparagine on an opposing helix. This provides a design rule: *helices with asparagine will preferentially assemble with other helices possessing asparagine at the same point in the helix.*

Second, the positions on either side of the hydrophobic core (*e* and *g*) tend to be occupied by charged amino acids. Complimentary pairs of charges can be used to favor specific helix pairing and noncomplementary charge pairs can be used to disfavor unwanted helix pairing.

Third, there is the possibility of using the size of the side chains to produce complementary fitting in the hydrophobic core. Large hydrophobic amino acids forming "knobs" can be accommodated by smaller "hole" residues on the opposing helix.

Many aspects of this statistically derived data have been confirmed experimentally by the group of Vinson using point mutations to a heterodimeric system derived from the PAR family member VBP B-ZIP domain.[44] In a second more comprehensive study,[45] 10 pairs of coiled coils were made, with each pair having a different amino acid at the single mutated *a* position (I, V, L, N, A, K S, T, E, and R). The thermodynamic stability of all 100 combinations of peptides was then measured. The first conclusion to be drawn from this work is that the most stable homotypic interactions were for isoleucine, followed by

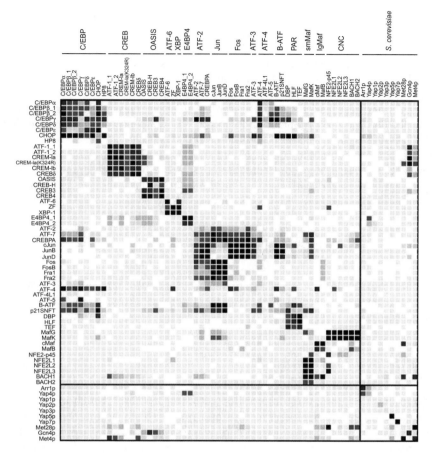

FIG. 6. The experimentally determined interactome for human bZIP proteins taken from Newman and Keating[41]. The darker the squares are the more statistically significant the interaction. Dark squares on the diagonal indicate strong homodimeric interactions, and off-diagonal dark squares heterodimeric interactions (surface-bound proteins are shown on the left and fluorescently labeled probes are shown across the top). Reproduced with permission from Newman and Keating[41], copyright The American Association for the Advancement of Science (2003). (For color version of this figure, the reader is referred to the web version of this chapter.)

valine, leucine, and asparagine. By looking at the coupling energies using double-mutant cycles, it was also deduced that the heterotypic interaction in which asparagine paired with isoleucine was the most repulsive, and that lysine and arginine paired with isoleucine, valine, or leucine, formed the most attractive set of heterotypic interactions.

Vinson's group also investigated the use of charged e and g positions within the same system, again using double-mutant cycles.[46,47] The amino acids alanine, lysine, arginine, glutamine, and glutamic acid were placed in e and g positions, and the stability of all the combinations of heterodimers was considered. The conclusions were that the interaction in which arginine at e paired with glutamic acid at g was the most stabilizing (with respect to alanine–alanine) and also had the highest coupling energy. This interaction was also found to be the least reduced by increased salt concentration, indicating that the charges may be partially buried. Other stabilizing interactions were lysine at e with glutamic acid at g, and the switched e for g versions of these interactions. As is expected, the like charged pairs are destabilizing, with glutamic acid paired with itself being the least stable, followed by arginine–arginine, lysine–arginine, and lysine–lysine pairs.

B. Designed Coiled Coils: Methods for Maximizing Specificity

The lessons learned in the previous section have been put to use. Many examples exist of coiled coils designed to be hetero, homo, dimer, and other oligomer states.

1. EXPLOITING NATURAL SPECIFICITY

Our first example of exploiting natural specificity is the work of Arndt's group in redesigning the transcription activator protein-1, which includes the coiled-coil heterodimer peptides Fos and Jun.[48] This is an interesting target, as it is implicated in various cancers where it can become upregulated or overexpressed. The idea is that a synthetic peptide that could outcompete either half of this interaction would be useful as a drug. In this work, the technique of protein-fragment complementation assays combined with growth competition was used to produce optimized binding partners for both the wild-type Fos and Jun (see Fig. 7). The library used in the assay was constructed semirationally using information from the families of Fos and Jun and knowledge of coiled-coil interactions. The winning peptides from the library had higher melting temperatures when mixed with their wild-type partners than the natural interaction, as well as an even tighter heterodimeric interaction with each other.

The two winning peptides along with the wild-type peptides and five other related peptides were screened for the melting temperatures of each of the homo or hetero dimeric combinations. The data from this were used to create an algorithm that predicts the melting temperature of coiled-coil dimers

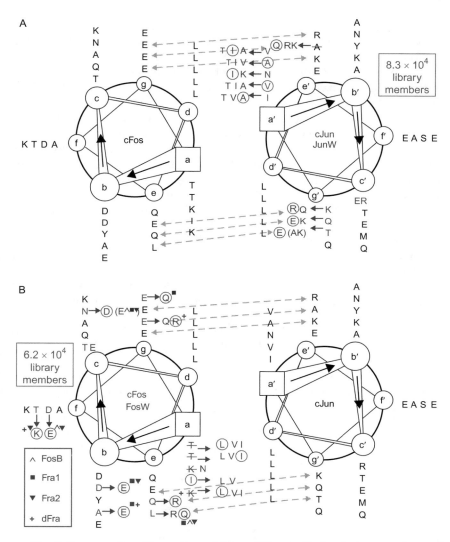

Fig. 7. Sequences for wild-type Fos and Jun are shown in black only. Amino acids in red indicate those used in the libraries for the mutant Fos and Jun. (Where the wild-type residue is not present in the library it is struck through in green.) The residues used in the winning peptides are circled in green. Dotted blue lines indicate potentials salt bridges. Reproduced with permission from Mason *et al.*[48] copyright National Academy of Sciences, USA (2006). (For interpretation of the references to color in this figure legend, the reader is referred to the Web version of this chapter.)

(see www.molbiotech.uni-freiburg.de/bCIPA). This algorithm uses empirically derived weights based on helical propensity, electrostatics, and packing of the hydrophobic core.

The designing of synthetic partners for bZIP proteins has also been explored by the Keating group. They have developed an algorithm called CLASSY, which uses a cluster expansion method to convert their structure-based interaction model into a sequence-based scoring function that is very fast to evaluate.[49] The algorithm begins by finding the sequence with the maximum interaction score with the target sequence. A value is then set for the difference between this interaction and the most favorable of the interactions with a set of competitor sequences (see Fig. 8). The introduction of competitor sequences is a key concept in protein design, the idea being that it is not enough to simply be stable in the desired fold but that other possible competing folds must be destabilized.

Once a low-energy sequence is found, it is then mutated until the difference (the specificity) between the energy of the target and of the competitors is maximized. This study demonstrates most vividly the trade-off between stability and specificity of interactions.

Experimentally, 48 peptides designed against 20 natural targets were tested for interaction with 33 representative human bZIP coiled coils and for self-association. Within in this dataset, many designs were found to outcompete the native partners for the targets, and furthermore, several of the designs also exhibited their strongest interaction with their target.

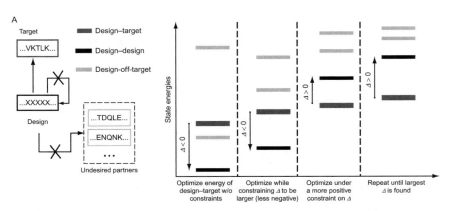

FIG. 8. Schematic indicating the operation of the CLASSY algorithm. The design–target interaction is desired, with the design–design and design–off-target interactions being disfavored. Initially, the design–target interaction energy is minimized, and then the energy of the competing interactions is raised until they are higher than that of the design–target interaction. This energy gap is then maximized. Reproduced with permission from Grigoryan et al.[49] copyright Nature Publishing Group (2009). (For color version of this figure, the reader is referred to the web version of this chapter.)

From this work, it was also found that while the designed peptides were somewhat limited in the sequence space they covered: they produced many new interaction profiles indicating that the original bZIP families have only explored a small part of the possible interaction space.

2. DEVELOPING A COILED-COIL TOOL BOX

With the information required to design sequences that will specifically interact with their target partners in hand, it is possible to address the problem of developing a tool box of synthetic coiled coils. Once such a set of coiled coils exists, their orthogonal interaction profiles can be used to generate more complex systems in which many more self-assembling components can be mixed at will.

Toward this goal, the Woolfson group has used a computer algorithm to find the maximum specificity that can be generated given a set of amino acid choices for positions a, e, and g.[50] This system was limited to coiled coils of only three heptads in length but nevertheless generated a set of six peptides that associated into the three targeted coiled coils preferentially out of all of the 21 possible homo and hetero dimeric possibilities (see Fig. 9). The success of this system was tested in two ways. The peptides were all labeled with terminal cysteine residues so that the coiled coils formed in solution could be trapped and examined. First, each of the designed coiled coils was checked for folding preferentially as a parallel dimer, using a combination of analytical ultracentrifugation (under reducing conditions designed to allow nondimeric species to form if desired) and comparing thermal denaturation under both oxidizing and reducing conditions. Second, the mixture of all six peptides was incubated under conditions where exchange of terminal disulfide bonds could occur. The reaction was then quenched, and mass spectrometry was used to identify that only the three desired coiled coils had formed.

This approach of pulling out selective sets of coiled coils has recently been generalized by Keating's group in a study in which they measured the interactions of 48 designed peptides and 7 natural bZIP proteins with no strong homodimerizing properties.[51] Within this dataset, 27 different hetero-specific pairings were found using 26 different peptides, with each peptides being involved in upto 7 different dimers. From this information, 10 different types of subnetworks of interactions were found including orthogonal pairs, orthogonal triplets (as in the previous system), and more complicated hub-type networks in which one peptide interacts with many. The behavior of two sets of four peptides predicted by the interactome to form orthogonal pair systems was experimentally demonstrated as matching the prediction.

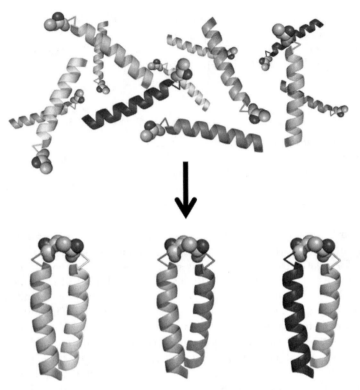

FIG. 9. Schematic showing how a mixture of six thiol-labeled peptides assembles into just 3 of the 21 possible disulfide-linked dimers under redox-buffered conditions. (For color version of this figure, the reader is referred to the web version of this chapter.)

One interesting observation from this work was that all of the networks found were sparsely connected, and it was hypothesized that this is due to the peptides having either been designed or selected for a lack of homospecificity. This may, in turn, lead to the desirable outcome that the peptides lack promiscuity, making them ideal for using in synthetic biological applications (Fig. 10).

3. NONNATIVE AMINO ACIDS

One way to extend the specificity available to designers is to incorporate nonnative amino acids. These are simply amino acids with side chains not found in nature, and it is often the case that the interactions between natural and synthetic amino acids show a significant level of specificity. Synthetic amino

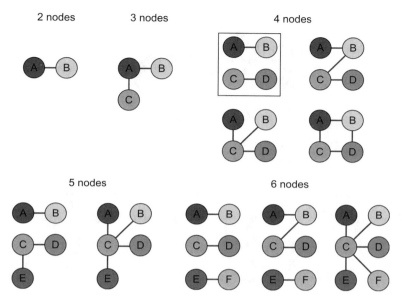

FIG. 10. Diagram indicating the subnetworks of interactions found within the combined design and natural bZIP protein interactome. Lines indicate interactions, and all peptides interacting as homodimers were excluded from the analysis. The boxed network corresponds to the system chosen for further characterization. Adapted with permission from Reinke *et al*.[51] copyright American Chemical Society (2010). (For color version of this figure, the reader is referred to the web version of this chapter.)

acids can be incorporated during peptide synthesis and, indeed, commonly used ones may be obtained with all of the protecting groups necessary for routine synthesis.

Several examples of this have come from the Kennan group who have used a variety of nonnatural amino acids to influence coiled-coil assembly (see Fig. 11). First, the effect of side-chain length in derivatives of glutamic acid and lysine used at *e* and *g* positions was investigated.[54] In this study, the number of methylene units separating the amine and carboxylic acid groups from the backbone was varied from 1 to 4. The general trend was a dramatic increase in stability as the total number of methylene units increased.

The second example from this group involved using guanidinium-functionalized side chains at *a* positions.[52] Again, the effect of chain length was investigated using arginine derivatives with one or two methylene groups removed and looking at interactions with asparagine, glutamic acid, and aspartic acid. The conclusions were that shorter chains in this core position confer more *specificity*. The combination of positive interactions between aspartic

A

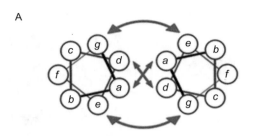

B   bcdefg abcdefg abcdefg abcdefg abc
      AQLXKX LQALXKX ZAQLXKX LQALXKX LAG

C   (i)   (ii)   D   (i)   (ii)   (iii)   (iv)

      a-position   e/g-position

E

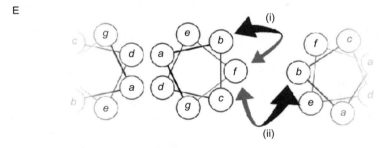

FIG. 11. Coiled-coil designs that incorporate nonnative amino acids. (A) Helical wheel illustrating the main interactions available to designers of coiled-coil structures; the core and flanking salt-bridge interactions. (B) A "typical" base sequence for the investigation of nonnative amino-acid interactions in coiled coils.[52–54] Residues in positions marked Z were substituted with amino acids with side chains as shown in C, while those marked X have been substituted with amino acids with side chains shown in D. (C) Guanidinium- and urea-based side chains with differing lengths, used

acid and the guanidinium side chains, combined with the negative interaction of asparagine with aspartic acid, allowed the creation of two dimer pairs that assembled with no cross-reactions. This set of coiled coils was extended to three (Section II.B.2) by the addition of peptides with a positions containing urea-functionalized side chains, specifically citrulline.[53]

Finally from this group, the nonnatural amino acids discussed above were explored in the context of a trimer-forming system.[56] In this work, the trimeric nature of the interactions was favored by the use of isoleucine at all a and d positions other than a single mutated a position. It was found that while urea-based side chains allowed trimer formation, guanidinium-based side chains did not—thus providing another rule in the protein design toolbox.

A final example of the utility of nonnative amino acids is the use of fluorinated side chains. Fluorine is not generally found in natural systems because of a lack of bioavailability; however, it has the property that it is immiscible with both water and most organic solvents. This makes it a target for incorporation into biological molecules, as it can provide another orthogonal interaction for increasing the complexity of designed systems.[57]

To demonstrate the utility of fluorinated side chains, the group of Kumar (and others) has used trifluoroleucine and trifluorovaline substitutions in the a and d positions of the leucine zipper GCN4.[58] They found that the substituted peptide formed a dimeric coiled coil which had a higher melting temperature than the original wild-type GCN4. Following on from this, they produced two variants of this system, one with an all leucine core and the other with an all hexafluoroleucine core.[59] These two peptides were shown to be mutually exclusive in their self-assembly, with only homodimers forming from mixed solutions. This is attributed to the vastly higher melting temperature of the hexafluoroleucine homodimers.

More recent work in this area by the Koksch group has focused on heterodimeric systems in which the fluorous side chains are partnered with natural amino acids. From their studies, it is found that in this environment fluorous chains pair best with the same hydrophobic amino acids normally found in the core.[60] A second study looking at a variety of fluorous side chains in the core

---

to investigate core interactions. (i) guanidinylated diaminopropionic acid, (ii) guanidinylated diaminobutyric acid, (iii) pUr, a urea-terminated side chain, and (iv) pUr°, a urea-terminated side chain with an additional methylene group in the "linking" region to those in pUr. (D) Amino- and carboxyl-based side chains used to probe salt-bridge interactions. (i–iv) Positively charged, amine-terminated side chains with increasing length (and thus hydrophobic contact area): (iv) forms the naturally occurring lysine. (v–viii) Negatively charged, carboxyl-terminated side chains with increasing length. (vi) and (vii) form the natural amino-acids aspartic acid and glutamic acid, respectively. (E) Intra- and interhelical cation–π interactions investigated using the nonnative amino-acid *norleucine*.[55] (For color version of this figure, the reader is referred to the web version of this chapter.)

a and d positions has found that, similar to the natural hydrophobic amino acids, the specific geometry of the side-chain packing is important in determining stability.[61] This suggests that fluorinated side chains may also be capable of specifying helix orientation and the coiled-coil oligomeric state.

C. Applications of Designs

So far, the rules of self-assembly of coiled-coil tectons have been explored. Next, some selected applications in protein design are examined, which are by no means an exhaustive list but should give a flavor of the possibilities.

1. MORE COMPLEX DISCRETE ASSEMBLIES

The first application examined is simply the building of more complex architectures, involving peptides with more than one coiled-coil domain and hence more than one coiled-coil interaction.

One of the first examples of this was the "belt and braces" system from the Woolfson lab, where a "belt" peptide, six heptads in length, is paired with two "brace" peptides, each three heptads long.[62] Specificity was achieved by having the two brace peptides carry only positively charged e and g positions and the belt carrying only negatively charged e and g positions. The two brace peptides were distinguished by one of them possessing an asparagine at a that matched with an asparagine at a in one half of the belt. The external ends of the peptide braces were functionalized with cysteine to enable interactions with gold. The success of the design was demonstrated by a combination of biophysical techniques and by incubation with gold nanoparticles (Fig. 12). In the absence of the belt peptide, the gold particles are not seen to assemble by transmission electron microscopy (TEM); however, once the belt is added, the nanoparticles collect together in groups spaced by the expected 6-nm length of the belt peptide.

This concept was extended by the Woolfson group using the set of six helices discussed previously.[50] In this case, two sets of two peptides were linked together using flexible glycine residues to create two different belt peptides. By combining the two belt peptides, which assembled in an offset manner, with the remaining two of the six peptides, a 9-nm long construct with four components was formed. Again, this assembly was demonstrated through biophysical techniques. While this increase in the complexity of the constructs seems modest, it opens the way for more complicated nanostructures to be designed in the future.

2. VESICLE FUSION

One recent application of coiled-coil design is the production of a system capable of fusing liposomes.[63] The system is based on a simplified model of the action of SNARE proteins, which are responsible for fusing vesicles in the cell (see Fig. 13). In nature, three types of SNARE proteins come together to form

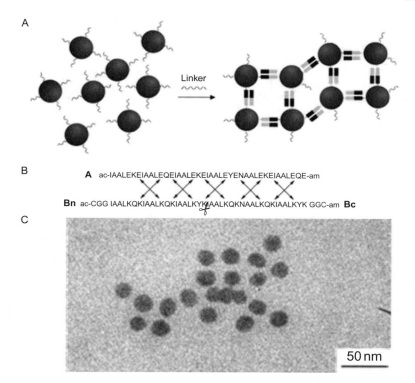

FIG. 12. (A) Schematic showing the "belt and braces" principle: dark and light indicate the presence or absence of the specifying asparagine residues. (B) The sequences for the belt and braces. (C) TEM image of resulting gold nanoparticle arrays. Reproduced with permission from Ryadnov et al.[62] copyright American Chemical Society (2003).

a coiled coil. One type of SNARE is associated with a transport vesicle; the second is associated with the destination membrane; and the third is present in solution. Once the three types are in proximity with each other, a stable tetrameric coiled-coil forms which colocalizes the membranes, allowing fusion to proceed.

The designed system uses just two types of peptides, each one anchored to a separate liposome via a PEG linker attached to a lipid domain (DOPE). In this system, a dimeric coiled coil is formed between the two peptides that serves the purpose of colocalizing the membranes to promote fusion. The success of this design has been shown by fusing liposomes that contain one each of a forster resonance energy transfer (FRET) pair of fluorophores.

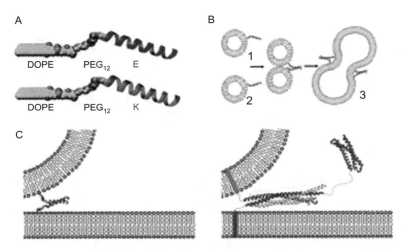

FIG. 13. Liposomes labeled with one of two interacting peptides fuse on mixing. (A) The two peptides are linked to DOPE via a PEG linker. (B) The two peptides associate to form a coiled coil which colocalizes the liposomes leading to fusion. (C) Comparison between the reduced SNARE model on the left and the natural process on the right. Reproduced with permission from Robson Marsden *et al.*[63] copyright Wiley-VCH Verlag GmbH & Co. KGaA (2009). (For color version of this figure, the reader is referred to the web version of this chapter.)

As the contents of the vesicles mix, the FRET becomes measurable, indicating the fusion of oppositely labeled liposomes. Conversely, liposomes labeled with the same peptide do not fuse.

3. SELF-REPLICATING PEPTIDES

Designed coiled coils have also been used to template peptide bond formation, creating a self-replicating system. Ghadiri's group developed a system based on the leucine zipper GCN4 in which the full-length peptide assembles two peptide fragments one of which has been preactivated as a thiobenzyl ester and the other terminated with cysteine.[64,65]

Once these two chemical functionalities are colocalized, they react to produce a peptide bond between the two fragments, making a copy of the full-length peptide (see Fig. 14).

In this system, a dimeric coiled coil is formed between three components, that is, a full-length peptide and two peptide fragments corresponding to the N- and C-terminal halves of the full-length peptide. Once the peptide bond has formed, the full-length peptide must dissociate in order to leave the catalyst free for the next cycle. The concentration of the full-length peptide has been

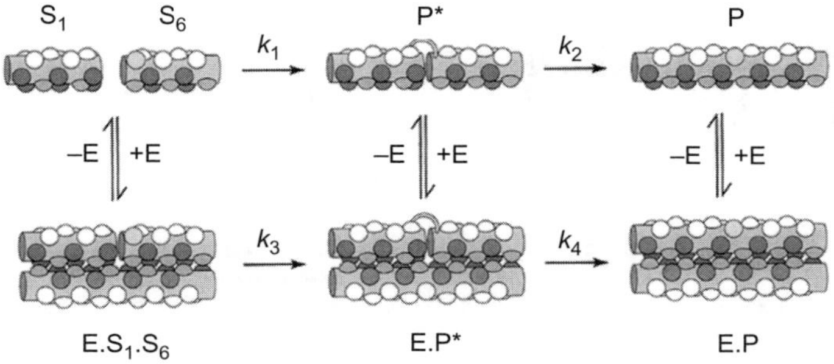

FIG. 14. Schematic diagram of the uncatalyzed and catalyzed routes to ligation for a self-replicating peptide. The condensation reaction of the substrate molecules S_1 and S_6 is greatly facilitated by their colocalization by the catalyst E (rate k_3). Residues critical for molecular recognition and ligation are shown colored (red, glutamic acid; blue, lysine and arginine; yellow, cysteine; gray, leucine and valine). Reproduced with permission from Severin et al.[66] copyright Nature Publishing Group (1997). (For interpretation of the references to color in this figure legend, the reader is referred to the Web version of this chapter.)

shown to increase exponentially as expected for this self-catalyzed reaction. This work was extended to using a heterodimeric coiled coil via manipulation of the charged residues at e and g positions.[66]

The group of Chmielewski has added an interesting design principle to their self-replicating peptide: specifically, the addition of a proline residue close to the ligation site.[67] Once ligation has occurred, this proline kinks the resulting full-length helix, destabilizing the coiled coil and leading to an increase in the efficiency of self-replication.

D. Switching and Dynamic Coiled-Coil Systems

As was stated in the introduction, coiled coils are involved in many dynamic processes in the cell and, in many cases, this results from structural switching either between folded and unfolded (in the case of signaling pathways) or more subtly in the case of motor proteins. In this section, some of the ways in which coiled-coil systems have had structural duality introduced is discussed, to bring to bear the function of switching.

1. SOLVENT CONDITIONS

To begin with, coiled-coil systems that respond to changes in the bulk solvent condition including temperature, pH, and redox potential are discussed.

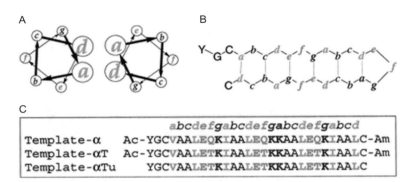

FIG. 15. Schematic of design principles for peptides that can adopt both coiled-coil and β-sheet structures. (A) Sequences aligned against the coiled-coil heptad repeat. (B) Sequences aligned against the β-sheet structure. Proposed structure from Ciani *et al.*[68] (left) and Kammerer *et al.*[69] (right).

Some examples for temperature-based switches are peptides that switch from coiled-coil folds to β-sheet-rich folds on heating. Woolfson and colleagues have found that the addition of β-strand favoring threonine at the surface-exposed *f* position of a dimeric coiled-coil-forming peptide (see Fig. 15) produces a switch to β-structure at elevated temperatures.[68] This switch is associated with the formation of amyloid fibers and is not therefore reversible. Kammerer also produced a series of peptides that were coiled coils at room temperature but switched to amyloid structures at high temperatures.[69] It was found that the coiled-coil structure can tolerate well the addition of several β-strand favoring residues without significant loss of stability.

An example of a pH-driven switch comes from the Kennan lab in which a heterotrimeric system is employed.[70] The assembly of the three different peptides in the trimer is driven by the use of cyclohexylalanine side chains at *a* positions. At each core *a* layer, one of the helices supplies a cyclohexylalanine which the other two complement with alanine. Each peptide has either all lysine or all glutamic acid at *e* and *g* positions, meaning that one of the *e/g* interfaces is forced to be mismatched (see Fig. 16). At high pH, the system is more tolerant of an all-lysine interface, whereas at low pH, an all glutamic acid interface is preferred. The pH switch was created in the form of a competition between two peptides with the same core pattern but with *e* and *g* positions with either all lysine or all glutamic acid. These peptides were shown to exchange with each other when the pH was cycled.

This work was extended to apply to a system that switched helices and orientation, between a parallel and antiparallel heterotrimer.[71]

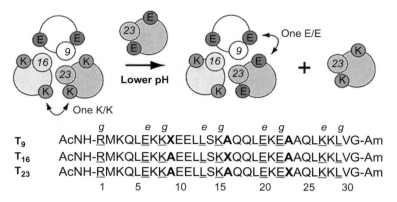

	g	e g	e g	e g	e g
T_9	AcNH-RMKQLEKKXEELLSKAQQLEKEAAAQLKKLVG-Am				
T_{16}	AcNH-RMKQLEKKAEELLSKXQQLEKEAAAQLKKLVG-Am				
T_{23}	AcNH-RMKQLEKKAEELLSKAQQLEKEXAQLKKLVG-Am				
	1	5	10	15	20 25 30

FIG. 16. pH-triggered helix exchange. At higher pH, the mismatched lysine interface is favorable and so T23K is incorporated. At lower pH, the mismatched glutamic acid interface becomes favorable and helix exchange occurs. Heptad repeat positions e and g shown only. Reproduced with permission from Schnarr and Kennan[70], copyright American Chemical Society (2003). (For color version of this figure, the reader is referred to the web version of this chapter.)

As an example of a redox-based switch, the Woolfson group produced a system that switched between a monomeric helix–turn–helix hairpin and a dimeric coiled coil.[72] The interactions in the monomeric form are those of an antiparallel coiled coil and can be made compatible with the dimeric form by switching the charge pattern at e and g half way along the helix. For the turn region of the monomeric form, a series of sequences were tested, from the flexible to the more rigid. The monomeric form is held together by an intramolecular disulfide bond between the N- and C-termini, and switching to the dimeric form is accomplished by reducing this bond. Although the various design iterations exhibited different stabilities in the two states, the dimeric unstrained state always took precedence once the disulfide bond was reduced, making the switch irreversible.

2. METAL-DRIVEN SWITCHING

In this section, two types of metal-driven switching are discussed: first, switches in which metal binding produces the folded state from the unfolded, and second, reversible switches in which the metal-bound form is different to the unbound but also folded.

As an example of the first type of switch, Tanaka's lab produced a parallel trimeric coiled coil which was adapted to contain metal binding.[73] In this design, two core positions were mutated from isoleucine to histidine. This destabilizes the coiled coil until Co(II), Ni(II), or Zn(II) is added, at which point the trimeric coiled-coil forms.

Moving on to the second type of switch, the Ogawa lab has produced a designed dimeric coiled coil which incorporates nonnative metal-binding 4-pyridylalanine side chains in solvent-exposed positions.[74] On the addition of a platinum center, this structure rearranges to form a four-helix bundle, and hence this system constitutes an oligomeric state switch.

As a more dramatic structural switch, two groups (Woolfson and Kuhlman) have developed peptides that switch from trimeric coiled coils to a zinc-binding motif known as the "zinc finger" on addition of zinc.[75,76] Although the two states of this switch are similar, the two studies differ in both the approach taken to design and in the details of how the two functionalities are overlaid in the peptide sequence. The Woolfson group switch uses an $HX_2HX_{12}HX_5H$ zinc-binding motif aligned with the heptad repeat such that none of the core residues of the coiled coil is affected. The sequence was designed manually using knowledge of both structural motifs. In contrast, the Kuhlman sequence uses the $CX_2CX_{12}HX_3H$ zinc-binding motif in an alignment, which puts the second cysteine residue at a d position in the core of the coiled coil (see Fig. 17). This sequence was designed using the ROSETTADESIGN program,[77] which in this case was used to thread target sequences onto the backbone structures for the two desired states simultaneously. The algorithm then mutates selected side chains to find a minimum-energy sequence for both states. Both these designed sequences produce a structural switch on the addition and removal of zinc.

As a final example of metal-driven switching, the Koksch lab has produced a metal-driven α- to β-structural switch.[78] A peptide has been designed that shares coiled-coil and β-sheet characteristics, which is found to be β-structured under aqueous conditions but will switch to helical with the addition of 40% helix-promoting trifluoroethanol (TFE). A mutation of this peptide, which contains a number of histidine residues, behaves similarly in water and TFE but will switch back to being β-structured on the addition of Zn or Cu. This switch is reversible on the addition of EDTA.

3. LIGHT-SENSITIVE SWITCHING

Switching using light is desirable because of the difficulty of rapidly changing other solution conditions such as pH or metal ion concentration. One illustration of a possible route to light-controlled conformational switching is the work of Woolley's group using the dye azobenzene as an activating switch within the transcription activator protein-1 system.[79] In this work, an azobenzene linker is used to connect two consecutive f positions in a coiled-coil peptide. While the end-to-end distance of the azobenzene linker is compatible with an α-helical fold in the *cis* form, it is incompatible in the *trans* form.

A

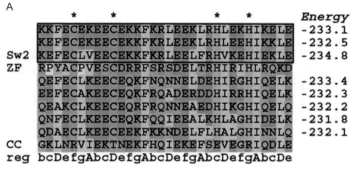

		Energy
	KKFECEKEECEKKFKRLEEKLRHLEKHIKELE	-233.1
	KEFECEKEECEKKFKRLEEKLRHLEEHIKKLE	-232.5
Sw2	KEFECLVEECEKKFKRLEELFRHVKEHIEKLE	-234.8
ZF	RPYACPVESCDRRFSRSDELTRHIRIHLRQKD	
	QEFECLKEECEQKFNQNNELDEHIRGHIQELK	-233.4
	EEFECAKEECEQKFRQADERDDHIRRHIQELK	-232.3
	QEAKCLKEECEQRFRQNNEAEDHIKGHIQELQ	-232.2
	QNFECLKEECEQKFQQIEEALKHLAGHIDELK	-231.8
	QDAECLKEECEEKFKKNDELFLHALGHINNLQ	-232.1
CC	GKLNRVIEKTNEKFHQIEKEFSEVEGRIQDLE	
reg	bcDefgAbcDefgAbcDefgAbcDefgAbcDe	

B

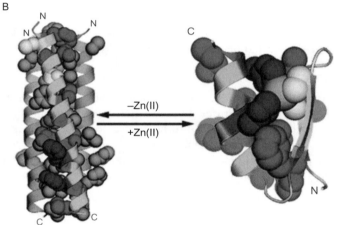

−Zn(II)

+Zn(II)

FIG. 17. Computational design of a zinc-driven switch. (A) A sequence alignment of the lowest energy sequences from the design process with a zinc finger sequence (residues 3–33 of Zif268) and a coiled-coil sequence (residues 13–44 of hemagglutinin) for comparison. The heptad repeat is also shown along with stars, indicating the zinc-binding residues. (B) The backbone structure of Sw2 (the sequence of the successfully produced switch) threaded onto the two target structures. Hydrophobic residues are colored red, cysteine ligands are colored yellow, histidine ligands are colored blue, and the Zn(II) ion is represented as an exaggerated sphere colored in violet. Reproduced with permission from Ambroggio and Kuhlman[76], copyright American Chemical Society (2006). (For interpretation of the references to color in this figure legend, the reader is referred to the Web version of this chapter.)

Switching from the *cis* to the *trans* form is achieved through irradiation with light at 460 nm, while switching from the trans back to the *cis* form requires light at 365 nm.

In order to produce a test system, the linker has been inserted into wFos, the winning peptide partner to wild-type Jun discovered in the work described in Section II.B.1. The idea is that, while the linker is in the *trans* form, wFos will be unavailable for binding and the native interaction of Fos–Jun will prevail. Once the *cis* form of the linker is made, wFos will be available to outcompete wild-type Fos and the cellular function of the Fos–Jun dimer will be disrupted. The success of this system was demonstrated via control of DNA-binding activity in cell treated with the modified wFos (Fig. 18).

III. Designing Higher-Order Helical Assemblies

Having discussed the way in which coiled coils have been exploited to create new discrete self-assembling systems, the next stage is to look at coiled coils in designed fiber and higher dimensionality assemblies.

A. Lessons from Nature

There are many different fibers found in cells performing various functions using a wide variety of structures; however, there are common themes connecting them. Fibers in nature tend to be nucleated; in other words, several monomers must be assembled before a stable growing fiber is formed. This gives the cell spatial and temporal control over the location of fiber growth by disfavoring the spontaneous formation of growing small oligomers in the

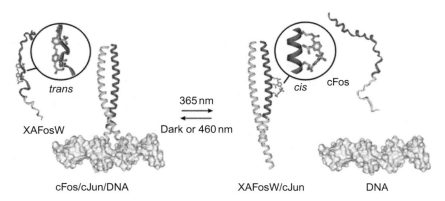

FIG. 18. Schematic showing how switching of the azobenzene from *trans* to *cis* allows the synthetic peptide to outcompete the natural binding partner of Jun. This in turn prevents binding to DNA. Reproduced with permission from Zhang *et al.*[79] copyright Wiley-VCH Verlag GmbH & Co. KGaA (2010). (For color version of this figure, the reader is referred to the web version of this chapter.)

bulk. Natural fibers are uniform in their width (and sometimes length): that is to say, all fibers formed from the same protein have approximately the same dimensions. This means that physical parameters such as flexibility, persistence length, probability of breakage, and the number of monomers per unit length are controlled. This enables the cell to produce fibers that are fit for purpose and are efficient in their use of protein. Natural fibers are dynamic: they can be switched from assembly to dissociation by changes in chemical potential. Many fibers achieve this property in an active fashion using adenosine triphosphate (ATP) hydrolysis as a method to manipulate the energy landscape; however, others are passively responsive to solvent conditions. Exchange of subunits can usually occur at the free ends of the fiber but can also occur along the length of the fiber. Finally, natural fibers are functional—they possess binding sites on their ends and surfaces that are used to recruit other protein machinery including protein motors.

Examples of naturally occurring coiled-coil fibers are the intermediate filaments, which are a broad class of related proteins containing a rod domain formed from a parallel dimeric coiled coil. The intermediate filaments form one of the primary elements determining the shape of both the nucleus and the cell in general. They have an extraordinarily hierarchical assembly pathway which results in fibers of a determined width and whose mechanical properties allow large deformations.[26]

The assembly of vimentin is a specific, well-studied example of an intermediate filament,[80] in which the first step is association of the parallel dimeric coiled coil (see Fig. 19). This leaves non-coiled-coil head and tail domains at either end, which are responsible for the onward assembly of two of these dimers to form a tetrameric species which is antiparallel and sticky-ended (meaning that, while there is a substantial overlap between the dimers, a large fraction of the coiled-coil domains are hanging over either end). Two of these tetramers assemble (via interactions between rod domains) to form an octomer, which can be trapped and examined under specific pH and ionic strength constraints. These octomers will further assemble to form an asymmetric 32-subunit-containing species,[81] which represents the unit-length building block of the filaments. Until this species is formed, elongation of the filament is disfavored. Once the elongation has progressed substantially, one final structural rearrangement occurs in which the fiber condenses to a more compact width. The final mature fibers exhibit exchange both at free ends and also in coiled-coil units along the length of the fiber.

As such, this example fiber displays nucleation, control of width, and hence mechanical properties and dynamic exchange of subunits along its length. It also displays onward binding motifs for interactions with, amongst other things, cell adhesion proteins.

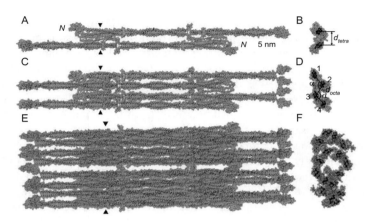

FIG. 19. The structural hierarchy of unit-length vimentin. (A, B) The tetramer state, or antiparallel dimer of parallel dimeric coiled coils. (C, D) The dimer of these tetramers. (E, F) The asymmetric octomer of tetramers. Reproduced with permission from Parry *et al.*[80] copyright Elsevier (2007). (For color version of this figure, the reader is referred to the web version of this chapter.)

B. Designed Fiber Assemblies

Armed with a combination of natural examples of fiber assemblies and rules for the design of coiled coils, it is possible to begin to build self-assembling systems that mimic natural fibers.

1. HOMOTYPIC COILED-COIL-BASED SYSTEMS

The earliest example of a designed coiled-coil fiber was a three-heptad-long homotypic system using the peptide sequence LETLAKA corresponding to heptad positions *abcdefg*.[82] The use of leucine at *a* and *d* positions and alanine at *e* and *g* makes this a rather generic coiled-coil sequence in which the oligomer state is not designed in. Indeed, this peptide forms tetrameric bundles under some conditions, but also assemblies into micrometer-long fibrils with widths of 5–10 nm. The morphology of the fibers is found to be dependent on the ionic strength.

The next system to be developed used a similar design principle with just under five-heptad repeats of the sequence QQLAREL at *bcdefga*.[83] This system was designed to assemble into a pentamer initially and then to undergo longitudinal assembly based on the predicted slippage this geometry induces providing staggered ends. This system produced fibrils at pH 6 and below, although this range was later extended by reducing the charge per heptad by replacing glutamic acid with glutamine or serine.

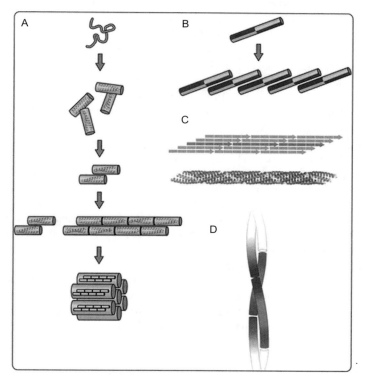

FIG. 20. A variety of possible coiled-coil fiber assembly schemes. (A) Blunt-ended coiled-coil assembly, as demonstrated by Dong *et al.*[84] (B) Coiled-coil assembly using a "staggered" hydrophobic core.[85] (C) A five-heptad coiled-coil system with a single heptad overlap.[83] (D) A palindromic charge pattern that produces sticky ends and allows fibrilization.[55] (For color version of this figure, the reader is referred to the web version of this chapter.)

In this design trend, the Hartgerink group has also produced fibers from short, blunt-ended coiled-coil-forming peptides.[84] This system has been explored in detail, with control of fiber width being demonstrated via the choice of amino acids used at b, c, and f positions, with higher charge resulting in thinner fibers. This study proposes a general mechanism for this type of assembly in which bundles of helices form first, and then slippage of the alignment allows longitudinal assembly, usually only at higher peptide concentrations (see Fig. 20).

Both the above systems rely on the slippage between the largely nonspecific hydrophobic cores. It is, however, also possible to promote a staggered assembly (and disfavor the entropically favorable blunt-ended configuration) using the design rules discussed earlier. Namely, this can be done by the insertion of strategically located polar residues into the hydrophobic core and manipulating the charged e and g positions. This was demonstrated in a homotypic system by the Conticello group with a sequence that comprised six heptads.[86] Specifying asparagine residues in the third and sixth heptad promote either the blunt-ended dimer or the three-heptad overlapped sticky-ended dimer, while all negative e and g positions in the first three heptads and all positive e and g positions in the last three heptads promote only the sticky-ended dimer. This system indeed forms the α-helical structure which appears as long fibers under TEM.

Another, more recent, implementation of this type is the magic wand system from the Woolfson lab.[55] This system demonstrated that the use of charge patterning at e and g positions was sufficient to produce sticky-ended assembly. By removing the polar core residue, this system was able to achieve fibrillogenesis from a shorter four-heptad peptide, with an increase in stability with respect to the previous example. This study also investigated a series of mutations to external coiled-coil positions, and began to highlight the importance of electrostatics and cation–pi interactions as the source of thickening in fiber systems.

A different way to force staggered ends is to break the hydrophobic core. This technique was demonstrated by the Fairman group using a peptide based on the GCN4 leucine zipper but with two key alanine residues inserted into the middle of the sequence.[87] This has the effect of locally breaking the heptad repeat and causing the hydrophobic seam of the peptide to be displayed on opposite faces at either end of the helix. This discontinuity of the hydrophobic interface prevents the blunt-ended dimer from forming, and indeed fibers are formed instead. Once again, the morphology of these fibers is dependent on salt concentration.

2. HETEROTYPIC COILED-COIL-BASED SYSTEMS

Taking one step further forward in complexity, there are fiber systems made from heterotypic coiled-coil designs. These have the primary advantage over single-peptide systems of being able to control the onset of fibrillogenesis— fibers will only form once the components are mixed.

One of the best studied heterodimeric systems is the SAF system that came out of the Woolfson lab. This system consisted originally of two peptides, each four heptads in length. The design utilizes offset asparagine side chains at a positions and charge patterning to promote staggered heterodimeric assembly. The original peptide design was found to be helical in structure and to produce long, thickened fibers.[88]

In later work, the stability of the fibers was improved by the inclusion of oppositely charged side chains in surface-exposed positions in the coiled coil.[89] For this second iteration of the system, the helices were found to be packed in a helical array with the helix axis parallel to the fiber axis,[90] and a mechanism for fiber formation was determined using a combination of kinetic experiments using TEM, circular dichroism (CD), linear dichroism (LD), and nuclear magnetic resonance (NMR).[91] The mechanism was found to involve an initial folding in which small oligomers are formed, which are approximately half helical, followed by a nucleation step and by simultaneous thickening and elongation. By adding preformed fiber fragments to fresh peptide solutions (seeding), it was shown that at later times elongation dominates. This method was also used to demonstrate control over the length distribution of the fibers, with high seeding density leading to shorter overall fiber lengths (see Fig. 21).

A later generation of the SAF system has been made in which the exposed b, c, and f positions have been altered to be alanine and glutamine. These systems produce thinner, more flexible fibers that form system-spanning networks in the form of hydrogels.[92] In particular, the alanine-based system has been shown to form a robust gel capable of supporting mammalian cell growth.

C. More Complex and Higher Dimensionality Assemblies

1. HELIX–TURN–HELIX MOTIF

A fiber system based on a helix–turn–helix motif forms the first example of a more complex system. The added complexity of the turn region is of interest, especially as the turn regions are notoriously difficult to understand and design. Lazar et al. produced a peptide that comprises two 18-amino-acid long helical segments joined by four different turn regions taken from apolipoprotein A-I.[93] Three of these designs, which include a helix-breaking proline in the turn, are found to form fibers, at least some of which have been shown to assemble with the helicies perpendicular to the axis of longitudinal growth. In general, having the helicies running across the fiber axis is considered to be an advantage in terms of being able to functionalize fibers. This is due to the fact that the labels can be added in a position that should be displayed on the outside of the fiber rather than interfering with the end-to-end stacking required by helices running parallel to the fiber axis.

2. POLYNANOREACTORS

The move from one-dimensional fiber systems up to two-dimensional assemblies is made with the creation by Ryadnov of polynanoreactors.[94] The key design principle used in this work is the addition of interactions patterned onto the outer surface of the coiled coil. Ryadnov uses arginine in c positions such that an interaction can take place with both the neighboring g positions on the same helix or with an e position on a helix incorporated into a different

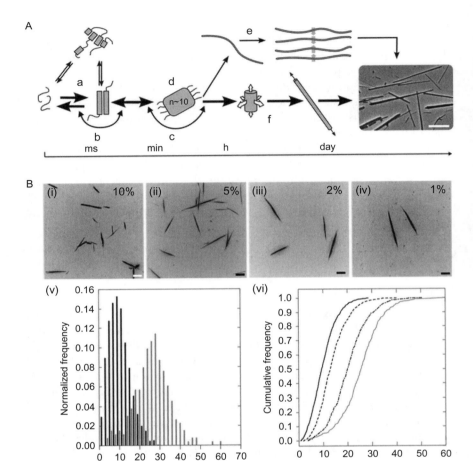

FIG. 21. Summary of the model proposed in Bromley et al.[50] for the assembly of SAF-type peptide fibers. (A) The proposed assembly mechanism proceeds from an initially unfolded state (a), through dimerization (b), and nucleation (d) steps to form a species that first grows in three dimensions (f) before ending with a species that grows by extension only. Several other possible paths are also shown, but these are found to be not present in this system. (B) The model predicts that control of fiber length should be possible through "seeding". Figures (i–iv) show representative fluorescence micrographs of SAF samples seeded with varying concentration of seeds. (v) Length distribution of fibers in samples seeded with 10% (black bars) and 1% (gray bars) preformed fibers. (vi) Cumulative length distributions for the four samples imaged, 10% (solid, black), 5% (dashed, black), 2% (dash-dot, black), and 1% (solid, gray). Reproduced with permission from Bromley et al.[50] copyright Elsevier (2009).

dimer (see Fig. 22). Arginine is chosen for this, as its guanidinium group can form two salt bridges at the same time (though with half of the strength), while all the e and g positions were filled with glutamic acid side chains. This homodimeric system produces very dense mesoscopic assemblies of a hexagonal paracrystalline phase, which was visible by TEM.

The system was then extended to a heterodimeric format in order to increase the size of the cavities. This was done by the introduction of a second peptide with lysine at all e and g positions and which was covalently triplicated into a starburst molecule. On mixing this peptide with the original, circular dense aggregates are seen by TEM. At higher resolution, these exhibit a surface patterning of rings with an average diameter of around 4.5 nm.

In order to bring function to this system, an f position of the second peptide was mutated to cysteine with the ambition of using the cavities to convert metal ions to colloidal metal. This was demonstrated via the creation of silver nanoparticles, which (once the peptide had been removed) could be seen by TEM to have an average diameter of around 5 nm (see Fig. 22C and D).

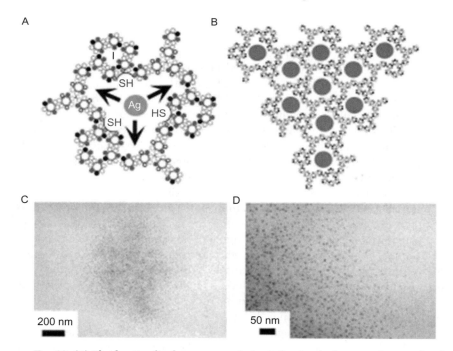

FIG. 22. (A) The functionalized nanoreactor design indicating the location of cavities lined with cysteine residues. (B) The network of silver deposits expected to be formed by the design. (C and D) Clusters of colloidal silver as seen by TEM. Reproduced with permission from Ryadnov[94], copyright Wiley-VCH Verlag GmbH & Co. KGaA (2007).

A   B

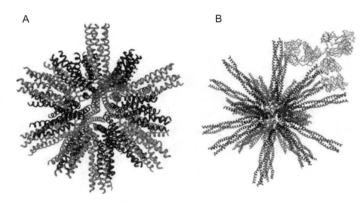

FIG. 23. (A) The original peptide nanoparticle structure showing the pentameric and the trimeric coiled coils. (B) The same particle with the trimeric coiled coil extended using the HIV surface protein gp41. Multiple copies are displayed, allowing a strong interaction with antibodies. Reproduced with permission from Raman *et al.*[95] copyright Elsevier (2006). (For color version of this figure, the reader is referred to the web version of this chapter.)

3. Peptide Nanoparticles

This example of peptide design links together some of the concepts used earlier in joining together functional tectons, with onwards assembly in three dimensions creating a sphere. Specifically, a peptide designed to form a homo-pentamer was linked to a homotrimeric sequence via a flexible diglycine linker.[95] This system assembles into a sphere containing 60 monomeric units (12 pentamers, 20 trimers) with icosahedral symmetry, via an intermediate with 15 monomeric units (see Fig. 23). The final assembly has been shown to contain 60 monomers in solution by analytical ultracentrifugation (AUC) and regular spherical particles by TEM. The particles have a diameter of around 16 nm and an internal cavity of around 6 nm. Both the internal cavity and the free ends of both the trimeric and pentameric peptides are available for functionalization.

One way in which this has been exploited is by appending a trimeric coiled-coil sequence taken from the severe acute respiratory syndrome (SARS) virus to the trimeric domain of the nanoparticle.[96] Once the particle assembles, the SARS sequence is displayed multiple times on the surface of the particle in its native trimeric state. As such, the nanoparticles have been shown to be a very promising platform for vaccine design.

Most recently, the nanoparticles have been appended with a tandem repeat of the B cell immunodominant repeat epitope of the malaria parasite *Plasmodium berghei* circumsporozoite protein.[97] The self-assembled nanoparticle was administered as a vaccine to mice and was found to confer a long-lasting, protective immune response.

D. Functionalization of Fiber Assemblies

There are many ways in which one can approach the functionalization of fiber systems.[98] In the first approach (coassembly), one may covalently label the peptides before assembly such that they possess both the assembly information and also some other functional domain or moiety. In the second approach, the fibers can be labeled either covalently or noncovalently after assembly. In the third approach (templated assembly), the fibers are used as a scaffold to condense some other functional material.

Each of these cases will be examined with a few examples from the α-helical assembly literature, but it should be noted that the techniques involved have been exploited to a much larger extent in the β-based assembly field and that many of the concepts are transferable. For a recent review of the functionalization of amyloid-based systems, see Ref. 99.

The covalent labeling of peptides has been used extensively in the case of peptides designed for tissue culture scaffolds. Many systems have therefore been modified to include cell recognition motifs including the integrin-dependent cell adhesion motif RGD. Systems modified in this way include the pentameric fibers produced by Potekhin et al.[100,101] Once the fibers are assembled, it is hoped that the RGD motif will be displayed on the surface.

Another form of covalent labeling involves the addition of fluorophores. This was used in the case of the SAF system discussed earlier, with one of the two co-assembling peptides covalently labeled with a fluorescent label.[102] Using this technique, it was possible to observe two previously unknown functions of this system. First, polar growth was shown, in that the added labeled peptide associated preferentially to one end of the fibers. This was demonstrated by adding the peptide labeled with rhodamine followed by a peptide labeled with fluorescein, giving fibers with multicolored tips (see Fig. 24). Second, a small level of exchange along the length of the fibers was seen, potentially drawing a parallel with the intermediate filament systems. Finally, within the SAF system, various assembling subunits have been covalently linked together, giving systems that produced fibers with branches, waves, and kinks.[103,104]

It should be noted that, in many of these systems, it is not desirable, necessary, or even tolerated that every peptide in the assembly carries a function, and that in many cases, only a small amount of functional peptide is doped into the assembly.

Postassembly modification has also been attempted in the SAF system, using biocompatible click chemistry.[105] This involves a component of the previous route, in that peptides are labeled with either azide- or alkene-modified side chains. These are small modifications and do not significantly alter the behavior of the peptide system and are incorporated at high levels into

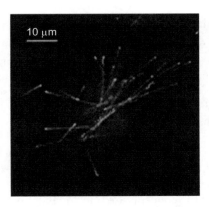

FIG. 24. Fibers from the SAF system labeled sequentially with rhodamine (red) and then fluorescein (green) labeled peptide. Reproduced with permission from Smith et al.[101] copyright Wiley-VCH Verlag GmbH & Co. KGaA (2005). (For interpretation of the references to color in this figure legend, the reader is referred to the Web version of this chapter.)

the fibers. Once the fibers are assembled, the function bearing click partner is added and labeled fibers are created. This has been demonstrated with gold nanoparticles and fluorphores (Fig. 25A).

Noncovalent postassembly labeling of the SAF system has also been achieved,[106] using short peptide tags with negatively or positively charged or neutral sequences that are incubated with assembled fibers. Given the net positive charge of the SAFs, it is the DEDEDE tag that is found to interact strongly. Again, this has been demonstrated to recruit function in the form of gold nanoparticles and fluorophores to the surface of the fibers (Fig. 25B).

The final route of templated assembly has also been demonstrated in this system, by the deposition of silica onto preformed fibers.[107] The fibers themselves can be removed by proteolysis leaving silica tubes, which appear hollow by TEM. Another example of this approach was demonstrated by the Conticello group in which a trimeric coiled-coil-based fiber containing histidine side chains was assembled.[108] The peptides could recruit silver, which formed a nanowire within the fiber core (Fig. 25C).

IV. Overview and Future Outlook

As we have seen in the course of this chapter, the design of α-helical tectons for self-assembly is maturing as a science. We have now reached the point where many different coiled-coil topologies can be reliably produced, using design rules that have been extracted from natural systems and validated in synthetic systems. The field is now moving on towards more complex, discrete

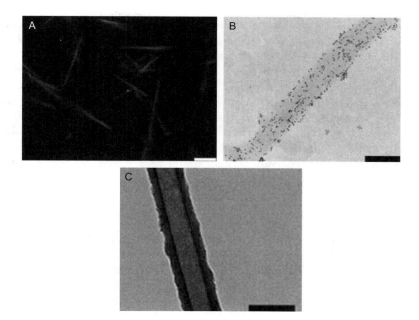

FIG. 25. Three examples of fiber functionalization applied to the SAF system. (A) Noncovalent attachment of dye to fibers using charged peptides. Scale bar 10 μm. Reproduced with permission from Mahmoud and Woolfson[106], copyright Elsevier (2010). (B) *In situ* functionalization of azide-containing SAFs functionalized with biotin, followed by streptavidin–nanogold. The fibers contained purely azide-labeled P1, scale bar 200 nm. Reproduced with permission from Mahmoud *et al.*[105] copyright Elsevier (2011). (C) TEM images of silica tubes left behind after digestion of the peptide fibers. Scale bar 200 nm. Reproduced with permission from Holmstrom *et al.*[107] copyright American Chemical Society (2008). (For color version of this figure, the reader is referred to the web version of this chapter.)

structures and applications. As we move along, we will encounter increasingly the challenges of creating systems in which multiple self-assembling components must interact. While several of the works discussed here have attempted to address the issues surrounding the orthogonality of tectons for mixed self-assembling systems, to date our success in producing applications using this information has been limited to small proofs of concept. As we move beyond this, we can expect to encounter greater difficulty in maintaining fidelity of interactions—which should in itself provide another rich source of basic research into the way in which natural systems behave.

The level and sophistication of functionalization of discrete, self-assembling structures is also rapidly increasing. In particular, the creation of responsive, structurally switching systems will be of great future interest. The study of natural systems has been moving away from the static and towards the dynamic for some time, and it is encouraging to see that the protein design field is following this trend. With the increase in functional building blocks of this nature, it will not be long before we can realize the ambition of designed protein-based molecular motion.

The story from the perspective of infinite or fiber assemblies is similar: that assemblies can now be designed, modified, and functionalized in a variety of ways. In this case, the future issue is likely to be the *control* of assembly. Although some control of width and length in fiber systems has emerged, it has mostly been of an observational nature, and developing more complex fiber systems that display the level of width and length control exhibited by nature will indeed be a challenge.

Spatial control of fibrillogenesis will also be a goal for the future, both in terms of controlling the location and direction of nucleation and polymerization, and in the control of labeling such that multiple ordered functionalities can be presented on fibers.

For many of the challenges listed above, it will be necessary to go back to the natural systems to extract more information and to perform more basic science and proof-of-concept experiments. However, the vast increase in the possible applications in both medicine and nanoscience that such research could bring should make the effort well worthwhile.

References

1. Szybalski W. In vivo and in vitro initiation of transcription. In: Kohn A, Shatkay A, editors. *Control of gene expression.* New York: Plenum Press; 1974.
2. Bromley EHC, Channon K, Moutevelis E, Woolfson DN. Peptide and protein building blocks for synthetic biology: from programming biomolecules to self-organized biomolecular systems. *ACS Chem Biol* 2008;3:38–50.
3. Channon K, Bromley EHC, Woolfson DN. Synthetic biology through biomolecular design and engineering. *Curr Opin Struct Biol* 2008;18(4):491–8.
4. Vickers CE, Blank LM, Kroemer JO. Chassis cells for industrial biochemical production. *Nat Chem Biol* 2010;6(12):875–7.
5. Gibson DG, Glass JI, Lartigue C, Noskov VN, Chuang RY, Algire MA, et al. Creation of a bacterial cell controlled by a chemically synthesized genome. *Science* 2010;329:52–6.
6. Keasling JD. Manufacturing molecules through metabolic engineering. *Science* 2010;330 (6009):1355–8.
7. Ro DK, Paradise EM, Ouellet M, Fisher KJ, Newman KL, Ndungu JM, et al. Production of the antimalarial drug precursor artemisinic acid in engineered yeast. *Nature* 2006;440:940–3.
8. Endy D. Foundations for engineering biology. *Nature* 2005;438(7067):449–53.
9. Sole RV, Munteanu A, Rodriguez-Caso C, Marcia J. Synthetic protocell biology: from reproduction to computation. *Philos Trans R Soc B* 2007;362(1486):1727–39.
10. Luisi PL. *The emergence of life: from chemical origins to synthetic biology.* Cambridge: Cambridge University Press; 2006.

11. Ramussen S, Bedau MA, Chen L, Deamer D, Krakauer DC, Packard NH, et al. *Protocells: bridging the nonliving and living matter.* Cambridge and London: The MIT Press; 2009.

12. Andersen ES, Dong M, Nielsen MM, Jahn K, Subramani R, Mamdouh W, et al. Self-assembly of a nanoscale DNA box with a controllable lid. *Nature* 2009;**459**:U73–5.

13. Gu HZ, Chao J, Xiao SJ, Seeman NC. A proximity-based programmable DNA nanoscale assembly line. *Nature* 2010;**465**:202–6.

14. Jaeger L, Chworos A. The architectonics of programmable RNA and DNA nanostructures. *Curr Opin Struct Biol* 2006;**16**:531–43.

15. Lund K, Manzo AJ, Dabby N, Michelotti N, Johnson-Buck A, Nangreave J, et al. Molecular robots guided by prescriptive landscapes. *Nature* 2010;**465**:206–10.

16. Rothemund PWK. Folding DNA to create nanoscale shapes and patterns. *Nature* 2006;**440**:297–302.

17. Seeman NC. Nanomaterials based on DNA. *Annu Rev Biochem* 2010;**79**:65–87.

18. Crick FHC. The packing of-helices: simple coiled-coils. *Acta Crystallogr* 1953;**6**:689.

19. Testa OD, Moutevelis E, Woolfson DN. CC plus: a relational database of coiled-coil structures. *Nucleic Acids Res* 2009;**37**(SI):D315–22.

20. Moutevelis E, Woolfson DN. A periodic table of coiled-coil protein structures. *J Mol Biol* 2009;**385**(3):726–32.

21. Rackham OJL, Madera M, Armstrong CT, Vincent TL, Woolfson DN, Gough J. The evolution and structure prediction of coiled coils across all genomes. *J Mol Biol* 2010;**403** (3):480–93.

22. Hurst HC. Transcription factors 1: bZIP proteins. *Protein Profile* 1995;**2**(2):101–68.

23. Rose A, Meier I. Scaffolds, levers, rods and springs: diverse cellular functions of long coiled-coil proteins. *Cell Mol Life Sci* 2004;**61**:1996–2009.

24. Lin RC, Scheller RH. Mechanisms of synaptic vesicle exocytosis. *Annu Rev Cell Dev Biol* 2000;**16**:19–49.

25. Shu W, Liu J, Ji H, Lu M. Core structure of the outer membrane lipoprotein from Escherichia coli at 1.9 A resolution. *J Mol Biol* 2000;**299**(4):1101–12.

26. Herrmann H, Bar H, Kreplak L, Strelkov SV, Aebi U. Intermediate filaments: from cell architecture to nanomechanics. *Nat Rev Mol Cell Biol* 2007;**8**(7):562–73.

27. Margolin W. Bacterial shape: concave coiled coils curve caulobacter. *Curr Biol* 2004;**14**(6): R242–4.

28. Mizuno N, Narita A, Kon T, Sutoh K, Kikkawa M. Three-dimensional structure of cytoplasmic dynein bound to microtubules. *Proc Natl Acad Sci USA* 2007;**104**:20832–7.

29. Vale RD, Milligan A. The way things move: looking under the hood of molecular motor proteins. *Science* 2000;**288**(5463):88–95.

30. Walshaw J, Woolfson DN. SOCKET: a program for identifying and analysing coiled-coil motifs within protein structures. *J Mol Biol* 2001;**307**:1427.

31. Woolfson DN. The design of coiled-coil structures and assemblies. *Adv Prot Chem* 2005;**70**:79–112.

32. Harbury PB, Zhang T, Kim PS, Alber T. A switch between two-, three-, and four-stranded coiled coils in GCN4 leucine zipper mutants. *Science* 1993;**262**:1401.

33. Meier M, Stetefeld J, Burkhard P. The many types of interhelical ionic interactions in coiled coils—an overview. *J Struct Biol* 2010;**170**(2):192–201.

34. Armstrong CT, Boyle AL, Bromley EHC, Mahmoud ZN, Smith L, Thomson AR, et al. Rational design of peptide-based building blocks for nanoscience and synthetic biology. *Faraday Discuss* 2009;**143**:305–17.

35. Steinmetz MO, Stock A, Schulthless T, Landwehr R, Lustig A, Faix J, et al. A distinct 14 residue site triggers coiled-coil formation in cortexillin I. *EMBO J* 1998;**17**(7):1883–91.

36. Kammerer RA, Kostrewa D, Progias P, Hannappa S, Avila D, Lustig A, et al. A conserved trimerization motif controls the topology of short coiled coils. *Proc Natl Acad Sci USA* 2005;**102**(39):13891–6.

37. Ciani B, Bjelic S, Honnappa S, Jawhari H, Jaussi R, Payapily A, et al. Molecular basis of coiled-coil oligomerization-state specificity. *Proc Natl Acad Sci USA* 2010;**107** (46):19850–5.

38. Apgar JR, Gutwin KN, Keating AE. Predicting helix orientation for coiled-coil dimers. *Proteins Struct Funct Bioinformatics* 2008;**72**(3):1048–65.

39. Hadley EB, Testa OD, Woolfson DN, Gellman SH. Preferred side-chain constellations at antiparallel coiled-coil interfaces. *Proc Natl Acad Sci USA* 2008;**105**:530–5.

40. Oakley MG, Hollenbeck JJ. The design of antiparallel coiled coils. *Curr Opin Struct Biol* 2001;**11**:450–7.

41. Newman JRS, Keating AE. Comprehensive identification of human bZIP interactions with coiled-coil arrays. *Science* 2003;**300**(5628):2097–101.

42. Fong JH, Keating AE, Singh M. Predicting specificity in bZIP coiled-coil protein interactions. *Genome Biol* 2004;**5**(2):R11.

43. Grigoryan G, Keating AE. Structure-based prediction of bZIP partnering specificity. *J Mol Biol* 2006;**355**(5):1125–42.

44. Acharya A, Ruvinov SB, Gal J, Noll JR, Vinson C. A heterodimerizing leucine zipper coiled coil system for examining the specificity of a position interactions: amino acids I, V, L, N, A, and K. *Biochemistry* 2002;**41**(48):14122–31.

45. Acharya A, Rishi V, Vinson C. Stability of 100 homo and heterotypic coiled-coil a–a′ pairs for ten amino acids (A, L, I, V, N, K, S, T, E, and R). *Biochemistry* 2006;**46**(38):11324–32.

46. Krylov D, Mikhailenko I, Vinson C. A thermodynamic scale for leucine-zipper stability and dimerization specificity: -e and g-interhelical interactions. *EMBO J* 1994;**13**(12):2849–61.

47. Krylov D, Barchi J, Vinson C. Inter-helical interactions in the leucine zipper coiled coil dimer: ph and salt dependence of coupling energy between charged amino acids. *J Mol Biol* 1998;**279** (4):959–72.

48. Mason JM, Schmitz MA, Muller K, Arndt KM. Semirational design of Jun-Fos coiled coils with increased affinity: universal implications for leucine zipper prediction and design. *Proc Natl Acad Sci USA* 2006;**103**(24):8989–94.

49. Grigoryan G, Reinke AW, Keating AE. Design of protein-interaction specificity affords selective bZIP-binding peptides. *Nature* 2009;**458**(7240):895.

50. Bromley EHC, Sessions RB, Thomson AR, Woolfson DN. Designed α-helical tectons for constructing multicomponent synthetic biological systems. *J Am Chem Soc* 2009;**131**:928–30.

51. Reinke AW, Grant RA, Keating AE. A synthetic coiled-coil interactome provides heterospecific modules for molecular engineering. *J Am Chem Soc* 2010;**132**(17):6025–31.

52. Diss ML, Kennan AJ. Orthogonal recognition in dimeric coiled coils via buried polar-group modulation. *J Am Chem Soc* 2008;**130**(3):1321–7.

53. Diss ML, Kennan AJ. Simultaneous directed assembly of three distinct heterodimeric coiled coils. *Org Lett* 2008;**10**(17):3797–800.

54. Ryan SJ, Kennan AJ. Variable stability heterodimeric coiled-coils from manipulation of electrostatic interface residue chain length. *J Am Chem Soc* 2007;**129**(33):10255–60.

55. Gribbon C, Channon KJ, Zhang WJ, Banwell EF, Bromley EHC, Chaudhuri JB, et al. MagicWand: a single, designed peptide that assembles to stable, ordered α-helical fibers. *Biochemistry* 2008;**47**(39):10365–71.

56. Diss ML, Kennan AJ. Heterotrimeric coiled coils with core residue urea side chains. *J Org Chem* 2008;**73**(24):9752–5.

57. Akcay G, Kumar K. A new paradigm for protein design and biological self-assembly. *J Fluor Chem* 2009;**130**(12):1178–82.

58. Bilgicer B, Fichera A, Kumar K. A coiled coil with a fluorous core. *J Am Chem Soc* 2001;**123** (19):4394–9.
59. Bilgicer B, Xing X, Kumar K. Programmed self-sorting of coiled coils with leucine and hexafluoroleucine cores. *J Am Chem Soc* 2001;**123**(47):11815–6.
60. Vagt T, Nyakatura E, Salwiczek M, Jackel C, Koksch B. Towards identifying preferred interaction partners of fluorinated amino acids within the hydrophobic environment of a dimeric coiled coil peptide. *Org Biomol Chem* 2010;**8**(6):1382–6.
61. Salwiczek M, Samsonov S, Vagt T, Nyakatura E, Fleige E, Numata J, et al. Position-dependent effects of fluorinated amino acids on the hydrophobic core formation of a heterodimeric coiled coil. *Chem A Eur J* 2009;**15**(31):7628–36.
62. Ryadnov MG, Ceyhan B, Niemeyer CM, Woolfson DN. "Belt and braces": a peptide-based linker system of de novo design. *J Am Chem Soc* 2003;**125**:9388.
63. Robson Marsden H, Elbers NA, Bomans PHH, Sommerdijk N, Kros A. A reduced SNARE model for membrane fusion. *Angew Chem Int Ed* 2009;**48**(13):2330–3.
64. Lee DH, Granja JR, Martinez JA, Severin K, Ghadiri MR. A self-replicating peptide. *Nature* 1996;**382**(6591):525–8.
65. Severin K, Lee DH, Martinez JA, Ghadiri MR. Peptide self-replication via template-directed ligation. *Chem A Eur J* 1997;**3**(7):1017–24.
66. Severin K, Lee DH, Kennan AJ, Ghadiri MR. A synthetic peptide ligase. *Nature* 1997;**389** (6652):706–9.
67. Li X, Chmielewski J. Peptide self-replication enhanced by a proline kink. *J Am Chem Soc* 2003;**125**:11820–1.
68. Ciani B, Hutchinson EG, Sessions RB, Woolfson DN. A designed system for assessing how sequence affects α to β conformational transitions in proteins. *J Biol Chem* 2002;**277**:10150–5.
69. Kammerer RA, Kostrewa D, Zurdo J, Detken A, Garcia Echeverria C, Green JD, et al. Exploring amyloid formation by a de novo design. *Proc Natl Acad Sci USA* 2004;**101**:4435–40.
70. Schnarr NA, Kennan AJ. pH-triggered strand exchange in coiled-coil heterotrimers. *J Am Chem Soc* 2003;**125**(21):6364–5.
71. Schnarr NA, Kennan AJ. pH-switchable strand orientation in peptide assemblies. *Org Lett* 2005;**7**(3):395–8.
72. Pandya MJ, Cerasoli E, Joseph A, Stoneman RG, Waite E, Woolfson DN. Sequence and structural duality: designing peptides to adopt two stable conformations. *J Am Chem Soc* 2004;**126**(51):17016–24.
73. Suzuki K, Hiroaki H, Kohda D, Nakamura H, Tanaka T. Metal ion induced self-assembly of a designed peptide into a triple-stranded alpha-helical bundle: a novel metal binding site in the hydrophobic core. *J Am Chem Soc* 1998;**120**(50):13008–15.
74. Tsurkan MV, Ogawa MY. Metal-mediated peptide assembly: use of metal coordination to change the oligomerization state of an alpha-helical coiled-coil. *Inorg Chem* 2007;**46**:6849.
75. Cerasoli E, Sharpe BK, Woolfson DN. ZiCo: a peptide designed to switch folded state upon binding zinc. *J Am Chem Soc* 2005;**127**(43):15008–9.
76. Ambroggio XI, Kuhlman B. Computational design of a single amino acid sequence that can switch between two distinct protein folds. *J Am Chem Soc* 2006;**128**(4):1154–61.
77. Kuhlman B, Baker D. Native protein sequences are close to optimal for their structures. *Proc Natl Acad Sci USA* 2000;**97**(19):10383–8.
78. Pagel K, Vagt T, Kohadja T, Koksch B. From α-helix to β-sheet—a reversible metal ion induced peptide secondary structure switch. *Org Biomol Chem* 2005;**3**(14):2500–2.
79. Zhang F, Timm KA, Arndt KM, Woolley GA. Photocontrol of coiled-coil proteins in living cells. *Angew Chem Int Ed* 2010;**49**(23):3943–6.
80. Parry DA, Strelkov SV, Burkhard P, Aebi U, Herrmann H. Towards a molecular description of intermediate filament structure and assembly. *Exp Cell Res* 2007;**313**(10):2204–16.

81. Sokolova AV, Kreplak L, Wedig T, Mucke N, Svergun DI, Herrmann H, et al. Monitoring intermediate filament assembly by small-angle x-ray scattering reveals the molecular architecture of assembly intermediates. *Proc Natl Acad Sci USA* 2006;**103**(44):16206–11.

82. Kojima S, Kuriki Y, Yoshida T, Yazaki K, Miura K. Fibril formation by an amphipathic alpha-helix-forming polypeptide produced by gene engineering. *Proc Jpn Acad Ser B Phys Biol Sci* 1997;**73**:7–11.

83. Potekhin SA, Melnik TN, Popov V, Anina NF, Vazina AA, Rigler P, et al. De novo design of fibrils made of short alpha-helical coiled coil peptides. *Chem Biol* 2001;**11**(8):1025–32.

84. Dong H, Paramonov SE, Hartgerink JD. Self-assembly of α-helical coiled coil nanofibers. *J Am Chem Soc* 2008;**130**(41):13691–5.

85. Zimenkov Y, Dublin SN, Ni R, Tu RS, Breedveld V, Apkarian RP, et al. Rational design of a reversible pH-responsive switch for peptide self-assembly. *J Am Chem Soc* 2006;**128**:6770.

86. Zimenkov Y, Conticello VP, Guo L, Thiyagarajan P. Rational design of a nanoscale helical scaffold derived from self-assembly of a dimeric coiled coil motif. *Tetrahedron* 2004;**60**:7237.

87. Wagner DE, Phillips CL, Ali WM, Nybakken GE, Crawford ED, Schwab AD, et al. Toward the development of peptide nanofilaments and nanoropes as smart materials. *Proc Natl Acad Sci USA* 2005;**102**(36):12656–61.

88. Pandya MJ, Spooner GM, Sunde M, Thorpe JR, Rodger A, Woolfson DN. Sticky-end assembly of a designed peptide fiber provides insight into protein fibrillogenesis. *Biochemistry* 2000;**39**:8728–34.

89. Smith AM, Banwell EF, Edwards WR, Pandya MJ, Woolfson DN. Engineering increased stability into self-assembled protein fibers. *Adv Funct Mater* 2006;**16**:1022–30.

90. Papapostolou D, Smith AM, Atkins EDT, Oliver SJ, Ryadnov MG, Serpell LC, et al. Engineering nanoscale order into a designed protein fiber. *Proc Natl Acad Sci USA* 2007;**104**:10853–8.

91. Bromley EHC, Channon KJ, King PJS, Mahmoud ZN, Banwell EF, Butler MF, et al. Flow linear dichroism of some prototypical proteins. *Biophys J* 2010;**98**:1668–76.

92. Banwell EF, Aberlardo ES, Adams DJ, Birchall MA, Corrigan A, Donald AM, et al. Rational design and application of responsive alpha-helical peptide hydrogels. *Nat Mater* 2009;**8**:596–600.

93. Lazar KL, Miller-Auer H, Getz GS, Orgel J, Meredisth SC. Helix-turn-helix peptides that form α-helical fibrils: turn sequences drive fibril structure. *Biochemistry* 2005;**44**:12681.

94. Ryadnov MG. A self-assembling peptide polynanoreactor. *Angew Chem Int Ed* 2007;**46**(6):969–72.

95. Raman S, Machaidze G, Lustig A, Aebi U, Burkhard P. Structure-based design of peptides that self-assemble into regular polyhedral nanoparticles. *Nanomed Nanotechnol Biol Med* 2006;**2**:95.

96. Pimentel TAPF, Yan Z, Jeffers SA, Holmes KV, Hodges RS, Burkhard P. Peptide nanoparticles as novel immunogens: design and analysis of a prototypic severe acute respiratory syndrome vaccine. *Chem Biol Drug Des* 2009;**73**(1):53–61.

97. Kaba SA, Brando C, Guo Q, Mittelhozer C, Raman S, Tropel D, et al. A nonadjuvanted polypeptide nanoparticle vaccine confers long-lasting protection against rodent malaria. *J Immunol* 2009;**183**(11):7268–77.

98. Woolfson DN, Mahmoud ZN. More than just bare scaffolds: towards multi-component and decorated fibrous biomaterials. *Chem Soc Rev* 2010;**39**:3464–79.

99. Channon KJ, MacPhee CE. Possibilities for smart materials exploiting the self-assembly of polypeptides into fibrils. *Soft Matter* 2008;**4**:647–52.

100. Melnik TN, Villard V, Vasiliev V, Corradin G, Kajava AV, Potehkin SA. Shift of fibril-forming ability of the designed α-helical coiled-coil peptides into the physiological pH region. *Protein Eng* 2003;**16**:1125.
101. Villard V, Kalyuzhniy O, Riccio O, Potehkin SA, Melnik TN, Kajava AV, et al. Synthetic RGD-containing alpha-helical coiled coil peptides promote integrin-dependent cell adhesion. *J Pept Sci* 2006;**12**:206.
102. Smith AM, Acquah SFA, Bone N, Kroto HW, Ryadnov MG, Stevens MSP, et al. Polar assembly in a designed protein fiber. *Angew Chem Int Ed* 2005;**44**:325–8.
103. Ryadnov MG, Woolfson DN. Introducing branches into a self-assembling peptide fiber. *Angew Chem Int Ed* 2003;**42**:3021–3.
104. Ryadnov MG, Woolfson DN. Engineering the morphology of a self-assembling protein fibre. *Nat Mater* 2003;**2**:329–32.
105. Mahmoud ZN, Gunnoo SB, Thomson AR, Fletcher JM, Woolfson DN. Bioorthogonal dual functionalization of self-assembling peptide fibers. *Biomaterials* 2011;**32**(15):3712–20.
106. Mahmoud ZM, Woolfson DN. The non-covalent decoration of self-assembling protein fibers. *Biomaterials* 2010;**31**:7468–74.
107. Holmstrom SC, King PJS, Ryadnov MG, Butler MF, Mann S, Woolfson DN. Templating silica nanostructures on rationally designed self-assembled peptide fibers. *Langmuir* 2008;**24**:11778–83.
108. Dublin SN, Conticello VP. Design of a selective metal ion switch for self-assembly of peptide-based fibrils. *J Am Chem Soc* 2008;**130**:49–51.

Nanobiotechnology with S-Layer Proteins as Building Blocks

Uwe B. Sleytr, Bernhard
Schuster, Eva M. Egelseer,
Dietmar Pum, Christine M.
Horejs, Rupert Tscheliessnig,
and Nicola Ilk

*Department of NanoBiotechnology,
University of Natural Resources and
Life Sciences, Vienna, Austria*

One of the key challenges in nanobiotechnology is the utilization of self-assembly systems, wherein molecules spontaneously associate into reproducible aggregates and supramolecular structures. In this contribution, we describe the basic principles of crystalline bacterial surface layers (S-layers) and their use as patterning elements. The broad application potential of S-layers in nanobiotechnology is based on the specific intrinsic features of the monomolecular arrays composed of identical protein or glycoprotein subunits. Most important, physicochemical properties and functional groups on the protein lattice are arranged in well-defined positions and orientations. Many applications of S-layers depend on the capability of isolated subunits to recrystallize into monomolecular arrays in suspension or on suitable surfaces (e.g., polymers, metals, silicon wafers) or interfaces (e.g., lipid films, liposomes, emulsomes).

Progress in Molecular Biology
and Translational Science, Vol. 103
DOI: 10.1016/B978-0-12-415906-8.00003-0

277

S-layers also represent a unique structural basis and patterning element for generating more complex supramolecular structures involving all major classes of biological molecules (e.g., proteins, lipids, glycans, nucleic acids, or combinations of these). Thus, S-layers fulfill key requirements as building blocks for the production of new supramolecular materials and nanoscale devices as required in molecular nanotechnology, nanobiotechnology, biomimetics, and synthetic biology.

I. Introduction

An important area in nanobiotechnology concerns "bottom-up" routes to fabricate supramolecular structures at subnanometer precision.[1] Presently, the most complex functional nanoscale structures are efficiently built from biomolecules that self-assemble into nanoscale three-dimensional (3D) shapes or two-dimensional (2D) crystals on surfaces and interfaces. So far, a variety of biomolecules optimized through evolution, including proteins, nucleic acids, and lipids, are used as molecular building blocks in nanobiotechnology. Owing to their unique morphogenetic potential, crystalline bacterial cell surface layer (S-layer) proteins or glycoproteins represent prime candidates as patterning elements for nanobiotechnological applications.[2]

S-layers represent an almost universal feature in archaeal cell envelopes and have been identified in hundreds of different species of bacteria. Moreover, since the biomass of prokaryotic organisms by far prevail eukaryotic biomass and S-layers represent 10% of cell proteins, the crystalline protein arrays can be considered as one of the most abundant biopolymers on earth.

As outermost envelope layer in prokaryotic organisms, S-layers also represent an evolutionary adaption of the organisms to a broad spectrum of selection criteria. Most of the presently known S-layers are composed of single protein or glycoprotein species endowed with the intrinsic ability to assemble into monomolecular arrays. Thus, S-layers can be regarded as the simplest protein membrane developed during evolution. Since archaea that dwell under most extreme environmental conditions (e.g., up to 120 °C, pH 0, concentrated salt solutions, high hydrostatic pressure) possess S-layers, these protein meshworks can also exhibit remarkable chemical stability.

Studies on the structure, chemistry, genetics, morphogenesis, and function have led to a broad spectrum of applications (for reviews, see Refs. 2–20). Up to now, most applications developed for using S-layer lattices depend on the *in vitro* self-assembly capabilities of native or recombinant S-layer proteins on the surface of solids and interfaces (e.g., silicon wafers, polymers and metals, lipid films, liposomes, and nanoparticles). Most importantly, because S-layers are periodic structures composed of a single (glyco)protein species and possess pores of identical size and morphology, they exhibit identical physicochemical properties on each molecular unit, down to the subnanometer scale. This unique feature enables very precise chemical and/or genetic modifications. Particularly, the possibility of modifying and changing the natural properties of S-layer proteins and glycoproteins by genetic engineering techniques and to incorporate specific functional domains has led to a broad spectrum of applications including ultrafiltration membranes, affinity structures, enzyme membranes, microcarriers, biosensors, diagnostic devices, biocompatible surfaces, as a matrix for biomineralization, and vaccines, as well as targeting, delivery, and encapsulating systems. It is now evident that S-layers can be used as structural basis for a versatile biomolecular construction kit involving all major species of biological molecules (proteins, lipids, glycans, nucleic acids, and combinations of these).

In this contribution, we provide a survey of the general principles of S-layers and their broad application potential in nanobiotechnology.

II. General Principles

A. Occurrence, Location, and Structure of S-Layers

Almost 60 years ago, Houwink and Le Poole described the presence of a "macromolecular monolayer" in the cell wall of *Spirillum* sp.,[21] but it took 15 more years until regular arrays of macromolecules were identified as common cell surface layers on living *Bacillus* and *Clostridium* cells.[22,23]

Crystalline cell surface layers now referred to as S-layers[24,25] have so far been identified in hundreds of different strains belonging to all major phylogenetic groups of the domain bacteria and represent an almost universal feature in archaea.[20,26–29] The location and ultrastructure of S-layers were investigated by different electron microscopy procedures, including thin sectioning, freeze-etching, and freeze-drying in combination with heavy metal shadowing and negative staining.[30–38]

Nevertheless, the most suitable procedure for identifying the presence of S-layer lattices on intact cells is still electron microscopy of freeze-etched preparations.[24,34,37,39] High-resolution studies on the mass distribution of the lattice are generally performed on negatively stained S-layer fragments or thin frozen foils.[31,40,41] High-resolution images of S-layer lattices can also be obtained by scanning force microscopy under aqueous conditions.[34,42,43]

Freeze-etching preparations (Fig. 1) of a great variety of bacteria and archaea have clearly demonstrated that S-layer lattices completely cover the cell surfaces at all stages of cell growth and cell division. Although considerable variation exists in the supramolecular structure and chemistry of cell envelopes

S-layer lattice types

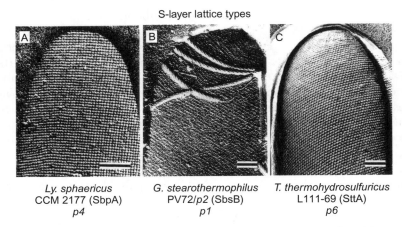

Ly. sphaericus	G. stearothermophilus	T. thermohydrosulfuricus
CCM 2177 (SbpA)	PV72/p2 (SbsB)	L111-69 (SttA)
p4	p1	p6

FIG. 1. Electron micrographs of freeze-etched preparations of intact cells from (A) *Lysiniba-cillus sphaericus* CCM 2177 showing square lattice symmetry (bar, 150 nm), (B) *Geobacillus stearothermophilus* PV72/p2 exhibiting an oblique lattice type (bar, 50 nm), and (C) *Thermoanaer-obacter thermohydrosulfuricus* L111-69 showing a hexagonal S-layer lattice (bar, 100 nm).

of prokaryotic organisms, S-layers must have coevolved with these diverse structures.[34] In most archaea, the S-layer (glyco)protein lattices are, as the only wall component, closely associated or integrated into the plasma membrane (Fig. 2). In Gram-positive bacteria and in certain archaea, S-layers assemble on the surface of the rigid wall matrix, which is composed mainly of peptidoglycan and covalently attached secondary cell wall polymers (SCWPs) or pseudomurein, respectively. In Gram-negative bacteria, the lattice is attached to the lipopolysaccharide component of the outer membrane.

Most remarkably, some Gram-positive and Gram-negative bacteria assemble two or even more superimposed S-layers which generally are composed of a different subunit species.[24,27] S-layer subunits can be aligned with oblique (*p1*, *p2*), square (*p4*), or hexagonal (*p3, p6*) symmetry (Fig. 3), with center-to-center spacings of the morphological units (composed of one, two, three, four, or six identical monomers) of approximately 3.5–35 nm. Hexagonal symmetry is predominant among archaea.[20] Most S-layer lattices are 5–25 nm thick and reveal a rather smooth outer and a more corrugated inner surface. As S-layers are monomolecular assemblies of identical subunits, they exhibit pores of identical size and morphology. In many S-layers, two or even more distinct classes of pores could be observed. Pore sizes were determined to be in the range of approximately 2–8 nm, and pores can occupy approximately 30–70% of the surface area.[31,38,40] Most S-layers of archaea exhibit pillar-like structures on the inner surface, which are involved in anchoring the arrays in the

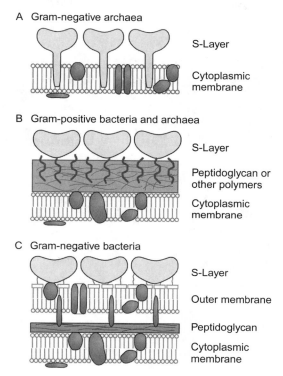

FIG. 2. Schematic illustration of the major classes of prokaryotic cell envelopes with crystalline bacterial cell surface layers (S-layers). (A) Cell envelope structure of Gram-negative archaea with S-layers as the only cell wall component external to the cytoplasmic membrane. (B) Cell envelope as observed in Gram-positive archaea and bacteria. (C) Cell envelope profile of Gram-negative bacteria composed of a thin peptidoglycan layer and an outer membrane. If present, the S-layer is closely associated with the lipopolysaccharide of the outer membrane. (For color version of this figure, the reader is referred to the web version of this chapter.)

underlying plasma membrane.[14,34,44,45] In many species of bacteria, the S-layers of individual strains exhibit great diversity with respect to lattice symmetry and center-to-center spacing of the morphological units.

B. Isolation, Chemical Characterization, Molecular Biology, and Function

S-layers constitute the outermost cell wall component facing the environment and are part of quite different supramolecular prokaryotic envelope structures. The latter can be classified into three main groups on the basis of the biological origin (bacteria or archaea) and their response to the so-called

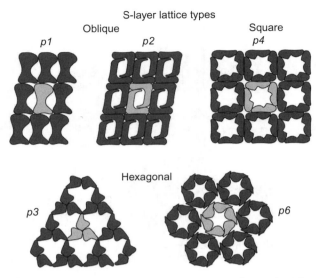

FIG. 3. Schematic drawing of different S-layer lattice types. The regular S-layer arrays show either oblique (*p1*, *p2*), square (*p4*), or hexagonal lattice symmetry (*p3*, *p6*). The morphological units are composed of one, two, three, four, or six identical subunits. Modified after Ref. 17; Copyright (1999) with permission from Wiley–VCH.(For color version of this figure, the reader is referred to the web version of this chapter.)

Gram-staining.[46,47] This bacteriological laboratory technique is frequently used to differentiate bacterial species into two large groups based on the physical properties of their cell walls. Gram-positive bacteria have a thick mesh-like cell wall made of peptidoglycan or pseudomurein (50–90% of cell wall), which stains purple, and the S-layer proteins assemble on the surface of this rigid wall matrix. In contrast, Gram-negative bacteria have a thinner polymer layer (10% of cell wall), which stains pink.[48] Gram-negative bacteria also have an additional outer membrane comprising lipids and lipopolysaccharides, and is separated from the cell wall by the periplasmic space.[44] In these more complex cell envelopes, the S-layers are linked to specific lipopolysaccharide fractions.[17,19,34,49–51] Gram-staining is not used to classify archaea, as these microorganisms yield widely varying responses that do not follow their phylogenetic groups.[48] In many archaea, S-layers represent the only cell wall component and can be closely associated with the plasma membrane so that it is partially integrated in the lipid layer.[44,52]

Owing to the diversity in the supramolecular structures of prokaryotic cell envelopes, S-layer proteins differ considerably in their susceptibility to disruption into constituent subunits. S-layers are usually not covalently attached to the cell surface[50]; hence, these proteins can be isolated in the presence of dissociating agents such as lithium chloride[53] or metal-chelating agents such as ethylenediaminetetraacetic acid (EDTA) and ethyleneglycoltetraacetic acid (EGTA).[54] Further, chaotropic denaturants such as guanidine hydrochloride and urea[55,56] or detergents at a pH-value lower than 4 can be applied to isolate S-layer subunits. In certain cases, even washing cells with deionized water can lead to dissociation of the S-layer lattice.[57–59] Extraction and disintegration experiments revealed that the inter-subunit bonds in the S-layer are stronger than those binding the crystalline array to the supporting envelope layer.[39] However, in the halophilic archaea *Halobacterium halobium* and *Haloferax volcanii*, the S-layer protomers are anchored by C-terminally located membrane-spanning domains to the cytoplasmic membrane.[60,61]

Chemical and genetic analyses on many bacterial and archaeal S-layers have shown that they are generally composed of a single protein or glycoprotein species with molecular masses ranging from 40 to 170 kDa.[15,17,20,27,62,63] Amino acid analysis of S-layer proteins of organisms from all phylogenetic branches revealed a rather similar overall composition.[40,64,65] Sequencing of genes encoding the S-layer proteins and isoelectric focusing provide evidence that, with a few exceptions (e.g., *Lactobacillus* and *Methanothermus*), S-layers are composed of an acidic protein or glycoprotein species with an isoelectric point between 3 and 6. Accordingly, S-layer proteins have a high amount of glutamic and aspartic acids (approximately 15 mol%) and the lysine content of S-layer proteins is in the range of 10 mol%. Thus, approximately one-quarter of the amino acids is charged, indicating that ionic bonds play an important role in inter-subunit bonding and/or in attaching the S-layer subunits to the underlying cell envelope layer. S-layer proteins have no or only a low content of sulfur-containing amino acids and a high proportion of hydrophobic amino acids (between 40 and 60 mol%). Interestingly, hydrophilic and hydrophobic amino acids do not form extended clusters but, instead, the hydrophobic and hydrophilic segments alternate with a more hydrophilic region at the very N-terminal end.[66] Information regarding the secondary structure of S-layer proteins is either derived from the amino-acid sequence or from circular dichroism measurements. In most S-layer proteins, 40% of the amino acids are organized as β-sheet, 10–20% occur as α-helix, whereas aperiodic foldings and β-turn content may vary between 5% and 45%.

Few posttranslational modification including cleavage of amino- or carboxy-terminal fragments, phosphorylation, and glycosylation of amino-acid residues are known to occur in S-layer proteins. The latter is a remarkable characteristic of many archaeal and some bacterial S-layer proteins. The glycan chains and types of linkage differ significantly from those of eukaryotes.[27,63,67,68]

During the past two decades, numerous S-layer genes from bacteria and archaea of quite different taxonomical positions have been sequenced and cloned.[17,20,49,69,70] Recently, these informations were summarized including a complete coverage of GenBank accession numbers of S-layer structural genes and presently known data on surface-layer glycosylation (*slg*) gene clusters.[27] The S-layer gene sequence itself provides valuable information especially for pathogenic organisms such as *Bacillus anthracis*, *Clostridium difficile*, and *Campylobacter fetus*, for which specific identification and discrimination is vital for the accurate treatment of afflicted persons.[71–75]

The accumulation of S-layer gene sequences made it possible to screen for putative sequence identities and to elucidate the structure–function relationship of distinct segments of S-layer proteins by the production of N- and/or C-terminally truncated forms.[76–78] Moreover, the assembly-negative, water-soluble, N- or C-terminally truncated forms of the S-layer protein SbsC of *Geobacillus stearothermophilus* ATCC 12980 turned out to be well suited for 3D crystallization studies.[79,80] Based on the C-terminally truncated form rSbsC$_{31-844}$, the first high-resolution structure of the bacterial S-layer protein SbsC could be obtained[80] (see also the chapter: "The structure of bacterial S-layer proteins").

In the case of the S-layer protein SbsB of *G. stearothermophilus* PV72/p2, the tertiary structure was predicted by using molecular dynamic simulations based on the amino-acid sequence using the mean force (MF) method.[81] This approach has led to a thermodynamically favorable atomic model of the tertiary structure of the S-layer protein, which could be verified by both the MF method and the lattice model[81] (see also Section III.B.4).

Kinns and coworkers pursued the approach of using epitope insertion mutagenesis to identify surface-located mutations of the S-layer protein SbsB that block assembly without affecting the protein's overall tertiary structure.[82] SbsB was chosen because previous studies using cysteine scanning mutagenesis and targeted chemical modification had identified 23 amino-acid positions that are surface accessible in the monomer.[83,84] The insertion mutagenesis screen yielded several assembly-compromised mutants that represent an important step toward structure elucidation of an S-layer protein by NMR or X-ray crystallography.[82] Cysteine-scanning mutagenesis and targeted chemical modification were also applied for SlpA, the S-layer protein of the potentially probiotic bacterium *Lactobacillus brevis* ATCC 8287, in order to distinguish amino-acid residues located in the outer and inner surfaces of the lattice, protein interior, and interface/pore regions.[85]

Common structural organization principles have been identified at least for S-layer proteins of Gram-positive bacteria. A cell wall targeting domain was found either at the N-terminal or C-terminal region of these S-layer proteins. The existence of an N-terminal cell wall targeting domain was sustained by the identification of the so-called S-layer homology (SLH) motifs, consisting of 50–60 amino acids each, which are mostly found in triplicate at the N-terminus of S-layer proteins.[86] If present, SLH motifs are involved in cell wall anchoring of S-layer proteins by recognizing a distinct type of SCWP, which carry pyruvic acid residues and belong to group I SCWPs.[76,87–95] Results obtained by May and coworkers indicated that a highly conserved motif termed TRAE based on the sequence of the four amino-acids threonine, arginine, alanine, and glutamic acid is necessary in all three SLH motifs forming the functional SLH domain, and at least one (preferentially positively) charged amino acid in the TRAE motif is required for the activity.[96] For SLH-mediated binding, the construction of knock-out mutants in *B. anthracis* and *Thermus thermophilus* in which the gene encoding a putative pyruvyl transferase was deleted demonstrated that the addition of pyruvic acid residues to the peptidoglycan-associated cell wall polymer was a necessary modification.[87,92] The need for pyruvylation was also confirmed by surface plasmon resonance (SPR) spectroscopy measurements using the S-layer protein rSbsB of *G. stearothermophilus* PV72/p2 and the corresponding SCWP[97] for interaction studies.[91] The interaction proved to be highly specific for the carbohydrate component, and the exclusive and complete responsibility of a functional domain formed by the three SLH motifs for SCWP recognition was clearly confirmed.[91] In addition, by using optical spectroscopic methods and electron microscopy, rSbsB could be characterized by its two functionally and structurally separated parts, namely, the SLH domain which is responsible for "cell wall targeting" by recognizing the SCWP, and the larger C-terminal part which corresponds to the self-assembly domain.[95] Interestingly, the C-terminal part of SbsB was highly sensitive against deletions, and the removal of even less than 15 amino acids led to water-soluble S-layer protein forms.[83,98] In contrast to SbsB, in SbpA, the S-layer protein of *Lysinibacillus sphaericus* CCM 2177, an additional 58-amino-acid long SLH-like motif located behind the third SLH-motif is required.[76] In the C-terminal part of this S-layer protein, up to 237 amino acids could be deleted without interfering with the formation of the square lattice structure.

A further main type of binding mechanism between S-layer proteins and SCWPs has been described for the *G. stearothermophilus* wild-type strains PV72/p6, NRS 2004/3a, and ATCC 12980[78,99–102] as well as for a temperature-derived variant of the latter.[103] The N-terminal part of these S-layer proteins contains a surplus of arginine and tyrosine, which typically occur in carbohydrate-binding proteins such as lectins.[104] In this binding mechanism, nonpyruvylated group II SCWPs interact with a highly conserved N-terminal region of the S-layer proteins without SLH motifs.[78,99,100,103] First affinity studies using different N- or C-terminally truncated forms of the S-layer protein SbsC from *G. stearothermophilus* ATCC 12980 indicated that the N-terminal part comprising amino acids 31–257 is exclusively responsible for cell wall binding.[78] This result was also corroborated by SPR studies using the S-layer protein SbsC and the corresponding nonpyruvylated SCWP of *G. stearothermophilus* ATCC 12980 as the model system.[105]

A cell wall targeting domain is not necessarily located in the N-terminal region of S-layer proteins. In the S-layer proteins SlpA of *Lactobacillus acidophilus* ATCC 4356 and CbsA of *Lactobacillus crispatus* JCM 5810, a cell wall binding domain has been identified in the C-terminal one-third of these S-layer proteins, and sequence alignment studies revealed a putative carbohydrate-binding repeat comprising approximately the last 130 C-terminal amino acids which were suggested to be involved in cell wall binding.[106,107] Although SLH motifs have not been found in *Lactobacillus* S-layer proteins, their attachment to the cell wall seems to involve SCWPs in several lactobacilli, too.[70] The S-layer proteins from *L. brevis* and *L. buchneri* are reported to bind to a neutral polysaccharide moiety of the cell wall,[70,108,109] but the location of the cell wall binding domain of these proteins is currently unknown.

In contrast to Gram-positive bacteria, no general S-layer anchoring motifs have been identified in Gram-negative organisms. Available evidence in each case so far implicates the involvement of the N-terminus of S-layer proteins. Concerning the S-layer of *Aeromonas salmonicida*, the majority of the trypsin-inaccessible residues were identified in the N-terminal 301 amino acids of the A-protein, suggesting cell surface anchoring via the N-terminus of the S-layer subunits.[110] Similarly, the *C. fetus* S-layer SapA is cell surface-anchored via its conserved N-terminal region comprising at least 189 amino acids, as N-terminal deletions disrupt SapA anchoring,[111] while C-terminal truncations do not.[112] The S-layer protein RsaA of the Gram-negative bacterium *Caulobacter crescentus* also follows this general pattern since mutations[113] and truncations[114] in the extreme N-terminus of RsaA lead to an S-layer shedding phenotype. By using reattachment assays, it became evident that the anchoring region of the *C. crescentus* S-layer protein lies within the first ∼225 amino acids and that RsaA anchoring requires a smooth lipopolysaccharide species found in the outer membrane.[115]

The observation of phenotypic S-layer variation was not surprising, as cell surface components can generally be considered as nonconservative structures that have to respond to changing environmental conditions in the course of evolution. S-layer variation leads to the synthesis of alternate S-layer proteins, either by the expression of different S-layer genes or by recombination of partial coding sequences, and has been described to occur in pathogens as well as in nonpathogens.[116–126] In pathogens, altered cell surface properties most probably protect the cells from the lytic activity of the immune system,[120] whereas in nonpathogens, the S-layer variation is frequently induced in response to altered environmental conditions, such as increased oxygen supply.[70,124,125,127] In *G. stearothermophilus* strain variants, expression of a completely new type of S-layer protein is accompanied by synthesis of a different type of SCWP, and S-layer variation may also lead to a change in the lattice type.[124,125] These results indicated that, in the course of variant formation, changes in S-layer protein and SCWP synthesis were strictly coordinated and that the type of SCWP must have been changed prior to or simultaneously with the expression of a different type of S-layer gene. Genetic studies revealed that variant formation was caused by recombinational events between a naturally occurring megaplasmid and the chromosome.[128] Regarding the development of S-layer-deficient strain variants, the importance of insertion sequence elements has been demonstrated for various organisms.[129,130]

In a nutshell, multiple mechanisms leading to S-layer protein variation, modification, or complete loss of the S-layer indicate the importance of diversification of the surface properties even of closely related organisms for their survival in a competitive habitat.

Synthesis of a coherent S-layer lattice on a cell surface requires a considerable biosynthetic effort. When bacteria are no longer subject to the natural environmental selection pressure, S-layers can be lost. This was often observed during continuous culture under optimal growth conditions, and, usually, the S-layer-deficient variant outgrows the S-layered parent.[34,49] So far, a general, all-encompassing natural function for S-layers has not been found and many of the functions assigned to S-layers still remain hypothetical and not based on firm experimental data (see Table I modified after Refs. 20,27).

However, for assignment of functions, S-layers must be considered as part of complex supramolecular structures of diverse prokaryotic cell envelopes rather than an isolated monomolecular (glyco)protein lattice. The fact that no structural models at atomic resolution of S-layer proteins are available until now makes detailed interpretations of functional aspects even more difficult (for reviews, see Refs. 2,20,27,34,49,51).

III. Assembly and Morphogenesis of S-Layers

A. Self-Assembly *In Vivo*

Owing to the fact that S-layers possess a high degree of structural regularity and are composed in most cases of a single protein or glycoprotein species, they represent ideal model systems for studying the morphogenesis of a supramolecular layer during cell growth.[33,39] It can be calculated that approximately 5×10^5 S-layer protein monomers are needed to cover an average-sized rod-shaped prokaryotic cell. Consequently, at a generation time of about 20 min, at least 500 copies of a single polypeptide species with a molecular

TABLE I
GENERAL AND SPECIFIC FUNCTIONS OF S-LAYERS (MODIFIED AFTER REFS. 20,28)

General function	Specific function
Determination and maintenance of cell shape	Determination of cell shape and cell division in archaea that possess S-layers as exclusive wall component
Isoporous molecular sieves	Molecular sieves in the ultrafiltration range
Adhesion zones for exoenzymes	High molecular weight amylase of *Geobacillus stearothermophilus* wild-type strains
	Pullulanase and glycosyl hydrolases of *Thermoanaerobacter thermosulfurigenes*
Protective coats	Prevention of predation by *Bdellovibrio bacteriovorus* in Gram-negative bacteria
	Phage resistance by S-layer variation
	Prevention or promotion of phagocytosis
	Adaption of *Bacillus pseudofirmus* to alkaline environment
Templates for fine grain mineralization	Induction of precipitation of gypsum and calcite in *Synechococcus* and shedding of mineralized S-layer
Pathogenicity and cell adhesion	Virulence factor in pathogenic organisms
	Important role in invasion and survival within the host
	Specific binding of host molecules
	Protective coat against complement killing
	Ability to associate with macrophages and to resist the effect of proteases
	Production of immunologically non-cross-reactive S-layers (S-layer variation)
Surface recognition and cell adhesion to substrates	Physicochemically and morphologically well-defined matrices
	Masking the net negative charge of peptidoglycan-containing layer in Bacillacea
Antifouling layer	Prevention of nonspecific adsorption of macromolecules
	Maintaining the permeability properties through the S-layer pores
Delineating a periplasmic space	Entrapping molecules between the cytoplasmic membrane and the S-layer lattice

mass of approximately 100,000 have to be synthesized, translocated to the cell surface, and incorporated into the preexisting S-layer lattice per second in order to maintain a closed protein meshwork.[34,131]

The rate of synthesis of S-layer proteins appears to be strictly controlled, as only small amounts are detectable in the growth medium at optimal growth conditions. Nevertheless, a few species shed considerable amounts of S-layers.[132,133]

It was also demonstrated in different Bacillaceae that a pool of S-layer subunits is accumulated in the peptidoglycan-containing layer in an amount at least sufficient for generating one coherent lattice on the cell surface.[134]

Determination of the half-lives of S-layer protein mRNAs from *C. crescentus* (10–15 min), *A. salmonicida* (22 min), and *L. acidophilus* (15 min) revealed exceptionally stable transcripts compared to typical half-lives of prokaryotic mRNAs.[135–137] Secondary-structure formation in the long, untranslated leader sequences found for many S-layer protein mRNAs might contribute to such unusually long half-lives.[138] As they mediate the synthesis of a major structural component of the cell, the high stability of S-layer mRNAs is not unexpected.

Most S-layer proteins contain an N-terminal signal peptide that allows for their secretion by the Sec-dependent general secretion pathway. These signal sequences, with a length ranging from 21 to 34 amino acids for *A. salmonicida* and *H. volcanii*, respectively, are generally processed by signal peptidases, resulting in the mature S-layer proteins.[138] In some cases, such as the S-layers from *Deinococcus radiodurans*[139] and *T. thermophilus* HB8,[140] the N-terminus of the protein seems to be chemically modified. Further processing also occurs in the S-layer protein of *Rickettsia prowazekii*.[141] The best studied models of signal peptide-containing S-layer proteins are *A. salmonicida* and *Aeromonas hydrophila*. In the case of *A. salmonicida*, secretion of the S-layer protein (VapA) through its outer membrane implicates a substrate-specific main terminal branch of the general secretion pathway.[142] Interestingly, S-layer proteins from other Gram-negative bacteria do not contain an N-terminal signal peptide. Examples of signal peptide-less S-layer proteins are those from *C. crescentus*,[143] *Serratia marcescens*,[144] and *C. fetus*.[54] Secretion of these S-layer proteins involves specific type I secretion systems.[145,146]

Once secreted, the subunits interact with each other and with the supporting envelope layer through noncovalent forces. Different physicochemical surface properties at the inner and outer surface of the S-layer lattice have been shown to be responsible for the proper orientation of the S-layer subunits and their insertion in the course of lattice growth.[24,37,39,147] Depending on the supramolecular cell envelope design, S-layer subunits may interact with different components (e.g., the plasma membrane in most archaea, the rigid wall component in Gram-positive bacteria, or the lipopolysaccharide layer of the outer membrane in Gram-negative bacteria). Electron microscopy of negatively stained, thin sectioned or freeze-etched preparations and labeling experiments using fluorescent antibodies or immunogold have demonstrated that different patterns of S-layer lattice extension exist for Gram-positive and Gram-negative bacteria. In Gram-positive bacteria, lattice growth occurs primarily by insertion at multiple bands on the cylindrical part of the cell.[148] By contrast, in Gram-negative bacteria, the insertion of new subunits occurs at random.[149]

As predicted by Harris and coworkers,[150] in "closed surface crystals" dislocations and disclinations could serve as sites for incorporation of new subunits in closed lattices that grow by "intussusceptions."[2] As both types of lattice faults can be observed on high-resolution freeze-etching images of intact cells, it can be assumed that the rate of growth of S-layer lattices by the mechanism of nonconservative climb of dislocations depends on the number of dislocations present and the rate of incorporation of new subunits at these sites.[2,19,37]

Different to organisms with S-layers attached to rigid wall components (Fig. 2), in archaea with S-layers as exclusive wall component it can be expected that the protein meshwork has the intrinsic potential to fulfill cell shape determining functions and to be involved in cell fission.[151,152] Cell division in these organisms appears to be determined by the ratio of the increase in protoplast volume and S-layer surface area extension.[152]

These observations and *in vitro* self-assembly studies (see below) have led to the speculation that S-layer-like membranes could have fulfilled barrier and support functions for self-reproducing systems at the origin of life.[34,147]

B. Self-Assembly *In Vitro*

The attractiveness of S-layer proteins for a broad range of applications in nanobiotechnology lies in their capability to form 2D arrays in suspension and at interfaces without the bacterial cell envelope from which they have been removed (Fig. 4).[20,33] Most techniques for isolation and purification of S-layer proteins involve a mechanical disruption of the bacterial cells and subsequent differential centrifugation in order to isolate the cell wall fragments.[24,37] A complete solubilization of S-layers into their constituent protein monomers and their release from the supporting cell envelope layers can be achieved with high concentrations of chaotropic agents (e.g., guanidine hydrochloride, urea), by lowering or raising the pH, or by applying metal-chelating agents (e.g., EDTA, EGTA) or cation substitution.[37] The reassembly of isolated S-layer proteins into extended self-assembly products or, alternatively, the preparation of solubilized S-layer protein monomers occurs upon dialysis of the disintegrating agent.[37] The dialysis protocol (presence of divalent cations, pH, or dialysis time) determines the final product. It has to be noted that the formation of S-layer lattices is determined only by the amino-acid sequence of the polypeptide chains where a specific part is responsible for the reassembly property.[33,76] As S-layer proteins have a high proportion of nonpolar amino acids, most likely hydrophobic interactions are involved in the assembly process. Studies on the distribution of functional groups on the surface have shown that free carboxylic acid groups and amino groups are arranged in close proximity and thus contribute to the cohesion of the proteins by electrostatic interactions.[20] Further, some S-layer proteins require divalent cations in order to form stable lattices.[153–156]

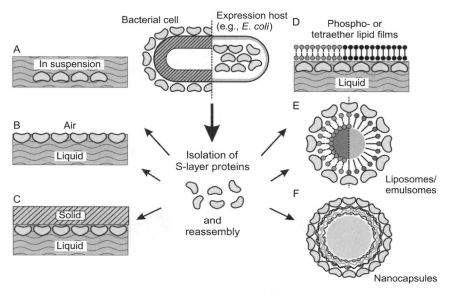

FIG. 4. Schematic drawing of the reassembly of native or recombinantly generated S-layer proteins in (A) suspension, (B) at the air–water interface, (C) on solid supports, (D) at lipid films, and (E, F) on spherical surfaces, such as liposomes, emulsomes, and nanocapsules. (See Color Insert.)

1. SELF-ASSEMBLY IN SUSPENSION

Depending on the specific bonding properties and the tertiary structure of the S-layer proteins, flat sheets, open-ended cylinders, or vesicles are formed (Fig. 5).[24,37,147,157,158] Both protein concentration and temperature determine the extent and rate of association. The assembly starts with a rapid initial phase and continues with a slow consecutive rearrangement step leading to extended lattices.[159] Further, depending on the S-layer proteins used and the environmental conditions (e.g., ionic content and strength in the buffer solution), the self-assembly products consist either of monolayers or double layers. In a systematic study with the S-layer protein SgsE from *G. stearothermophilus* NRS 2004/3a, it was shown that two types of monolayered and five types of double-layered assembly products with back-to-back orientation of the constituent monolayers were formed.[2,157] The double layers differed in the angular displacement of their constituent S-layer sheets. As the monolayers had an inherent inclination to curve along two axes, cylindrical or flat double-layer assembly products were formed depending on the degree of neutralization of the inherent "internal bending strain."[157]

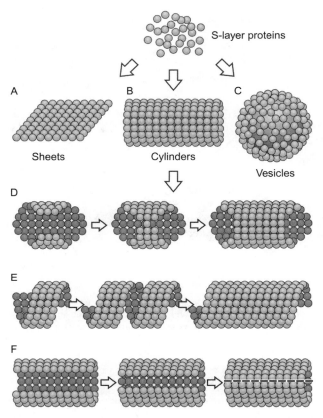

FIG. 5. Schematic drawing of the different assembly routes of S-layer proteins into (A) flat sheets, (B, D–F) cylinders, and (C) vesicles.(For color version of this figure, the reader is referred to the Web version of this chapter.)

2. SELF-ASSEMBLY ON SURFACES AND AT INTERFACES

Crystal growth on surfaces and at interfaces is initiated simultaneously at many randomly distributed nucleation points and proceeds in plane until the crystalline domains meet, thus leading to a closed seamless mosaic of individual S-layer domains of several micrometers long.[153,156,160] S-layer protein monolayer formation at the liquid–air interface was studied by transmission electron microscopy (TEM) (Fig. 6).[160] In this work, electron microscopy grids were deposited on and removed from the water surface by a Langmuir–Schäfer transfer at regular time intervals. After staining with uranyl acetate, the samples were inspected in the transmission electron microscope. In a recent study, it was demonstrated that atomic force microscopy (AFM) is most suitable to

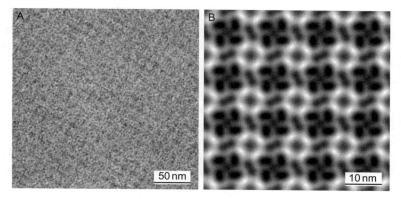

FIG. 6. Low-dose transmission electron microscopical image of a negatively stained preparation of S-layer proteins from *Lysinibacillus sphaericus* CCM 2177. (A) shows the original micrograph and (B) an image-processed zoom region.

image the lattice formation in real time.[161,162] Approximately 10 min after injection of the protein solution into the fluid cell, the first small crystalline patches became visible, and about 30 min later, the silicon surface was completely covered and only small holes remained free which were closed in due course. Extremely low loading forces of the AFM tip were necessary in order to minimize the influence of the scanning tip on the reassembly of the proteins. Nevertheless, the formed S-layer lattices showed a perfect long-range order and a resolution of the molecular details down to approximately 1 nm (as determined from the digital diffraction pattern).

The importance of controlling the hydrophobicity of the surface had been demonstrated by adsorbing disulfides on gold substrates prior to reassembling S-layer proteins.[163] Disulfides are OH- and CH_3-terminated thiolates in an inclined position.[164] By changing the length stepwise by one CH_2 unit, the hydrophobicity could be adjusted very precisely. It was shown that the length of the alkyl chains determined the lattice parameters and the domain sizes of the reassembled S-layer. An increase in hydrophobicity led to a transition from bilayer to monolayer formation.

In summary, the formation of coherent crystalline S-layer arrays depends strongly on the particular S-layer protein species, the environmental conditions of the bulk phase (e.g., temperature, pH, ion composition, and ionic strength), and, most importantly, on the surface properties of the substrate (hydrophobicity, surface charge).[153] For example, on hydrophobic silicon supports the S-layer protein SbpA form *Ly. sphaericus* CCM 2177 forms monolayers exposing the outer face of the S-layer lattices, while on silicon surfaces rendered hydrophilic the S-layer proteins form double layers facing each other with their inner faces and thus exposing the outer face as well. At incomplete double layers where the lower S-layer is larger than the upper one, the inner S-layer face becomes visible. Binding experiments with negatively charged gold nanoparticles demonstrated that only the inner positively charged S-layer faces could be labeled, confirming the AFM results.[165]

In addition, all these parameters have a direct effect on the mobility of the S-layer proteins on the surface and their probability to be trapped at the front lines of growing domains. This continuous rearrangement of subunits leads to a stable configuration which is characteristic of crystal growth in equilibrium where large and perfectly regular arrays are formed. By changing one of these critical parameters, such as the concentration of calcium ions in the reassembly of the S-layer protein SbpA from *Ly. sphaericus* CCM 2177 (see above), crystal growth is no longer in equilibrium and a broad spectrum of crystal morphologies ranging from tenuous, fractal-like structures to large monocrystalline patches is obtained.[155] The observed patterns indicate that diffusion-limited processes are dominating. For example, the borderline may advance so rapidly that the stable phase does not have time to reach its lowest energy state on the microscopic level, and a metastable microstructure with "extended fingers" results. The structures are tenuous and open because holes are formed and not filled up. From these data, it was concluded that S-layer crystal formation at interfaces is determined by a fast nucleation and assembly process involving subunits associated with the interface, and a slow incorporation of subunits from the subphase.

The possibility for recrystallizing isolated S-layer proteins at the air/water interface or on lipid films on the air/water interface and for handling such layers by standard Langmuir–Blodgett (LB) techniques opened a broad spectrum of applications in basic and applied membrane research (for review, see Refs. 165–167). Further, the reassembly of S-layer proteins on spherical lipid structures such as liposomes and nanocapsules has great technological and, in particular, medicinal importance (drug targeting and delivery, diagnostics, etc.).[168–171] A detailed summary on the S-layer protein–lipid interaction, the biophysical and electrochemical characterization of S-layer-supported planar and spherical lipid membranes, and the application potential of these biomimetic supramolecular architectures in nanobiotechnology and synthetic biology is given in Section VIII.

3. Patterning of S-layers on Solid Supports

For many technical applications of S-layers, spatial control over the reassembly is mandatory. For example, when using S-layers as affinity matrices in the development of biochips, or as templates in the fabrication of nanoelectronic devices, the S-layer must not cover the entire device area. Several approaches including optical lithography were used to pattern S-layer lattices on silicon wafers. Nevertheless, the soft lithographical method called *micro-molding in capillaries* has proven to be most suitable for patterning S-layer lattices, as it allowed restricting the reassembly of the S-layer proteins to certain areas on a solid support.[161] For this purpose, the S-layer protein solution was dropped onto the substrate in front of the channel openings of the attached mold. The solution was sucked in and the S-layer protein started to recrystallize. After removal of the mold, a patterned S-layer remained on the support, as demonstrated by AFM. Micromolding in capillaries offers the advantage that all preparation steps may be performed under ambient conditions. In contrast, patterning S-layers by optical lithography requires drying of the protein layer prior to exposure to (deep ultraviolet) radiation.[172] This is a critical step, as denaturation of the protein and consequently loss of its structural and functional integrity cannot be excluded.

4. MOLECULAR MODELING AND COMPUTER SIMULATIONS

Over the past decades, a large number of studies addressed the 3D structure determination of S-layer proteins, which led to a considerable knowledge about the distribution of amino acids on S-layer lattices, the structure–function relationship, molecular mechanisms of the self-assembly process, and even structural details of some S-layer species.[80,82,83,173–177] However, experimental structure determination techniques, for example, NMR and X-ray crystallography, pose problems due to the size and crystallization characteristics of S-layer proteins, as in solution they form crystallized monomolecular layers rather than isotropic 3D crystals. The dissolved proteins immediately interact to form small oligomers, which provide the nucleation seed for the formation of large layers.[178] Additionally, some S-layer proteins do not fold into their native tertiary structure as monomers in solution, but rather condense into amorphous clusters in an extended conformation. Only when assembled into the lattice structure, do they restructure into their native conformation.[173,179] Thus, 3D reconstructions were limited to truncated or mutated forms of the proteins.

The combination of molecular simulations and low-resolution experimental techniques, for example, small angle X-ray scattering (SAXS) and TEM, offers an alternative to determine the atomistic structure of unmodified native S-layer proteins and self-assembled lattices.

The folding of small protein domains and of entire proteins can be monitored by reverse and steered molecular dynamics simulations. However, in order to facilitate the equilibration process, secondary structure elements taken from homologous protein models have to be implemented first. The calculated 3D model of the entire protein can be consequently verified by a systematic exploration of the free energy, by a reverse Monte Carlo simulation based on scattering contrast data obtained by SAXS, or by 3D density distribution data as calculated by TEM. Following this approach, the structural models of the S-layer proteins SbsB from *G. stearothermophilus* pV72/p2 (*p1* lattice symmetry)[81,178] and of the unit cell of SbpA from *Ly. sphaericus* CCM 2177 (*p4* lattice symmetry)[179] could be calculated. Additionally, based on the model of the S-layer protein SbsB, the molecular mechanisms guiding the self-assembly into monomolecular sheets exhibiting a *p1* lattice symmetry could be analyzed using Monte Carlo simulations.[180]

The structure prediction of the S-layer protein SbsB is shown in Fig. 7. The protein is split into structurally meaningful domains based on homology searches, secondary structure, and domain predictions. To obtain 3D coordinates, a premodeling by fold recognition is performed. Molecular dynamics simulations were processed with each part, and consequently, the domains were joined and the whole structure was equilibrated in vacuum (Fig. 7A–D). The resulting structure was analyzed by pulling parts of the protein along a chosen reaction coordinate, and the protein was deformed to quantify the stability, which is expressed as the potential of MF (Fig. 7F). The final structural model of SbsB is shown in detail in Fig. 7E. The structure could be verified and additionally refined by SAXS studies, where the monomeric structure and self-assemblies were investigated. The analysis is based on a fractal mean potential, which describes best the behavior of S-layers in solution.[178]

The reconstruction of the 3D structure of an SbpA unit cell is based on a similar approach. An intermediate structural model was calculated by fold recognition and molecular dynamics simulations. The resulting 3D model of an SbpA tetramer is shown in Fig. 8A–C. In this case, 3D density data facilitated the modeling process, which were obtained by tilting studies and inverse Fourier transform using TEM. Regions of high and low electron density contrast were identified by SAXS studies (Fig. 8D), where those with high contrast were classified as noninteracting and those with low contrast as interacting with the other monomers in the tetramer. The resulting 3D model of an SbpA tetramer is shown in Fig. 8.

The structural model of the S-layer protein SbsB in combination with Monte Carlo simulations was used to study the functional protein self-assembly into S-layers.[180] Using a coarse-grained model, the specific interactions between two protein monomers in solution were investigated to determine the ground-state conformations, which led to a *p1* lattice symmetry. Consequently, the calculated energies of the interactions between two proteins were used to study the large-scale self-assembly by means of lattice Monte Carlo simulations, as schematically shown in Fig. 9. Only very few and mainly hydrophobic amino acids, located on the surface of the monomer, are responsible for the formation of the highly anisotropic protein lattice, which is in excellent agreement with known experimental results.

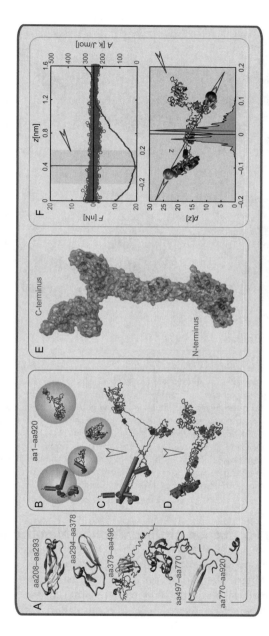

FIG. 7. (A) Three-dimensional models of the single domains of the S-layer protein SbsB created by fold recognition. Yellow arrows, beta-sheets; violet strands, alpha-helices; green line, turns; and red line, coils. Modeling method: (B) the individual domains were equilibrated in water spheres at 310 K, (C) joined in vacuum, and (D) the final structure was obtained by molecular dynamics simulations. (E) Structural model of SbsB. The protein is L-shaped, where the L is formed by the C-terminal domains. The N-terminus contains the SLH domains and is mainly made up of alpha-helices. (F) Structural analysis of the monomer structure by a calculation of the global free energy. The protein was deformed along the reaction coordinate z. Mean force values F are indicated by open blue circles. The red full line gives the potential of mean force A, which has a clear minimum at $z = 0$. Orange body gives the local density probability distribution $p[z]$. The model of the protein is given as an insert, the reaction coordinate is indicated, and the green and blue spheres indicate fixed regions. Figure modified after Ref. 81 with friendly permission of the American Institute of Physics. (See Color Insert.)

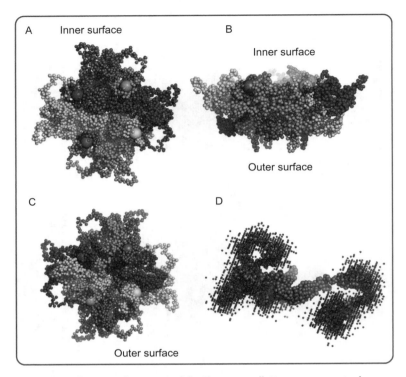

Fɪɢ. 8. Three-dimensional structure of the SbpA unit cell. Every monomer in the tetramer is illustrated in a different color. The proteins are interlocked into each other. (A) Inner surface of the tetramer, which anchors the protein on the cell surface. The N-termini are represented by magnified beads and are accessible on the surface. (B) Side view of the unit cell. (C) Outer surface of the tetramer, which is exposed to the surroundings of the cell. The C-termini are also accessible and marked as magnified beads. (D) Scattering clusters (red beads) of one SbpA monomer as determined by SAXS and a Monte Carlo algorithm. The scattering clusters represent regions of high electron density contrast, where those domains in the protein that do not show high contrast are related to interacting or overlapping parts in the tetramer. Figure modified after Ref. 179 with friendly permission of the American Institute of Physics. (See Color Insert.)

IV. S-Layers for the Production of Ultrafiltration Membranes

As S-layers are composed of identical subunits, they possess pores of identical size and morphology. Depending on the mass distribution of the constituent subunits, more than one type of pore may generate the protein meshwork. High-resolution electron microscopy in combination with digital image processing methods have indicated that pore diameters can range from 2 to 8 nm and thus function in the ultrafiltration range (Fig. 10).[64,181] More

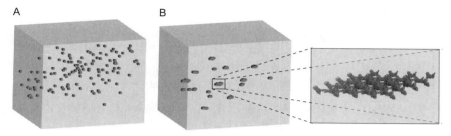

Fɪɢ. 9. Representation of the lattice Monte Carlo simulations of the large-scale self-assembly of the S-layer protein SbsB. Energy values were taken from Monte Carlo simulations of the interaction of two monomers. Proteins are represented as cubes. (A) Initial configuration: cubes are randomly distributed in the simulation box. (B) Competitive growth study. Multiple sheets start to grow during initial period. Magnified view of the corresponding S-layer sheet using a coarse-grained model. Figure modified after Ref. 180 with friendly permission of the American Institute of Physics. (For interpretation of the references to color in this figure legend, the reader is referred to the Web version of this chapter.)

precise data on the pore size could be derived from permeability studies on S-layer vesicles using the space technique[182,183] and S-layers that had been deposited on porous supports.[184] S-layers from thermophilic Bacillaceae (e.g., *G. stearothermophilus*) revealed sharp exclusion limits between molecular weights of 30,000 and 40,000 suggesting a limiting pore diameter of 3.5–5 nm. Carbonic anhydrase with a molecular weight of 30,000 and a molecular size of $4.1 \times 4.1 \times 4.7$ nm could still pass through the pores, whereas ovalbumin with a molecular weight of 43,000 was rejected to more than 90%. No significant difference in the rejection characteristics could be observed between native S-layers and S-layer protein lattices cross-linked with glutaraldehyde.[185,186]

For the production of S-layer ultrafiltration membranes (SUMs), cell wall fragments or isolated S-layers were deposited on microfiltration membranes (MFMs) using a pressure-dependent procedure and the S-layer protein was cross-linked with glutaraldehyde.

To increase the chemical stability of such cross-linked S-layer lattices, Schiff bases formed by the reaction of glutaraldehyde with ε-amino groups from lysine were reduced with sodium borohydride.[185–188] In these composite membranes, the active filtration layer consists of a coherent layer of super-imposed S-layer material, whereas the MFM provides the mechanical support. Generally, SUMs were made of S-layer material from different *G. stearother-mophilus* strains and *Ly. sphaericus* CCM 2120 (see Table II).

Even though S-layers of different crystallographic types were used, all molecular cut-off values resembled those determined for S-layer vesicles applying the space technique.[183]

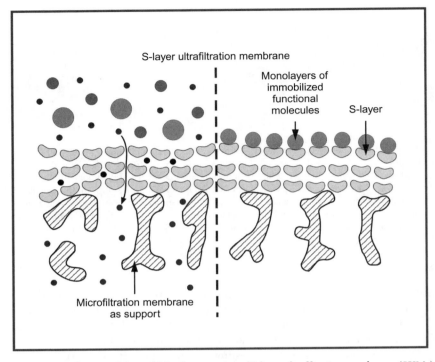

F_IG_. 10. Schematic drawing of the fine structure of S-layer ultrafiltration membranes (SUMs): (left) the active ultrafiltration layer consists of coherent S-layers deposited on open-celled foam-like microfiltration membranes. (right) SUMs can also be used for covalent attachment of functional molecules.(For color version of this figure, the reader is referred to the web version of this chapter.)

TABLE II

R_EJECTION_ C_HARACTERISTICS OF_ SUM_S_ P_REPARED OF_ S-L_AYER_ C_ARRYING_ C_ELL_ W_ALL_ F_RAGMENTS FROM_ *L_YSINIBACILLUS_ S_PHAERICUS_* CCM 2120 (M_ODIFIED_ A_FTER_ R_EF_. 19)

Protein	M_r	Molecular size (nm)	p*I*	%R	pH value of the protein solution
Ferritin	440,000	12	4.3	100	7.2
Bovine serum albumin (BSA)	67,000	$4.0 \times 4.0 \times 14.0$	4.7	100	7.2
Ovalbumin (OVA)	43,000	4.5	4.6	95	4.6
Carbonic anhydrase (CA)	30,000	$4.1 \times 4.1 \times 4.7$	5.3	80	5.3
Myglobin (MYO)	17,000	$4.4 \times 4.4 \times 2.5$	6.8	0	6.8

Cross-linking the S-layer protein with glutaraldehyde during the production of SUMs leads to net negatively charged membranes due to the reaction of a considerable proportion of free amino groups. As under physiological conditions most proteins in solution are negatively charged, it is advantageous for many ultrafiltration processes to use membranes that have a negative charge, too. Repulsive forces between the protein in solution and the membrane surface prevent unspecific adsorption and pore-blocking, which cause deterioration in membrane performance in the form of flux losses and decrease in selectivity. For example, SUMs with a net negative surface charge showed no or negligible flux losses after filtration of solutions of ferritin, bovine serum albumin, or ovalbumin, which had a net negative surface charge under the applied experimental conditions.[3,189–193]

As in S-layer lattices the physicochemical properties are determined by those of the single constituent (glyco)protein subunit, SUMs are ideal model systems for studying the effects of chemical modifications on the rejection and adsorption properties of ultrafiltration membranes. For example, after the cross-linking, carboxyl groups from the S-layer lattices were activated with 1-ethyl-3(3-dimethylaminopropyl) carbodiimide (EDC) and allowed to react with the free amino groups from nucleophiles of different molecular sizes, structures, hydrophobicity, and charge.

Contact angle measurements clearly demonstrated that covalent attachment of low molecular weight nucleophiles to S-layer lattices led to SUMs with more hydrophilic or hydrophobic surface properties.

Covalent attachment of low molecular weight nucleophile SUMs not only led to alterations of the surface properties and antifouling characteristics but were also responsible for an accurately controlled shift of the rejection curves to the lower molecular weight range.[19] Moreover, a correlation was observed between the molecular size of attached nucleophiles and the shift of the rejection curve to the lower molecular weight range.[189,194]

In conclusion, up to now, SUMs are the only ultrafiltration membranes that enable most precisely controlled modifications of the physicochemical and molecular sieving properties. This broad spectrum of potential modifications allows the properties of SUMs to be adapted to very specific requirements.[19,187,195]

Moreover, it has to be remembered that native S-layers reveal remarkable antifouling properties. This is considered an essential requirement to prevent plugging of pores in the protein meshwork and maintain free exchange of molecules up to a defined molecular weight between the cell and the environment.

V. S-Layer as Matrix for the Immobilization of Functional Molecules and Nanoparticles

Because S-layers are periodic structures, they exhibit repetitive physico-chemical and morphological properties down to the subnanometer scale and possess pores of identical size and morphology. Most importantly, S-layer recrystallization can be induced on flat surfaces and highly porous structures such as MFMs or porous beads. Owing to the high density of functional groups on the surface and their accessibility for chemical modifications, S-layers are well-defined matrices for controlled immobilization of functional molecules such as enzymes, antibodies, antigens, and ligands as required for affinity and enzyme membranes in the development of solid-phase immunoassays or in biosensors.

Chemical modification and labeling experiments revealed that S-layer lattices possess a high density of functional groups on the outermost surface. For covalent attachments of foreign (macro)molecules such as protein A, monoclonal antibodies, streptavidin, or various enzymes, the free carboxylic acid groups originating from either aspartic acid or glutamic acid in the S-layer protein were activated with carbodiimide and subsequently reacted with free amino groups of functional macromolecules leading to stable peptide bonds between the S-layer matrix and the immobilized macromole-cule.[168,193,196–199] From the amount, molecular mass, and size of the foreign proteins bound to the S-layer lattice, as well as from the molecular size of the S-layer subunits and the area occupied by one morphological unit, it was derived that most macromolecules formed a monomolecular layer on the surface of the S-layer lattice. Immobilization via spacer molecules such as 6-amino capronic acid was advantageous when the molecules were small enough to be entrapped inside the pores of the S-layer lattice, which, in the case of enzymes, was linked to a significant activity loss.[168] In general, the activity of enzymes immobilized to S-layer lattices was well preserved.[19,191,193]

The enzymes were either coupled to the hexagonally ordered S-layer lattices from *Thermoanaerobacter thermohydrosulfuricus* L111-69[200] or from *G. stearothermophilus* PV72.[201] The covalently bound carbohydrate chains of the S-layer glycoprotein from *T. thermohydrosulfuricus* L111-69[202,203] were also exploited for enzyme immobilization.[168,191,198,204] Independent of the type of S-layer protein used from different *Bacillaceae*, the large enzymes invertase ($M_r = 270,000$), glucose oxidase (GOD) ($M_r = 150,000$), glucuronidase ($M_r = 280,000$), and β-galactosidase ($M_r = 116,000$) formed a dense monolayer on the outer face of the S-layer lattice.[29] After direct coupling of the enzymes invertase, GOD, naringinase ($M_r = 96,000$), and β-glucosidase ($M_r = 66,000$) to the EDC-activated carboxylic acid groups of the S-layer protein from

T. thermohydrosulfuricus L111-69, the retained enzymatic activities were in the range of 70%, 35%, 60%, and 16%, respectively. By immobilization via spacer molecules, a significant increase in enzymatic activity could be achieved for GOD and naringinase of 60% and 80%, respectively. The most striking increase was observed for β-glucosidase, for which immobilization via spacers led to a 10-fold increase in activity to 160%. The significant increase in enzymatic activity indicated that immobilization via spacers most probably increased the distance between the enzyme molecules and the crystalline S-layer matrix.[205]

The S-layer immobilization matrix can be used with planar and curved supports. Affinity microparticles (AMPs) represent 1-μm large cup-shaped structures. They are produced from S-layer carrying cell wall fragments, in which the S-layer protein is cross-linked with glutaraldehyde and Schiff bases are reduced with sodium borohydride. Because of the applied preparation procedure, AMPs possess a complete outer and inner S-layer, which can be exploited for immobilization of foreign macromolecules. Protein A as an Fc-binding ligand was linked to the carbodiimide-activated carboxylic acid groups of the S-layer protein of *T. thermohydrosulfuricus* L111-69.[199,206] The Fc-binding ligand formed a monolayer on the exposed outer face of the S-layer lattice. AMPs based on S-layer-carrying cell wall fragments revealed excellent stability properties under cross-flow conditions. AMPs were used as escort particles in affinity cross-flow filtration and as novel immunoadsorbent particles in blood purification.[199,206] For both applications, the advantage of AMPs can be seen in the cup-shaped structure, leading to a high surface-to-volume ratio, as well as in the dense monolayer of protein A molecules on the outermost surface of the S-layer lattices.

SUMs were not only used as ultrafiltration membranes but also exploited as novel matrices for the development of dipstick-style solid-phase immunoassays and for the development of an amperometric glucose sensor (Fig. 10). Depending on the test system, the respective monoclonal antibody was covalently bound to the carbodiimide-activated carboxylic acid groups of the S-layer lattice. After immobilization of the monoclonal antibodies, disks of 3-mm diameter were punched out and sandwiched between Teflon foils, leaving the SUM exposed for further binding reactions. By immobilizing monolayers of either protein A or streptavidin onto SUMs, a universal biospecific matrix for immunoassays and dipsticks could be generated.[207] Matrices based on protein A as an immunoglobulin G (IgG)-specific ligand were obtained by immobilizing dense monolayers of this ligand to carbodiimide-activated carboxylic acid groups from the S-layer protein of SUMs.[199,207] Because of the high affinity of human IgG and rabbit IgG to protein A, the protein A-SUM was shown to be particularly suitable for generating dense monolayers of correctly aligned antibodies on the SUM surface. However, mouse IgG with lower affinity to protein

A than human IgG or rabbit IgG was either first biotinylated and subsequently bound to a streptavidin-coated SUM, or it was directly linked to carbodiimide-activated carboxylic acid groups exposed on the surface of the S-layer lattice. Proof of principle was demonstrated for different types of SUM-based dipsticks: for example, for diagnosis of type I allergies (determination of IgE against the major birch pollen allergen Bet v1 in whole blood or serum); for quantification of tissue type plasminogen activator (t-PA) in patients whole blood or plasma for monitoring t-PA levels in the course of thrombolytic therapies after myocardial infarcts; or for determination of interleukin 8 in supernatants of human umbilical vein endothelial cells (HUVEC) induced with lipopolysaccharides.[208–210] Further, a dipstick assay was developed for prion diagnosis based on a sandwich enzyme-linked immunosorbent assay (ELISA) specific for prion protein, exploiting S-layer lattices as an immobilization matrix. The sensitivity of the prion dipsticks were similar to that published for time-resolved fluorescence ELISA methods, which are among the most sensitive detection methods for prions.[211]

In addition, the feasibility of SUMs as a new type of immobilization matrix was already demonstrated many years ago by the development of an amperometric glucose sensor using GOD as the biologically active component.[204,212] Subsequently, a layer-by-layer technique was established allowing the fabrication of a multienzyme biosensor for sucrose.[213] Based on the demonstrated suitability of the S-layer protein self-assembly system for covalent enzyme immobilization, genetic approaches were pursued to construct fusion proteins between S-layer proteins of *Bacillaceae* and enzymes from extremophiles for the development of novel biocatalysts (see Section VI).[214,215]

The ultimately high requirements in positional control at the nanometer scale, the synthesis of molecular functional units (memory cells or switches), and their internal (nano-to-nano) and external (nano-to-micro) interconnection can only be met when novel concepts based on bottom-up approaches are developed. In this context, the broad base of knowledge about the S-layer-mediated binding of biological molecules has paved the way for investigating the potential of S-layer proteins and their self-assembly products as catalysts, templates, and scaffolds for the generation of ordered nanoparticle arrays for nonlife-science applications (e.g., nonlinear optics, nanoelectronics). In particular, the *in situ* precipitation and the controlled binding of metallic or semiconducting nanoparticles on S-layers laid the foundation for novel concepts in the field of molecular electronics and optics (Fig.11).

The first approach in using S-layers as lithographic templates in the formation of perfectly ordered nanoparticle arrays was based essentially on the deposition of a metal vapor onto S-layer fragments of *Sulfolobus acidocaldarius*. In a three-step parallel process, a 1-nm thick tantalum/tungsten film with holes (15 nm in diameter) periodically arranged according to the

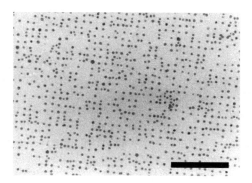

FIG. 11. TEM image of gold nanoparticles (mean diamter 5 nm) bound in a regular order on an S-layer exhibiting square lattice symmetry and a center-to-center spacing of 13.1 nm (bar, 100 nm).

center-to-center spacing (22 nm) of the hexagonal S-layer lattice was fabricated.[216] Although it has been demonstrated that nanoparticle arrays may be fabricated in this way, the real breakthrough was achieved by using S-layer lattices in the direct precipitation of metals from solution or by binding preformed nanoparticles. In the wet chemical approach, which was derived from mineral formation on bacterial surfaces,[217] self-assembled S-layer structures were exposed to metal salt solutions, such as tetrachloroauric (III) acid ($HAuCl_4$) solution, followed by slow reaction with a reducing agent such as hydrogen sulfide (H_2S) or by electron irradiation in an electron microscope.[218–223] As the precipitation of the metals was confined to the pores of the S-layer, nanoparticle arrays with prescribed symmetries and lattice geometries could be obtained. Although the wet chemical methods resulted in crystalline arrays of nanoparticles with spacings in register with the underlying S-layer lattice, they do not allow controlling of the particle size or composition, such as in core–shell nanoparticles. Thus, the binding of preformed standardized nanoparticles into regular arrays on S-layers has significant advantages in the development of biomolecule-driven assemblies of nanoscale electronic devices compared to vapor deposition or wet chemical methods.

Several studies have already demonstrated the outstanding performance of S-layer lattices as patterning elements.[165,218,224] The pattern of bound molecules and nanoparticles frequently reflects the lattice symmetry, the size of the morphological units, and the physicochemical properties of the array. For example, the distribution of net negatively charged domains on S-layers could be visualized by electron microscopic methods after labeling with positively charged topographical markers, such as polycationic ferritin (diameter, 12 nm).[19,151] Metal (Au) or semiconductor (CdSe) nanoparticles were either electrostatically or covalently bound onto solid-supported S-layer monolayers and self-assembly products of SbpA, the S-layer protein of *Ly. sphaericus* CCM 2177.[165] Upon activation of carboxyl groups in the S-layer lattice, a close-packed monolayer of 4-nm sized, amino-functionalized CdSe nanoparticles could be covalently established on the outer face of the solid-supported S-layer lattices. However, owing to electrostatic interactions, anionic citrate-stabilized 5-nm gold nanoparticles formed a superlattice at those sites where the inner face of the S-layer lattice was exposed.[165]

Further, S-layer protein lattices isolated from the Gram-positive bacterium *D. radiodurans* and the acidothermophilic archaeon *S. acidocaldarius* were investigated and compared for their ability to biotemplate the formation of self-assembled, ordered arrays of inorganic nanoparticles.[225,226] The authors demonstrated the possibility to exploit the physicochemical/structural diversity of prokaryotic S-layer scaffolds to vary the morphological patterning of nano-scale metallic and semiconductor NP arrays.[225]

More recently, genetic approaches were used for the construction of chimeric S-layer fusion proteins whereby precipitation of metal ions or binding of metal nanoparticles is confined to specific and precisely localized positions in the S-layer lattice (see Section VI).[98,227]

VI. S-Layer Fusion Proteins—Construction Principles and Applications

During the past 10 years, genetic approaches have been focused on the construction of chimeric S-layer fusion proteins aiming at a very controlled and specific way of making highly ordered functional arrays. The genetically engineered S-layer fusion proteins comprised (i) an accessible N-terminal cell wall anchoring domain, which can be exploited for oriented binding on supports precoated with SCWP; (ii) the self-assembly domain; and (iii) a functional sequence.[8,15,228] The most relevant advantages of the genetically engineered self-assembly system based on S-layers over less nanostructured approaches are (i) the alignment of functional domains at predefined distance in the nanometer range on the outermost surface of the S-layer lattice and, thus, availability for further binding reaction (e.g., substrate binding, antibody binding, enzymatic reactions); (ii) the requirement of only a simple, one-step incubation process for site-directed immobilization without preceding surface activation of the support; (iii) the general applicability of the "S-layer tag" to any fusion partner; (iv) the high flexibility for variation of the functional groups within a single S-layer array by cocrystallization of different S-layer fusion proteins to construct multifunctional arrays; and (v) the provision of a cushion for the functional group through the S-layer moiety preventing denaturation, and, consequently, loss of reactivity upon immobilization. Since these advantages opened up a broad spectrum of applications for S-layer fusion proteins particularly in the fields of biotechnology, molecular nanotechnology, and biomimetics (Fig.12),[3,5,7,15,228] a great variety of functional S-layer fusion proteins were cloned and heterologously expressed in *Escherichia coli* or used for surface display after homologous expression (overview given in Table III).

The S-layer protein SbpA of the mesophilic *Ly. sphaericus* CCM 2177 consists of total of 1268 amino acids, including a 30-amino-acid long signal peptide.[77] By producing various C-terminally truncated forms and performing surface accessibility screens, it became apparent that amino-acid position 1068 is located on the outer surface of the square lattice. The derivative rSbpA$_{31-1068}$ fully retaining its ability to self-assemble into a square S-layer lattice with a center-to-center spacing of the morphological units of 13.1 nm.[77] Therefore,

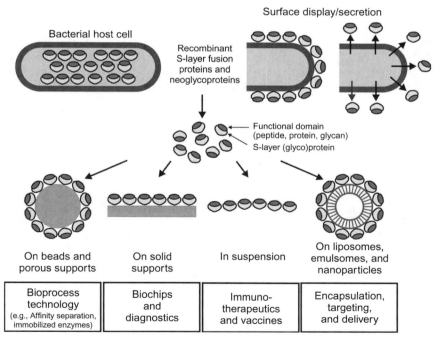

FIG. 12. Schematic drawing of technologies based on recombinant S-layer fusion proteins and their applications. (For color version of this figure, the reader is referred to the web version of this chapter.)

TABLE III

FUNCTIONAL RECOMBINANT S-LAYER FUSION PROTEINS AND THEIR APPLICATIONS

Recombinant S-layer protein	Functionality	Length of function	Application	References
SbpA, SbsB	Core streptavidin	118 aa	Binding of biotinylated ligands (DNA, protein), Biochip development	98,229
SbpA, SbsC	Major birch pollen allergen (Bet v1)	116 aa	Vaccine development, treatment of type 1 allergy	77,230
SbpA	*Strep*-tag II, affinity tag for streptavidin	9 aa	Biochip development	77
SbpA	ZZ, IgG-binding domain of Protein A	116 aa	Extracorporeal blood purification	231
SbpA	Enhanced green fluorescent protein (EGFP)	238 aa	Coating of liposomes, development of drug targeting and delivery systems	232
SbpA	cAb, heavy chain camel antibody	117 aa	Diagnostic systems and sensing layer for label-free detection systems	233
SbpA	Hyperthermophilic enzyme laminarinase (LamA)	263 aa	Immobilized biocatalysts	215
SbpA	Cysteine mutants	3 aa	Building of nanoparticle arrays	227
SbpA, SbsB	Mimotope of an Epstein–Barr virus (EBV) epitope (F1)	20 aa	Vaccine development	234
SbpA, SbsB	*M. tuberculosis* antigen (mpt64)	204 aa	Vaccine development	Tschiggerl H.° Nano-S°
SbpA	IgG-Binding domain of Protein G	110 aa	Downstream processing	214
SgsE	Glucose-1-phosphate thymidylyltransferase (RmlA)	299 aa	Immobilized biocatalysts	235,236
SgsE	Enhanced cyan fluorescent protein (ECFP)	240 aa	pH biosensors *in vivo* or *in vitro*, fluorescent markers for drug delivery systems	
	Enhanced green fluorescent protein (EGFP)	240 aa		
	Yellow fluorescent protein (YFP)	240 aa		
	Monomeric red fluorescent protein (RFP1)	225 aa		
SbsA	*H. influenzae* antigen, (*Omp*26)	200 aa	Vaccine development	237
SlpA	Antigenic poliovirus epitope (VP1)	11 aa	Development of mucosal vaccines	127
	Human c-myc proto-oncogene	10 aa		
SLH-EA1, SLH-Sap	Levansucrase of *B. subtilis*	473 aa	Vaccine development	238

S-layer protein	Fusion partner	Size	Application	Ref.
SLH-EA1	Tetanus toxin fragment C of C. tetani (ToxC)	451 aa	Development of live veterinary vaccines	239
RsaA	P. aeruginosa strain K pilin	12 aa	Vaccine development	240
RsaA	IHNV glycoprotein	184 aa	Development of vaccines against hematopoietic virus infection	241
RsaA	Beta-1,4-glycanase (Cex)	485 aa	Immobilized biocatalysts	242
RsaA	IgG-binding domain of Protein G	GB1$_{xs}$	Development of immunoactive reagent	243
RsaA	Domain 1 of HIV receptor CD4 MIP1α ligand for HIV coreceptor CCR5	81 aa 70 aa	Anti-HIV microbicide development	244
RsaA	His-tag, affinity tag	6 aa	Bioremediation of heavy metals (Cd) from aqueous systems, bioreactor	245

S-layer proteins: SbsB of *Geobacillus stearothermophilus* PV72/p2, ShpA of *Lysinibacillus sphaericus* CCM 2177, SbsC of *Geobacillus stearothermophilus* ATCC 12980, SgsE of *Geobacillus stearothermophilus* NRS 2004/3a, SbsA of *Geobacillus stearothermophilus* PV72/p6, SlpA of *Lactobacillus brevis* ATCC 8287, SLH (S-layer homology domain of EA1 or Sap) of *Bacillus anthracis*, RsaA of *Caulobacter crescentus* CB15A.
[a]personal communication.

this C-terminally truncated form was used as base form for the construction of various S-layer fusion proteins. For SbpA, the recrystallization process is dependent on the presence of calcium ions, thus allowing control over lattice formation,[155] which is of advantage for nanobiotechnological applications of the SbpA system.

Another S-layer protein that adopts C-terminal fusions without affecting self-assembly into lattices is SbsB of thermophilic G. stearothermophilus PV72/p2. SbsB consists of a total of 920 amino acids, including a 31-amino-acid long signal peptide.[98] As the removal of fewer than 15 C-terminal amino acids led to water-soluble rSbsB forms, the C-terminal part can be considered extremely sensitive to deletions. When the C-terminal end of full-length SbsB was exploited for linking a foreign functional sequence, water-soluble S-layer fusion proteins were obtained,[98] which recrystallized into the oblique (p1) lattice on solid supports precoated with SCWP of G. stearothermophilus PV72/p2. Alternatively, functional groups were fused toward the N-terminus of SbsB to construct self-assembling S-layer fusion proteins, which attached with their outer surface to, for example, liposomes and silicon wafers, so that the N-terminal region with the fused functional sequence remained exposed to the environment.[98]

The protein precursor of the S-layer protein SbsC from G. stearothermophilus ATCC 12980 includes a 30-amino-acid long signal peptide and consists of 1099 amino acids.[100] The investigation of the self-assembly properties of several truncated SbsC forms revealed that on the C-terminal part, 179 amino acids could be deleted without interfering with the self-assembling properties of the S-layer protein.[78] Thus, $SbsC_{31-920}$, which is the shortest C-terminal truncation still capable of forming self-assembly products, was used as base form for the construction of functional SbsC fusion proteins.[230]

To generate a universal affinity matrix for binding of any kind of biotinylated molecules, S-layer streptavidin fusion proteins have been constructed.[98,229,246] For that purpose, core streptavidin was either fused to N-terminal positions of the S-layer protein SbsB or to the C-terminal end of the truncated form $SbpA_{31-1068}$.[98,229,246] As biologically active streptavidin occurs as tetramer, heterotetramers consisting of one chain fusion protein and three chains of core streptavidin were prepared by applying a special refolding procedure (Fig. 13). A biotin binding capacity of about 75% could be determined for soluble heterotetramers, indicating that three of four biotin binding sites were active.[98] In addition, the use of S-layer–streptavidin fusion proteins allowed specific binding of biotinylated ferritin molecules into regular arrays.[98] The lattice formed by the fusion protein displayed streptavidin in defined repetitive spacing, capable of binding D-biotin and biotinylated proteins, in particular ferritin.[98]

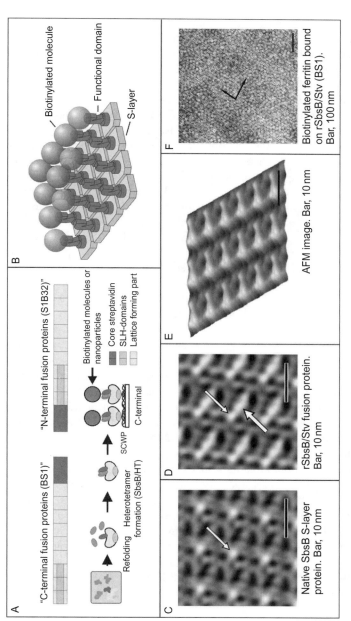

Fig. 13. S-layer/streptavidin fusion proteins as template for nanopatterned molecular arrays. (A) Refolding of heterotetrameric S-layer/streptavidin fusion protein. Core streptavidin and fusion protein were produced in *E. coli* and isolated independently, mixed in the molar ratio of 1:3, and refolded to heterotetramers by applying the rapid dilution protocol. (B) Cartoon illustrating self-assembled S-layer fusion protein carrying functional domains (e.g., streptavidin) in defined position and orientation with bound biotinylated molecules. (C and D) Digital image reconstruction by Fourier processing of electron micrographs showing self-assembly products of wild-type SbsB or rSbsB/Stv fusion protein, respectively. (C) The region of the highest protein mass in the SbsB lattice is the SLH domain. (D) In the lattice of the fusion protein, streptavidin showed up as additional protein mass. (E) Side-on view of the AFM image data of an rSbsB/Stv monolayer. (F) Electron micrograph of cell wall fragments carrying recrystallized S-layer/streptavidin fusion protein which was capable of binding biotinylated ferritin. Bound ferritin reflected the underlying oblique lattice symmetry. (See Color Insert.)

Hybridization experiments with biotinylated and fluorescently labeled oligonucleotides evaluated by surface plasmon-field-enhanced fluorescence spectroscopy indicated that a functional sensor surface could be generated by recrystallization of heterotetramers on gold chips.[229] Such promising structures could be exploited for the development of DNA or protein chips as required for many nanobiotechnological applications.

The Z-domain is a synthetic analog of the B-domain of protein A from *Staphylococcus aureus*, capable of binding the Fc part of IgG. For production of an antibody-binding matrix, the S-layer fusion protein $rSbpA_{31-1068}/ZZ$ carrying two copies of the 58-amino-acid long Fc-binding Z-domain on the C-terminal end was recrystallized on gold chips precoated with thiolated SCWP.[231] The binding capacity of the native or cross-linked $rSbpA_{31-1068}/ZZ$ monolayer for human IgG was determined by SPR measurements. On average, $\sim 66\%$ of the theoretical saturation capacity of a planar surface was covered by IgG aligned in upright position.[231] By recrystallization of this chimeric protein on microbeads, a biocompatible matrix for the microsphere-based detoxification system used for extracorporeal blood purification of patients suffering from autoimmune disease has been generated. To increase the IgG-binding variety for further immunoaffinity applications, current approaches focus on the construction of an S-layer fusion protein carrying the SPG1 domain of protein G from *Streptococcus* (NanoS, personal communication).

For the development of a sensing layer for label-free detection systems such as SPR, surface acoustic wave, or quartz crystal microbalance with dissipation monitoring (QCM-D), the S-layer fusion protein $rSbpA_{31-1068}/cAb$-PSA carrying the camel antibody sequence recognizing the prostate-specific antigen (PSA) was recrystallized on gold chips precoated with thiolated SCWP.[233] For determining the binding capacity, PSA-containing sera were conducted over the sensor surface. The fused ligands on the S-layer lattice showed a well-defined spatial distribution down to the subnanometer scale, which might reduce diffusion-limited reactions.[233,247]

Owing to their immunomodulating capacity, chimeric S-layer proteins comprising allergens are generally considered as a novel approach to specific immunotherapy (SIT) of allergic diseases.[77,230,248] For that purpose, two chimeric S-layer proteins, $rSbpA_{31-1068}/Bet$ v1 and $rSbsC_{31-920}/Bet$ v1, carrying the major birch pollen allergen Bet v1 at the C-terminus have been constructed.[77,230] In cells of birch pollen-allergic individuals, the histamine-releasing capacity induced by the S-layer fusion proteins was significantly reduced compared to stimulation with free Bet v1 and no Th2-like immune response was observed.[248,249] These data clearly supported the concept that genetic fusion of allergens to S-layer proteins is a promising approach to improve vaccines for specific treatment of atopic allergy.

A further promising application potential can be seen in the development of drug and delivery systems based on liposome–DNA complexes coated with functional S-layer fusion protein for transfection of eukaryotic cell lines. In this context, the S-layer fusion protein rSbpA$_{31-1068}$/enhanced green fluorescent protein (EGFP) incorporating the sequence of EGFP was recrystallized as a monolayer on the surface of positively charged liposomes. Because of its ability to fluoresce, liposomes coated with rSbpA$_{31-1068}$/EGFP represent a useful tool to visualize the uptake of S-layer-coated liposomes into eukaryotic cells.[232]

Another field of research deals with the production of S-layer fusion proteins between the S-layer proteins of *Ly. sphaericus* CCM 2177 or *G. stearothermophilus* PV72/p2 and peptide mimotopes such as F1 that mimics an immunodominant epitope of the Epstein–Barr virus (EBV). Diagnostic studies have been performed by screening 83 individual EBV IgM-positive, EBV-negative, and potential cross-reactive sera, which resulted in 98.2% specificity and 89.3% sensitivity as well as no cross-reactivity with related viral diseases. This result demonstrates the potential of these S-layer fusion proteins to act as a matrix for site-directed immobilization of small ligands in solid-phase immunoassays.[234]

In a recent study, C-terminally functionalization of the S-layer protein SbpA by introduction of cysteine residues combined with targeted chemical modification was used to identify amino acids that are located at the surface of the S-layer lattice.[227] Crystalline monolayers of these S-layer cysteine mutants offered free sulfhydryl groups for the activation with various heterobifunctional cross-linkers and covalent attachment of differently sized (macro)molecules. Finally, functionalized 2D S-layer lattices formed by rSbpA cysteine mutants exhibiting highly accessible cysteine residues in a well-defined arrangement on the surface were utilized for the template-assisted patterning of gold nanoparticles.[227]

On the basis of the remarkable intrinsic feature of S-layer proteins to self-assemble and the possibility for genetic modifications, S-layer proteins were exploited as component for the development of novel immobilized biocatalysts based on fusion proteins comprising S-layer proteins of Bacillaceae and enzymes from extremophilic organisms (extremozymes). By exploiting the self-assembly property of the S-layer protein moiety, the chimeric protein was used for spatial control over display of enzyme activity on planar and porous supports. As proof of principle, the enzyme β-1,3-endoglucanase LamA from the extremophilic *P. furiosus* was C-terminally fused to the S-layer protein SbpA$_{31-1068}$ of *Ly. sphaericus* CCM 2177.[215] The results obtained clearly demonstrate that S-layer-based bottom-up self-assembly systems for functionalizing solid supports with a catalytic function could have significant advantages over processes based on random immobilization of sole enzymes. In general, clear advantages for enzyme immobilization offered by the S-layer self-assembly system include the high flexibility for variation of enzymatic groups within a single S-layer array by cocrystallization of different enzyme/S-layer fusion proteins to construct multifunctional, nanopatterned biocatalysts, as well as the possibility for deposition of the biocatalysts on different supports with the additional option of cross-linking of individual monomers to improve robustness.[215] It is remarkable to note that the measured enzyme activities of the recrystallized S-layer/enzyme fusion proteins reach up to 100% compared to the native enzyme. The S-layer protein portion of the biocatalysts confers significantly improved shelf-life to the fused enzyme without loss of activity over more than 3 months, and also enables biocatalyst recycling.

In addition, recent research activities are focused on the production of immobilized biocatalysts based on fusion proteins comprising self-assembling S-layer proteins and multimeric enzymes of extremophiles (e.g., xylose isomerase of *Thermoanaerobacterium strain* JW/SL-YS 489) (Fig. 14).[250]

SgsE, the S-layer glycoprotein of the thermophilic Gram-positive bacterium *G. stearothermophilus* NRS 2004/3a, has a molecular weight of 93,684 Da and a *p*I of 6.1.[214] SgsE has the ability to form 2D crystalline arrays with oblique symmetry exhibiting nanometer-scale periodicity. Studies on the structure–function relationship of SgsE revealed that the N-terminal region is involved in anchoring the protein to the cell wall and the C-terminal region encodes the self-assembly information.[214] Concerning the biotechnological application of S-layer fusion proteins aiming at controllable display of biocatalytic epitopes, storage stability, and reuse, S-layer/enzyme fusion proteins comprising the glucose-1-phosphate thymidylyltransferase RmlA from *G. stearothermophilus*

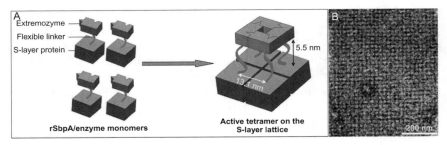

FIG. 14. Novel approach for site-directed immobilization of multimeric enzymes of extremophilic organisms via the S-layer self-assembly technique, allowing oriented and dense surface display of the extremozyme in its native confirmation and ensuring accessibility for the substrate. (A) Schematic drawing of an immobilized biocatalyst based on S-layer fusion proteins carrying a multimeric extremozyme. (B) Transmission electron micrograph of a negatively stained preparation of an S-layer/xylose isomerase fusion protein self-assembled in solution into a monomolecular array. (See Color Insert.)

NRS 2004/3a as well as the SgsE derivatives $SgsE_{31-773}$ or $SgsE_{31-573}$ from *G. stearothermophilus* NRS 2004/3a, respectively, were cloned and expressed in *E. coli*.[214]

To build up novel functional fluorescent architectures, the 903-amino-acid containing S-layer protein SgsE from *G. stearothermophilus* NRS 2004/3a was used for the production of four S-layer fusion proteins carrying different colored GFP mutants.[235,236] For this purpose, the nucleotide sequence encoding the EGFP, the enhanced cyan fluorescent protein, the yellow-shifted YFP 10C variant, as well as the yellow-shifted red fluorescent protein mRFP1 were fused to the C-terminus of the N-terminally truncated form $SgsE_{131-903}$. Results derived from investigation of the recrystallization properties, absorptions spectra, steady-state, and lifetime fluorescence measurements in different pH environments revealed that the assembling and fluorescence properties of the fusion proteins can be used for building up nanopatterned bifunctional surfaces that can be exploited as pH biosensors *in vivo* and *in vitro* or as fluorescent markers for drug delivery systems.[235,236]

The molecular masses of *Lactobacillus* S-layer proteins are among the smallest (43–46 kDa) known S-layer proteins and show sequence similarity at the C-terminus, which is responsible for binding of the S-layer subunits to the cell wall layer.[107,127] In *L. brevis* ATCC 8287, the C-terminal part of the 48-kDa S-layer protein SlpA is heterogeneous and lacks a cell wall anchoring domain whereas this strain strongly adheres to several human epithelial cell types, pig intestinal epithelial cells, and fibronectin via a receptor-binding site that is located at the N-terminus of SlpA.[127] In terms of immunological studies, an inducible *L. brevis* expression system for the production, secretion, and surface display of antigenic epitopes inserted at distinct sites of the S-layer protein SlpA was developed.[127] So, for surface display of foreign antigenic epitopes, *L. brevis* strains displaying a VP1 poliovirus epitope of 10 amino acids and the 11-amino-acid c-Myc epitopes from the human *c-myc* proto-oncogene as part inserted into the outermost proteinaceous S-layer of the cell were constructed.[127] For this purpose, the epitope insertion site allowing the best surface expression was used for construction of an integration vector, and a gene replacement system for the replacement of the wild-type *slpA* gene with the *slpA-c-myc* construct was developed. Electron microscopy investigation revealed that the S-layer lattice structure was not affected by the presence of the c-Myc epitope expressed in every SlpA subunit.[127]

G. stearothermophilus strain PV72/p6 contains the S-layer protein SbsA with a size of 1228 amino acids, which shows a hexagonal lattice symmetry.[251] In 2003, the antigen Omp26 was introduced at different positions of SbsA and the fusion protein was expressed in empty bacterial cell envelopes (ghosts) to deliver candidate antigens for the development of vaccines against nontypeable

Haemophilus influenzae (NTHi) infections.[237] The bacterial ghost system inducing Omp26-specific antibody response in mice is a novel vaccine delivery system endowed with intrinsic adjuvant properties.[237]

B. anthracis synthesizes two abundant surface proteins, EA1 and Sap, which form an S-layer.[252] Both proteins have the same modular organization, an N-terminal cell targeting domain consisting of three SLH motifs followed by a putative self-assembly domain.[93] Chimeric genes encoding the sequence of the SLH domains of Sap or EA1 as well as the levansucrase of *Bacillus subtilis* or the C-fragment of the tetanus toxoid of *Clostridium tetani* were cloned and expressed in *B. anthracis*.[238,239] The fusion proteins were secreted and could attach to the bacterial cell surface of *B. anthracis* and one resulting recombinant strain carrying the RPL-toxin fragment C (ToxC), was used for veterinary vaccinal purposes.[239]

C. crescentus is a Gram-negative, nonpathogenic bacterium that is covered by a hexagonal S-layer lattice composed of a single 98-kDa protein species termed RsaA.[253,254] A calcium binding domain that is located near the C-terminus likely mediates the calcium-dependent assembly process as well as surface attachment of RsaA.[253,255,256] The feasibility of the commercially available PurePro TM Caulobacter expression and secretion system (Invitrogen) for the surface display of functional epitopes (e.g., *Pseudomonas aeruginosa* strain K pilin peptide,[240] a 184-amino-acid segment of the infectious hematopoietic necrosis virus (IHNV),[241] the beta-1,4-glycanase from *Cellulomonas fimi*,[242] the IgG-binding domain of Streptococcal Protein G,[243] or Domain 1 of the HIV receptor CD4 and MIP1α ligand for HIV coreceptor CCR5[244]) could be demonstrated. In a recent study, the genetically engineered RsaA was employed as a delivery system for displaying hexahistidine peptides on the *Caulobacter* cell surface and to construct a recombinant bioremediation agent to remove heavy metals from aqueous solutions.[245]

In a biological context, protein glycosylation is often the key to protein function as well as for regulating and influencing many cellular processes.[257,258] Because of this, the development of tailor-made, bioactive glycoproteins (referred to as *neo*glycoproteins) by genetic engineering will drastically change the capabilities of influencing and controlling complex biological systems. Two different strategies are envisaged. The first one is the *in vivo* display of functional glycans on the surface of bacteria enabled by means of recombinant DNA technology with various applications in microbiology, nanobiotechnology, and vaccinology.[259,260] A second approach can be seen in the *in vitro* line of development that utilizes the recrystallization ability of the S-layer portion on a broad spectrum of supports. For both strategies, the S-layer "anchor" provides a crystalline, regular matrix for the display of functional glycosylation motifs.[27,68]

VII. S-Layers for Vaccine Development

Owing to their intrinsic adjuvant ability as well as their capability to surface display proteins and epitopes, S-layers are excellent candidates to be used as antigen carriers, either as self-assembly products coated on liposomes, or displayed on bacteria (overview given in Table IV).

B. anthracis causes lethal infections in mammals after cutaneous inoculation, inhalation, or ingestion.[273] As reported in Ref. 261, the gene bsIA encoding the S-layer protein BslA of B. anthracis is located on the pXO1 pathogenicity island and the expressed S-layer protein is necessary and sufficient for adhesion of the anthrax vaccine strain, B. anthracis Sterne, to host cells.[261,262] Surface localization and abundant expression of BslA make this polypeptide a candidate antigen for purified subunit vaccines against B. anthracis.[261] In an earlier study, the S-layer protein genes of B. anthracis were used to develop a cell surface display system for vaccination studies. For this purpose, a recombinant B. anthracis strain was constructed by integrating a translational fusion harboring DNA fragments encoding the cell wall targeting domain of the S-layer protein EA1 and tetanus ToxC into the chromosome. The humoral immune response was sufficient to protect mice against tetanus toxin challenge. Therefore, the expression system will be tested for the development of new live veterinary vaccines.[239]

A further important field of S-layer-based vaccine development is the investigation of the immunomodulating capacity as well as the adjuvant activity of the S-layer proteins of C. difficile.[274,275] This pathogenic organism is the major cause of antibiotic-associated diarrhea as well as pseudomembraneous colitis in hospitalized patients.[276] For active immunization, the surface-layer protein of C. difficile R13537 was tested as a vaccine component in a series of immunization and challenge experiments with Golden Syrian hamsters, combined with different systemic and mucosal adjuvants.[263]

For the development of fish vaccines to fight Aeromonas infections which can cause furunculosis in fish in freshwater and marine environments, the crystalline cell surface protein of the fish-pathogenic bacteria itself is essential for virulence and was considered a good vaccine candidate.[264] Although numerous attempts have been made to vaccinate salmon, trout, and catfish using outer membrane protein,[277] extracellular products, lipopolysaccharide preparations,[278] biofilms,[279] or whole cells of A. hydrophila,[280,281] there is still no commercial vaccine available.[265] A possible reason can be seen in the inability of these vaccines to cross-protect against different isolates of A. hydrophila, which is a biochemically as well as serologically heterogeneous bacterium.[265] To overcome this problem, a common antigen among different isolates of A. hydrophila that could serve as a vaccine candidate is required.

TABLE IV

IMMUNOGENIC S-LAYER (FUSION) PROTEINS

Antigen/Hapten	Conjugation	S-layer source (S-layer protein)	References
	Sole S-layer	*B. anthracis* Sterne	261,262
	Sole S-layer	*C. difficile* R13537	263
	Sole S-layer	*A. salmonicida* A449, *A. hydrophila* TF7	264,265
S. pneumoniae serotype 8 poly- and oligosaccharides	Chemically coupled hapten	*B. alvei* CCM 2051	266
Tumor marker T-disaccharide	Chemically coupled hapten	*T. thermohydrosulfuricus* L111-69, *G. stearothermophilus* NRS2004/3A, *B. alvei* CCM 2051	64
Tumor-associated Lewis Y (Le^y) tetrasaccharides	Chemically coupled hapten	*G. stearothermophilus* PV72	267
Birch pollen allergen, Bet v1	Chemically coupled antigen	*Ly. sphaericus* CCM 2177, *T. thermohydrosulfuricus* L111-69, *T. thermohydrosulfuricus* L110-69	268–270
Tetanus toxin fragment, ToxC	S-layer fusion protein[a]	*B. anthracis* RPL686	239
H. influenzae antigen, Omp26	S-layer fusion protein[b]	*G. stearothermophilus* PV72/p6	237
Hematopoietic necrosis virus glycoprotein segment	S-layer fusion protein[a]	*C. crescentus* JS 4011	241
Birch pollen allergen, Bet v1	S-layer fusion protein[b]	*Ly. sphaericus* CCM 2177, *G. stearothermophilus* ATCC 12980	77,230,248,249,271
M. tuberculosis protein, mpt64	S-layer fusion protein[b]	*Ly. sphaericus* CCM 2177, *G. stearothermophilus* PV72/p2	Tschiggerl H., unpublished
Adhesintope of the *P. aeruginosa* pilin	S-layer fusion protein[a]	*C. crescentus* JS 4011	272
Human *c-myc* proto-oncogene	S-layer fusion protein[a]	*L. brevis* ATCC 8287 (GRL1)	127

[a]Surface display of S-layer fusion proteins.
[b]*In vitro* formation of immunogenic S-layers.

In this context, in a recent study, the S-layer protein of *A. hydrophila* was produced recombinantly in *E. coli* and its ability to protect common carp *Cyrinus carpio* L. against six virulent isolates of *A. hydrophila* could be demonstrated.[265]

IHNV causes a hemorrhagic disease in young salmonid fish and is another severe threat for fish farming. A recombinant subunit model vaccine was developed by fusing a 184-amino-acid segment of IHNV glycoprotein to the C-terminal portion of the S-layer protein of *C. crescentus*.[241] The fusion protein was expressed by the *C. crescentus* S-layer secretion system and laboratory trials revealed a relative survival of 26–34% in rainbow trout fry.[241]

Another application of S-layers is their use as carrier for immunogenic antigens and haptens.[62,266,282,283] As common carriers for peptide epitopes are used as monomers in solution (e.g., tetanus or diphtheria toxoids) or as dispersions of unstructured aggregates on aluminum salts, a reproducible immobilization of ligands to the carrier protein cannot be achieved.[284,285] Consequently, the use of regularly structured S-layer self-assembly products as immobilization matrices represents a completely new approach. Investigations focused on the development of several model conjugate vaccines with S-layer (glyco) proteins of thermophilic bacilli and clostridia and weekly immunogenic carbohydrate antigens, for example, Streptococcus pneumoniae serotype 8 poly- and oligosaccharides, haptens or recombinant birch pollen allergen showed promising results in vaccination trials.[64,266–268,286–288]

A further approach was the use of recombinant S-layer fusion proteins and empty bacterial cell envelopes (ghosts) to deliver candidate antigens (Omp26) for a vaccine against the nontypeable *H. influenzae* (NTHi) infection. Immunization studies with the resulting SbsA/Omp26 in bacterial ghosts induced an Omp26-specific antibody response in BALB/c mice.[237]

S-layer self-assembly products and S-layer-coated liposomes[168,169] can be considered as particulate adjuvants with dimensions comparable to those of bacteria or viruses that the immune system evolved to combat. The mechanical and thermal stability of S-layer-coated liposomes[169,289,290] and the possibility for immobilization or entrapping biologically active molecules[168,170,291] introduced a broad application potential, particularly as carrier and/or drug delivery and drug targeting systems or in gene therapy, for example, as artificial viruses.[20]

Further, there is an urgent need for new vaccines that allow mucosal administration instead of intramuscular injections for the achievement of specific desired effects, such as adjuvant targeting, site-specific delivery, and controlled immune responses. S-layer–hapten conjugates induced significant vaccination responses even after oral/nasal administration.[267] One project was directed toward immunotherapy of cancer, where conjugates of S-layer with small, tumor-associated oligosaccharides were found to elicit hapten-specific DTH responses.[267]

Immunization experiments in mice have indicated that S-layers served not only as carriers but also as adjuvants.[267–269] Allergen–S-layer conjugates and S-layer/allergen fusion proteins have been prepared with the intention of suppressing the Th2-directed, IgE-mediated allergic responses to Bet v1, the major allergen of birch pollen.[77,230,269,270] These studies showed that the S-layer protein conjugate induced IFN-γ production, thus activating the phagocytotic cells and confirming that Th1-enhancing properties were clearly attributable to the S-layer protein. Further, the recombinant S-layer/Bet v1 fusion proteins altered an established Th2-dominated phenotype as well as the *de novo* cytokine secretion profile toward a more balanced Th1/Th0-like phenotype.[248,249,271] These data clearly confirm the immunomodulating properties of the S-layer moiety in S-layer fusion proteins and support the concept that recombinant fusion of allergens and S-layer proteins is a promising approach to improve vaccines for SIT of atopic allergy.

Current studies using immunogenic self-assembly products of S-layer fusion proteins comprising the antigen mpt64, a *Mycobacterium tuberculosis* protein, investigate the ability of S-layer proteins to serve as a carrier and adjuvant for the vaccination against tuberculosis (Tschiggerl, unpublished).

A number of vaccine approaches involve the development of vaccination vehicles, serving to potentiate the immune response to an antigen. For using S-layers to display foreign peptides on the *C. crecentus* cell surface in the dense, highly ordered S-layer structure was explored,[240,272] and the *C. crescentus* RsaA secretion apparatus was used to produce a fusion protein comprising RsaA and the adhesintope of the *P. aeruginosa* pilin. This presentation system could have many potential applications, such as the development of whole-cell vaccines, tumor suppressors, cellular adsorbents, and peptide display libraries.[292–294] Further, the 11-amino-acid long epitope *c-myc* from the human *c-myc* proto-oncogene was successfully expressed in every S-layer subunit of the *L. brevis* S-layer (SlpA) while maintaining the S-layer lattice structure.[127] Delivery of antigens to mucosal surfaces by lactic acid bacteria was considered to offer a safe alternative to live attenuated pathogens because of their food grade status.

VIII. S-Layers as Supporting Structure for Functional Lipid Membranes

One major topic in nanobiotechnology nowadays is the design, synthesis, and fabrication of supramolecular interfacial architectures comprising biomolecules (lipids, (membrane)-proteins, glycans, nucleic acids, and combinations thereof) and inorganic or organic materials of technological importance.[295,296]

Biological systems are prime candidates for controlled "bottom-up" production of defined nanostructures.[2,297,298] Although self-assembly of molecules is a ubiquitous strategy of morphogenesis in nature, research in the area of molecular nanotechnology, nanobiotechnology, and biomimetics are only beginning to exploit its potential for the functionalization of surfaces and interfaces as well as for the production of biomimetic membranes and encapsulation systems.

Biomimetic model membranes (free-standing, tethered, and supported lipid mono- or bilayers with associated or integral peptides or proteins) have attracted lively interest in recent years, as the advances in genome mapping revealed that approximately one-third of all the genes of an organism encode for membrane proteins such as pores, ion channels, receptors, and membrane-anchored enzymes.[299–302] These proteins are key factors in the cell's metabolism, for example, in cell–cell interaction, signal transduction, and transport of ions and nutrients, and, thus, in health and disease.[303] Owing to this important function, membrane proteins are a preferred target for pharmaceuticals (at present more than 60% of consumed drugs)[304] and have received widespread recognition for their application in drug discovery, protein–ligand screening, and biosensors.

The present section intends to give a survey on particular biomimetic planar and spherical lipid membranes, which consist, besides the lipid matrix, of a closely associated proteinaceous S-layer lattice. The concept of exploiting this supramolecular building principle for stabilizing planar or vesicular membrane systems[11,13,16,20] evolved from the observation that most archaea are exclusively composed of a cytoplasmic membrane and a closely associated or even integrated monomolecular crystalline S-layer (Fig. 15A). Moreover, these organisms dwell under extreme environmental conditions such as temperatures up to 120 °C, pH down to zero, high hydrostatic pressure, or high salt concentrations.[305–307] As suitable methods for disintegration of archaeal S-layer protein lattices and their reassembly into monomolecular arrays on lipid films are not yet available, S-layer proteins from Gram-positive bacteria were used for copying the archaeal building principle for the generation of S-layer-stabilized lipid membranes. S-layer proteins can be utilized as biofunctional surfaces[308] and constitute also a fascinating base structure for hosting functionalized planar lipid membranes.[12,13,16,20] The lipid membranes either consists of an artificial phospholipid bilayer (Fig. 15B) or a tetraetherlipid monolayer replaces the cytoplasmic membrane, and isolated bacterial S-layer proteins may be attached on one or even on both sides of the lipid membrane (Fig. 15E and F).

Further, S-layer lattices as the outermost envelope component covering spherical supramolecular structures such as liposomes or emulsomes constitute biomimetic "artificial virus-like particles" enabling both stabilization of, for example, liposomes and presenting addressor molecules in a well-defined

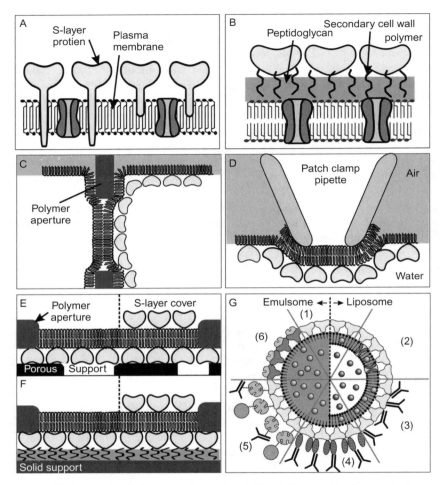

Fɪɢ. 15. Supramolecular structure of an archaeal (A) and Gram-positive bacterial cell envelope (B). Schematic illustrations of various S-layer-supported lipid membranes. In (C), a folded or
painted membrane spanning a Teflon aperture is shown. A closed S-layer lattice can be self-
assembled on either one or both (not shown) sides of the lipid membranes. (D) A bilayer lipid
membrane is generated across an orifice of a patch clamp pipette by the tip-dip method. Subsequently, a closely attached S-layer lattice is formed on one side of the lipid membrane. (E)
Schematic drawing of a lipid membrane generated on an S-layer ultrafiltration membrane
(SUM). Optionally, an S-layer lattice can be attached on the external side of the SUM-supported
lipid membrane (right part). (F) Schematic drawing of a solid support covered by a layer of
modified secondary cell wall polymer (SCWP). Subsequently, a closed S-layer lattice is assembled
and bound via the specific interaction between S-layer protein and SCWP. On this biomimetic
structure, a lipid membranes is generated. As shown in (E), a closed S-layer lattice can be
recrystallized on the external side of the solid-supported lipid membrane (right part). (G) Schematic drawing of (1) an S-layer-coated emulsome (left part) and S-liposome (right part) with

orientation and spatial distribution (Fig. 15G). The prerequisite for creating such supramolecular structures is given by the unique noncovalent interaction of S-layer proteins with lipid head groups within planar and spherical lipid mono- and bilayers.[11,13,166,167,309]

A. Planar Lipid Membranes

A broad range of techniques including TEM and AFM, Fourier transform infrared spectroscopy, dual-label fluorescence microscopy, and X-ray and neutron reflectivity measurements have been applied to characterize S-layer-supported lipid membranes[154,160,310–314] Formation of S-layer lattices covering the entire area of lipid films has been observed on zwitterionic phospholipids such as phosphatidyl ethanolamines and phosphatidyl cholines.[315] The addition of a small portion of positively charged surfactants[168,316] or lipid derivatives[317] facilitated the crystallization process, in particular, on phosphatidyl cholines. A systematic study has provided evidence that electrostatic interactions exist between exposed carboxyl groups on the S-layer lattice and zwitterionic lipid head groups. At least two to three contact points between the lipid film and the attached S-layer protein have been identified.[315] Hence, less than 5% of the lipid molecules of the adjacent monolayer are anchored to these contact points on the S-layer protein, whereas the remaining $\geq 95\%$ lipid molecules diffuse freely in the membrane between the pillars consisting of anchored lipid molecules.[318,319] This calculation is based on the S-layer lattice of SbpA from *Ly. sphaericus* CCM 2177 having a square unit cell with a spacing of 13.1 nm[153,171] and an area per lipid molecule of 0.65 nm.[320] These nanopatterned lipid membranes are also referred to as "semifluid membranes"[154] because of its widely retained fluid behavior.[317,321] Most importantly, although peptide side groups of the S-layer protein interpenetrate the phospholipid head group regions almost in its entire depth, no impact on the hydrophobic lipid alkyl chains has been observed.[312–314,322,323] Further, the S-layer lattice neither constitutes a significant barrier for the porcine pancreatic phospholipase A$_2$ (PLA$_2$; $M_r = 13.8$ kDa) nor induces lipid packing defects that would result in shorter enzymatic lag periods.[324] The alterations of the molecular-level organization of the lipid monolayer upon S-layer protein binding and

---

entrapped water-soluble (gray) or lipid-soluble (black) functional molecules and (2) functionalized by reconstituted integral membrane proteins. S-layer-coated emulsomes and S-liposomes can be used as immobilization matrix for functional molecules (e.g., IgG) either by direct binding (3), by immobilization via the Fc-specific ligand protein A (4), or biotinylated ligands can be bound to S-layer-coated emulsome and S-liposome via the biotin–streptavidin system (5). Alternatively, emulsomes and liposomes can be coated with S-layer fusion proteins incorporating functional domains (6). Modified after, Ref. 20 Copyright (2002) and, Ref. 16 Copyright (2004), with permission from Wiley–VCH. (See Color Insert.)

recrystallization observed by various biophysical methods do not cause serious inhibition of the PLA$_2$. Hence, there is much evidence that the recrystallized S-layer lattice did not modulate a large proportion of the head group region of the phospholipid monolayer to an extent that could dramatically impede the recognition of the phospholipids by the biological interplay with the PLA$_2$.[324]

In addition, a second S-layer acting as protective molecular sieve and further stabilizing scaffolding can be recrystallized on the top of S-layer-supported lipid membranes. With two recrystallized S-layers, nanopatterned fluidity determined by S-layer–lipid head group interactions from both sides are introduced. This is particularly the case whenever S-layer lattices differing in lattice constants and symmetry are used. Therefore, S-layer lattices constitute unique supporting scaffolding, resulting in lipid membranes with nanopatterned fluidity and considerably extended longevity.[11,166,167,309,325,326]

Schematic illustrations of S-layer-supported lipid membranes spanning the orifice of a bilayer lipid membrane (BLM) chamber and on the tip of a micropipette are shown in Fig. 15C and D, respectively. Whereas the impact of an attached S-layer lattice on the membrane capacitance, membrane resistance, and the boundary potential on free-standing BLMs is negligible, the mechanical properties of S-layer-supported lipid membranes are considerably altered. Hydrostatic pressure applied across painted BLMs caused them to bulge, resulting in an increase of the membrane capacitance due to area expansion. A significantly higher area expansion was observed for plain BLMs compared to S-layer-supported lipid membranes forced from the S-layer-facing side, demonstrating a protecting effect of the S-layer lattice against hydrostatic pressure.[327] Relaxation experiments revealed a considerably longer delay time between the applied voltage pulse and the appearance of an initial defect at S-layer-supported lipid membranes.[316] The membrane tension of BLMs upon the attachment of S-layer proteins has been determined by dynamic light scattering.[317] For BLMs, the collective motions of the lipid molecules are dominated by membrane tension rather than by membrane curvature energy. S-layer lattices on both sides of the BLM resulted in a considerable reduction of the membrane tension by a factor of approximately five. However, the membrane bending energy increased by three orders of magnitude.[317] Hence, the attached S-layer lattice facilitates the transverse shear motions of the lipid molecules. In accordance with voltage pulse experiments,[316] a significant increase of the previously negligible surface viscosity of the membrane has been observed during the S-layer protein attachment.[317]

The most challenging property of model lipid membranes is the feasibility to incorporate membrane-active peptides and, more important, (complex) integral membrane proteins in a functional state. Reconstitution of the staphylococcal pore-forming protein α-hemolysin (αHL),[328,329] the M2 segment that forms the ion-conducting channel of the nicotinic acetylcholine receptor,[330] and the ion carrier valinomycin into plain and S-layer supported lipid membranes has been successfully performed. S-layer-supported tetraetherlipid monolayers functionalized with valinomycin revealed a 10-fold higher life time compared to a membrane without an attached S-layer lattice.[323] Although no reconstitution of αHL could be achieved with tetraetherlipid membranes, lytic pores were formed in a membrane mainly composed of the branched phospholipid 1,2-diphytanosyl-sn-glycero-3-phosphocholine (DPhPC) by adding αHL to the lipid-exposed side of the S-layer-supported lipid membrane. No pore formation was detected upon addition of αHL monomers to the S-layer face of the S-layer-supported lipid membrane. Therefore, one can conclude that the intrinsic molecular sieving properties of the S-layer lattice do not allow passage of αHL monomers through the S-layer pores to the lipid surface.[322] In addition, these data represent a quality control for the existence of a closed S-layer lattice without any defects and a tight attachment to the BLM. Compared to plain BLMs, S-layer-supported lipid membranes have a decreased tendency to rupture in the presence of αHL, again indicating an enhanced stability due to the attached S-layer lattice.[322] Nevertheless, even single pore recordings have been performed with αHL reconstituted in free-standing S-layer-supported lipid membranes.[331]

S-layer-stabilized lipid membranes formed by the tip-dip methodology were functionalized with M2 ion channels.[332] The M2 ion channel characteristics were compared for BLMs of the same lipid composition with and without S-layers, and the attached S-layer lattices were found to be nonintrusive to the channel functionality and characteristics. The ability of S-layer proteins to stabilize BLMs and their nonintrusive character on ion channel activity make them attractive for biosensor applications, especially those that enhance the stability of BLMs beyond the use of tethers or polymer supports.[332]

However, although free-standing S-layer-supported lipid membranes revealed a higher mechanical stability (e.g., against hydrostatic pressure) and longevity in particular with reconstituted peptides or proteins, these membranes are up to now not stable enough for many practical applications.[13,167,318] Hence, a sophisticated long-term strategy is to attach BLMs to porous or solid supports to enhance their practical applicability.[333–338] Lipid membranes generated on a porous support combine the advantage of possessing an essentially unlimited ionic reservoir on each side of the BLM, individual excess to both membrane surfaces, and easy manual handling (Fig. 15E). This is seen as basic requirement of experiments copying the in vivo situation (e.g., plasmatic/exoplasmatic side). However, the surface properties of porous supports, such as roughness or great differences in pore size, have significantly impaired the stability of attached BLMs. Hence, the strategy to use SUMs with the S-layer as the stabilizing and smoothening biomimetic layer between the lipid membrane and the porous support is a straightforward approach.[325,339,340]

Composite SUM-supported DPhPC bilayers are structures that maintain its membrane resistance in the gigaohm range during their whole life time of up to 17 h.[325,339,340] In contrast, lipid membranes on plain MFMs revealed only a life time of approximately 3 h.[340] Interestingly, an additional monomolecular S-layer protein lattice recrystallized on the lipid-faced side, forming an S-layer–lipid membrane–S-layer sandwich-like structure, increased the life time significantly to about 1 day.[325,339] Stable membranes comprising tetraetherlipids, phospholipids, and their mixtures have also been generated on SUMs.[340] The capacitance of these electrically tight SUM-supported membranes increased continuously with increasing tetraetherlipid-to-phospholipid ratio. This result nicely demonstrated that the pure DPhPC membrane was thicker than membranes with a certain amount of tetraetherlipid and finally, the pure tetraetherlipid monolayer constituted the thinnest membrane.[340]

Incorporation of the membrane-active peptide gramicidin D resulted in high-resolution conductance measurements on single gramicidin D pores in all the above-mentioned S-layer-supported lipid membranes.[340] Functional reconstitution of αHL could be achieved with SUM-supported DPhPC bilayers, but no pore formation was observed with BLMs generated on pure MFMs.[339] However, even single pore recordings have been performed with αHL reconstituted in BLMs resting on an SUM.[339]

Solid-supported lipid membranes have been fabricated by several methods. A common feature of the S-layer lattice is not only to provide a stabilizing and defined tethering layer to decouple the BLM from the (inorganic) support but also to generate an ionic reservoir if desired. First, S-layer proteins have been self-assembled on glass and modified silicon surfaces before generating a BLM by the LB technique.[311,321] This composite structure has been compared with silane- and dextran-supported phospholipid bilayers.[321] Most probably due to the repetitive local interactions of the S-layer lattice with the lipid head groups, the nanopatterned fluidity of lipids was highest in the S-layer-supported lipid membranes compared to hybrid silane alkyl-phospholipid membranes or dextran-supported phospholipid bilayers, as determined by the fluorescence recovery after photobleaching. Phospholipid bilayers and tetraetherlipid monolayers have also been generated on S-layer-covered gold electrodes (Fig. 15F). The tetraetherlipid monolayer in between the S-layer covering the gold electrode and a second S-layer on the top revealed an exceptional long-term robustness of approximately 1 week.[13,167,318] If desired, a layer of thiolated SCWP may be chemisorbed on the gold surface prior recrystallization of the S-layer protein to enhance the long-range order and the smoothness of the S-layer lattice.[9,308,309] Hence, the nanopatterned anchoring of the membrane is a promising strategy for generating stable *and* fluid lipid membranes.

The functionality of these biomimetic membranes resting on solid supports has been investigated by the incorporation of the membrane-active peptides valinomycin, alamethicin, and gramicidin D.[325] S-layer-supported lipid membranes with incorporated valinomycin, a potassium-selective ion carrier, revealed a remarkable high resistance when bathed in a sodium buffer. In contrast, because of the valinomycin-mediated ion transport, a pronounced decrease in resistance by a factor of 500 was observed for the same membrane bathed in a potassium buffer.[325] Further, alamethicin channels could not only be incorporated in S-layer-supported lipid membranes on solid supports, but the conductive alamethicin channels could even be blocked as increasing amounts of inhibitor gave rise to a significantly increased membrane resistance.[325] Thus, proof of principle for the applicability of these composite structures for biosensing and screening purposes has been demonstrated. In future, the ability to reconstitute integral membrane proteins in defined structures on, for example, sensor surfaces is one of the most important concerns in designing biomimetic sensing devices.

B. S-Layer-Coated Liposomes and Emulsomes

Unilamellar liposomes are artificially prepared vesicles comprising of a phospholipid bilayer shell and an aqueous core.[341–343] The core can be filled with hydrophilic drugs, whereas the lipidic shell can be loaded with hydrophobic drugs. Emulsomes, however, are lipoidal vesicular systems with an internal solid fat core surrounded by a phospholipid mono- or bilayer.[344,345] Hence, emulsomes can be loaded with a much higher amount of lipidic drugs. Both spherical structures can be used for targeted drug delivery for cancer and other diseases.[345,346]

Isolated S-layer subunits were recrystallized on positively charged, unilamellar liposomes (Fig. 15G) composed of 1,2-dipalmitoyl-sn-glycero-3-phosphocholine (DPPC), cholesterol, and hexadecylamine in a molar ratio of 10:5:4.[168–170] The subunits attached to positively charged liposomes by their inner face (bearing a net negative charge) in an orientation identical to the lattice on intact cells. The S-layer protein, once recrystallized on liposomes, can be cross-linked with glutaraldehyde or bis(sulfosuccinimidyl)suberate to achieve stabilization of the whole supramolecular structure and can be utilized for covalent attachment of macromolecules.

Coating of the positively charged liposomes with the S-layer protein of G. stearothermophilus PV72/p2 resulted in inversion of the ζ-potential from $+29.1$ to -27.1 mV.[169] A similar behavior was also observed for liposomes coated with S-layer proteins from lactobacilli.[347] To study the influence of an S-layer lattice on the stability of liposomes, the hydrophilic marker carboxyfluorescein (CF) was encapsulated and its release was determined for plain and

S-layer-coated liposomes in the course of mechanical and thermal challenges. In comparison to plain liposomes, S-layer-coated liposomes released only half the amount of enclosed CF upon exposure to shear forces or ultrasonication as mechanical stress factors. Further, temperature shifts from 25 to 55 °C and vice versa induced considerably less CF release from S-layer-coated than from plain liposomes. A similar stabilizing effect of the S-layer lattice was observed after glutaraldehyde treatment of plain and S-layer-coated liposomes, although there was increased CF release in all glutaraldehyde-treated liposomes.[169] As chemical analysis revealed that almost all amino groups ($> 95\%$) from hexadecylamine in the liposomal membrane were involved in the cross-linking reaction, phase separation phenomena might be responsible for this observed behavior.[169,290]

The thermotropic phase behavior of S-layer-coated and uncoated liposomes was characterized by differential scanning microcalorimetry, indicating, for both preparations, a broad phase transition around 50 °C due to the chain-melting from a liquid-ordered gel-like to a liquid-ordered fluid phase as described for DPPC/cholesterol mixtures. The slightly higher phase transition temperature for the S-layer-coated liposomes was explained by increased intermolecular order. Covalent cross-linking of the S-layer subunits to hexadecylamine with glutaraldehyde induced phase separation within the liposomes. Based on deconvolution of the normalized excess heat capacity functions, it was proposed that the different lipid domains arise from phospholipids representing different degrees of mobility.[290] This is also in accordance with data of the CF release experiments.[169]

Sound velocity and density measurements have been used to study further physical properties of plain and S-layer-coated unilamellar liposomes.[289] It turned out that the adiabatic compressibility of S-layer-coated liposomes at $T < 20$ °C was higher and at $T > 20$ °C lower in comparison with that of plain liposomes. This provided evidence of an interesting phenomenon of softening and condensing effects of S-layer proteins on the liposomal lipid bilayer depending on the temperature.[289]

S-layer-coated liposomes have been investigated for their ability to act as a versatile system for entrapping and binding target molecules (Fig. 15G). A first study provided evidence that S-layer-coated liposomes constitute a proper matrix for the covalent attachment of macromolecules like ferritin.[168] The latter was used as a model system for demonstrating the suitability of S-layers attached to liposomes as immobilization matrices because ferritin does not penetrate into S-layer pores[184] and can be easily detected by electron microscopy procedures. Further, a targeted immobilization of immunoglobulins by bacterial S-layer proteins recrystallized on liposomes was exploited as immobilization matrix for antibody (Ab)-human IgG. The interaction of rabbit or swine anti-human IgG as antigens (Ag) was studied by measuring changes of the ultrasound velocity.[291] The ultrasound velocity decreased linearly with increase in Ag concentration. The decrease of ultrasound velocity was presumably caused by changes of hydration of the membrane due to the binding process. Finally, no substantial differences in the behavior of ultrasound velocity were observed for interactions of human IgG with rabbit or swine anti-human IgG.[291]

Another approach was the biotinylation of S-layer-coated liposomes. This was achieved by coupling p-diazobenzoyl biocytin, which preferably reacts with the phenolic residue of tyrosine or with the imidazole ring of histidine. By applying this method, two biotin residues accessible for subsequent avidin binding were introduced per S-layer subunit.[170] As visualized by labeling with biotinylated ferritin, an ordered monomolecular layer of streptavidin was formed on the surface of the S-layer-coated liposomes. As a second model system, biotinylated anti-human IgG was attached via the streptavidin bridge to the biotinylated S-layer-coated liposomes. The biological activity of the bound anti-human IgG was confirmed by ELISA.[170] Further, S-layer/streptavidin fusion proteins have been constructed, and hence biotinylated binding partners can be bound in a much better defined orientation and position. By applying this method, three biotin residues accessible for subsequent avidin binding were introduced per S-layer subunit.[98]

Another approach is to recrystallize functional chimeric S-layer fusion proteins carrying the sequence of EGFP on liposomes.[232] Because of the intrinsic EGFP fluorescence, the uptake of S-layer/EGFP fusion protein-coated liposomes into eukaryotic cells such as, for example, HeLa cells could nicely be visualized by confocal laser scanning microscopy. For instance, in HeLa cells the major part of the coated liposomes was internalized within 2 h of incubation by endocytosis.[232] With regard to further experiments, the most interesting advantage of these fusion proteins can be seen in recrystallization of this S-layer/EGFP fusion protein in combination with other S-layer fusion proteins such as the S-layer/streptavidin fusion proteins[98] on the same liposome surface. The uptake of these specially coated liposomes by target cells and the functionality of transported drugs could be investigated simultaneously without using any additional labeling.

Up to now, only preliminary studies have been performed with S-layer-coated emulsomes. Cationic emulsomes could be entirely covered with a crystalline S-layer lattice as demonstrated by transmission electron microscopical studies (Ücisik, M., unpublished data). Further, the coating of emulsomes with the S-layer proteins resulted in inversion of the ζ-potential from a positive value to a negative one. This shift in ζ-potential was in the same dimension as observed with liposomes (Ücisik, M., unpublished data).

To summarize, S-layer-protein-coated liposomes (Fig. 15G) are biomimetic structures with remarkably high mechanical and thermal stability.[169] Further, the possibility for entrapping and, most importantly, for immobilizing biologically active molecules[168,170] makes S-layer-coated liposomes and S-layer-coated emulsomes attractive for nanobiotechnological applications, particularly as carrier and/or drug delivery systems, as artificial virus envelopes in, for example, medicinal applications and in gene therapy.[13,170,319,348,349] These biomimetic approaches are exciting examples for synthetic biology mimicking structural and functional aspects of many bacterial and archaeal cell envelopes having S-layer lattices as the outermost component.[34]

IX. S-Layers as Matrix for Biomineralization

Currently, there is much interest in the synthesis of inorganic materials using biomimetic approaches. Unlike physical methods such as electron beam microlithography, there are relatively few reports of self-assembled organic templates being employed in the direct chemical synthesis of patterned arrays of inorganic nanoparticles.[220–223,350,351] Such materials are required as patterning and functional elements in molecular electronics and optics. But more than a decade ago and based on the investigation of mineral formation by bacteria in natural environments, S-layer lattices had already been used to generate periodic templates for the *in situ* nucleation of ordered arrays of uniform 5-nm sized cadmium sulfide (CdS) and gold (Au) nanoparticles.[222] Inorganic superlattices with either oblique or square lattice symmetry of approximately 10 nm repeat distance and particle sizes of 4–5 nm have been fabricated by exposing self-assembled S-layers to the respective metal salt solutions (e.g., Cd (II)) followed by either slow reaction with a reducing agent such as H_2S or by exposing the precipitated inorganic layer (attached to the S-layer) to the electron beam in a transmission electron microscope.[222,350] The latter approach is particularly interesting since nanoparticles were only formed at the exposed regions, allowing to gain spatial control over the array formation. Based on this approach, a broad range of further inorganic (mostly metallic) materials was used to fabricate nanoparticle arrays: palladium (Pd; salt: $PdCl_2$), nickel (Ni; $NiSO_4$), platinum (Pt; salt: $KPtCl_6$), lead (Pb; salt: $Pb(NO_3)_2$), and iron (Fe; salt: $(KFe(CN)_6)$).[220–223,350,351] As a general rule, the nanoparticles were mostly microcrystalline but not crystallographically aligned along the lattice lines of their superlattice. Further, the lattice spacing of the nanoparticles arrays resembled the lattice parameters of the underlying S-layer lattice (lattice symmetry and lattice spacings). As the precipitation of the metals was confined to the pores of the S-layer, the nanoparticles also resembled the morphology of the pores. Recently, small-spot X-ray photoelectron spectroscopy (XPS) was used to characterize the elemental composition of the nanoclusters.[219] XPS demonstrated that they consisted primarily of elemental gold. In addition to the precipitation of nanoparticles for applications in molecular electronics and optics, it must be noted here that recent investigations of the electronic structure of an S-layer protein revealed a semiconductor-like behavior with an energy gap value of ~ 3.0 eV and the Fermi energy close to the bottom of the lowest unoccupied molecular orbital.[352–354]

In a different work, S-layer protein lattices were used as scaffolds in the precipitation of biogenic silica and titania.[355] The current understanding of the key proteins (silicateins and sillafins), genes, and molecular mechanisms involved in the bioinspired formation of silica structures laid the foundation for investigating the biocatalytic activity of S-layer proteins and their self-assembly products as catalysts, templates, and scaffolds for the directed growth of silica and titania into novel nano- to micrometer sized structures. Based on established and published protocols for silica and titania synthesis, *in vitro* investigations were focused on native S-layer proteins and on genetically engineered S-layer fusion proteins involving silica and titania precipitating peptides as exposed surface functionalities.

Precipitation of tetramethoxysilane led to the formation of silica layers and of titanium(IV) bis(ammonium lactato)dihydroxide to titania layers on S-layer lattices (Göbel, C., personal communication). It was shown that the silicification of S-layer lattices led to a downsizing of the diameters of the S-layer pores (comparable to closing an iris diaphragm). This result is important for the development of nanoporous materials as, for example, used in fuel cells. Further, genetically engineered S-layer–rSilC (recombinant sillafin C (rSilC)) fusion protein lattices were coated with silica and titania.[356] In particular, due to its high refractive index, rutile titania would be a highly desirable material for applications in nano-optics.

X. Conclusion and Perspectives

The study of biological self-assembly systems is a new and rapidly growing scientific and engineering field that crosses the boundaries of different disciplines. Although self-assembly processes are common in biosystems, there are only a few examples where proteins possess the intrinsic capability to aggregate into monomolecular crystalline arrays.

S-layers composed of a single protein or glycoprotein species represent the simplest self-assembling membrane developed during biological evolution. Moreover, S-layers are now recognized as one of the most common cell surface component of prokaryotic organisms and consequently one of the most abundant biopolymers on earth. S-layers reveal different lattice types ($p1, p2, p3, p4, p6$), lattice constants (~ 5–30 nm), and a great diversity regarding their constituent (glyco)protein subunits. Moreover, S-layers that share almost identical lattice parameters can have dissimilar molecular sequences.[20,27] Those that possess similar physicochemical characteristics or functions might not be related at all to one another. This raises the question: are they an example of parallel evolution or a common structural theme, or an example of extreme divergence from a single ancient structure?[34]

Currently, most nanobiotechnological applications based on S-layers depend on the *in vitro* self-assembling capabilities of isolated S-layer subunits in suspension and on surfaces of solids, lipid films, liposomes, emulsomes, and nanoparticles. Most important, S-layer proteins also assemble as coherent layers on highly corrugated and porous structures.

The wealth of information accumulated on the structure, chemistry, morphogenesis, function, and genetics of S-layers has led to a broad spectrum of applications in areas of both life- and materials sciences. Presently, the most important area of development for S-layer technologies concerns changes of the natural properties of S-layer proteins or glycoproteins by genetic manipulation. Most important, S-layer proteins incorporating specific single- or multifunctional domains of proteins (e.g., antibodies, antigens, ligands, enzymes, fluorescent proteins, and peptide mimotopes) maintain the capability to assemble into coherent lattices on a great variety of solid supports. Such S-layer fusion proteins have already revealed a broad application potential for the production of affinity matrices, diagnostics, vaccines, biocompatible and antifouling surfaces, microcarriers, and matrices for biomineralization. It is expected that many other potentials for nanobiotechnological and biomimetic applications will emerge. A major advantage in using S-layer fusion proteins for functionalizing solids relates to the observation that functional domains associated with S-layers are more resistant to denaturation and thus exhibit a prolonged life time. Moreover, S-layer (fusion)proteins can be considerably strengthened by introducing intermolecular and/or intramolecular bonds. Up to now, fusion proteins have been produced with S-layers from organisms dwelling at moderate environmental conditions but even higher stability of fusion proteins may be achieved if S-layers of archaea are used that grow in extreme habitats (e.g., up to 120 °C, pH 0). An important line of development concerns strategies for copying the supramolecular principle of cell envelopes of archaea that possess S-layers as exclusive wall component and inhabit thermophilic and acidophilic environments. This biomimetic approach is expected to lead to new technologies for stabilizing functional lipid membranes and their use at the mesoscopic and macroscopic scale.[13] Such composite structures have the potential for generating very stable, long-lasting plane and vesicular membrane systems incorporating functional proteins as required for biosensors, photovoltaics, and high-throughput screening. Moreover, preliminary studies have clearly demonstrated that S-layer technologies have a great potential for nanopatterning of surfaces, biological templating, and functionalizing microfluidic devices. S-layer lattices may also enable the defined generation and deposition of metal or semiconductor nanoparticles across macroscopic surface areas as required for nonlife-science applications (e.g., nanoelectronics, nonlinear optics, or catalytic substrates).[6] Another area of future development of S-layers concerns their utilization as cell surface display system of rationally designed glycosylation motifs. Controlled S-layer glycosylation may add a new and very valuable component to an S-layer-based molecular construction kit as required for receptor mimics, vaccine design, and drug delivery involving carbohydrate recognition.[14,67]

Finally, a most challenging question remains unanswered: Could S-layer-like membranes fulfill barrier and supporting functions as required for self-reproducing systems at the beginning of life?[34,147]

ACKNOWLEDGMENTS

Financial support from the Austrian Science Fund (FWF, projects P18510-B12 and P20256-B11), the EU project MemS (Grant Agreement No. 244967), the Erwin-Schrödinger Society for Nanosciences, the Austrian NanoInitiative under the project SLAYSENS within the project cluster ISOTEC, and the Air Force Office of Scientific Research, USA (AFOSR, projects FA9550-06-1-0208, FA9550-07-1-0313, FA9550-09-1-0342, and FA9550-10-1-0223) is gratefully acknowledged.

REFERENCES

1. Papapostolou D, Howorka S. Engineering and exploiting protein assemblies in synthetic biology. *Mol Biosyst* 2009;**5**:723–32.
2. Sleytr UB, Sára M, Pum D, Schuster B. Crystalline bacterial cell surface layers (S-layers): a versatile self-assembly system. In: Ciferri A, editor. *Supramolecular polymers.* 2nd ed. Boca Raton: CRC Press LLC; 2005. pp. 583–612.
3. Egelseer EM, Ilk N, Pum D, Messner P, Schäffer C, Schuster B, Sleytr UB. S-layers, microbial, biotechnological applications. In: Flickinger MC, editor. *The encyclopedia of industrial biotechnology: bioprocess, bioseparation, and cell technology,*Vol. 7. Hoboken, USA: John Wiley & Sons, Inc.; 2010. pp. 4424–48.
4. Egelseer EM, Sára M, Pum D, Schuster B, Sleytr UB. Genetically engineered S-layer proteins and S-layer-specific heteropolysaccharides as components of a versatile molecular construction kit for applications in nanobiotechnology. In: Shoseyov O, Levy I, editors. *NanoBioTechnology.* Totowa, New Jersey: Humana Press; 2008. pp. 55–86.
5. Ilk N, Egelseer EM, Ferner-Ortner J, Küpcü S, Pum D, Schuster B, Sleytr UB. Surfaces functionalized with self-assembling S-layer fusion proteins for nanobiotechnological applications. *Colloids Surf A Physicochem Eng Aspects* 2008;**321**:163–7.
6. Pum D, Sleytr UB. Protein-based nanobioelectronics. In: Offenhäusser A, Rinaldi R, editors. *Nanobioelectronics for electronics, biology, and medicine.* Berlin, Germany: Springer; 2009. pp. 167–80.
7. Sára M, Egelseer EM, Huber C, Ilk N, Pleschberger M, Pum D, Sleytr UB. S-layer proteins: potential application in nano(bio)technology. In: Rehm B, editor. *Microbial bionanotechnology: biological self-assembly systems and biopolymer-based nanostructures.* New Zealand: Bernd Rehm Massey University, PN; 2006. pp. 307–38.

8. Sára M, Pum D, Schuster B, Sleytr UB. S-layers as patterning elements for application in nanobiotechnology. *J Nanosci Nanotechnol* 2005;**5**:1939–53.

9. Schuster B, Györvary E, Pum D, Sleytr UB. Nanotechnology with S-layer proteins. *Methods Mol Biol* 2005;**300**:101–23.

10. Schuster, B. & Sletyr, U. B. (in press). Nanotechnology with S-layer proteins. In *Protein nanotechnology: protocols, instrumentation and applications* (Gerrard, J., ed.). Humana Press, Heidelberg.

11. Schuster B, Sleytr UB. Biomimetic S-layer supported lipid membranes. *Curr Nanosci* 2006;**2**:143–52.

12. Schuster B, Sleytr UB. Fabrication and characterization of functionalized S-layer supported lipid membranes. In: Bernstein EM, editor. *Bioelectrochemistry research developments.* Hauppauge, NY: Nova Science Publishers Inc.; 2008. pp. 105–24.

13. Schuster B, Sleytr UB. Composite S-layer lipid structures. *J Struct Biol* 2009;**168**:207–16.

14. Sleytr UB, Egelseer EM, Ilk N, Messner P, Schäffer C, Pum D, Schuster B. Nanobiotechnological applications of S-layers. In: König H, Claus H, Varma A, editors. *Prokaryotic cell wall compounds—structure and biochemistry.* Heidelberg, Germany: Springer; 2010. pp. 459–81.

15. Sleytr UB, Egelseer EM, Ilk N, Pum D, Schuster B. S-layers as a basic building block in a molecular construction kit. *FEBS J* 2007;**274**:323–34.

16. Sleytr UB, Egelseer EM, Pum D, Schuster B. S-layers. In: Niemeyer CM, Mirkin CA, editors. *Nanobiotechnology: concepts, methods and perspectives.* Vol. 7. Weinheim, Germany: Wiley-VCH; 2004. pp. 77–92.

17. Sleytr UB, Messner P, Pum D, Sára M. Crystalline bacterial cell surface layers (S-layers): from supramolecular cell structure to biomimetics and nanotechnology. *Angew. Chem. Int. Ed.* 1999;**38**:1034–54.

18. Sleytr, U. B., Pum, D., Egelseer, E., Ilk, N. & Schuster, B. (in press). S-layer proteins. In *Handbook of biofunctional surfaces* (Knoll, W., ed.). Pan Stanford Publishing.

19. Sleytr UB, Sára M, Pum D, Schuster B. Molecular nanotechnology and nanobiotechnology with two-dimensional protein crystals (S-layers). In: Rosoff M, editor. *Nano-surface chemistry.* New York, Basel: Marcel Dekker, Inc.; 2001. pp. 333–89.

20. Sleytr UB, Sára M, Pum D, Schuster B, Messner P, Schäffer C. Self-assembly protein systems: microbial S-layers. In: Steinbüchel A, Fahnestock SR, editors. *Biopolymers.* 1st Ed. *Polyamides and complex proteinaceous materials I* Vol. 7. Weinheim: Wiley-VCH; 2002. pp. 285–338.

21. Houwink AL, Le Poole JB. Eine Struktur in der Zellwand einer Bakterie. *Physikalische Verhandlung* 1952;**98**.

22. Sleytr UB, Adam H, Klaushofer H. Die elektronenmikroskopische Feinstruktur von Zellwand, Cytoplasmamembran und Geißeln von *Bacillus stearothermophilus*, dargestellt mit Hilfe der Gefrierätztechnik. *Mikroskopie* 1967;**22**:233–42.

23. Sleytr UB, Adam H, Klaushofer H. Die Feinstruktur der Zellwandoberfläche von zwei thermophilen Clostridienarten, dargestellt mit Hilfe der Gefrierätztechnik. *Mikroskopie* 1968;**23**:1–10.

24. Sleytr UB. Regular arrays of macromolecules on bacterial cell walls: structure, chemistry, assembly, and function. *Int Rev Cytol* 1978;**53**:1–62.

25. Sleytr UB, Messner P, Pum D, Sára M, editors. *Crystalline bacterial cell surface layers.* Berlin: Springer; 1988.

26. Claus H, Akca E, Debaerdemaeker T, Evrard C, Declercq JP, Harris JR, Schlott B, König H. Molecular organization of selected prokaryotic S-layer proteins. *Can J Microbiol* 2005;**51**:731–43.

27. Messner P, Schäffer C, Egelseer EM, Sleytr UB. Occurence, structure, chemistry, genetics, morphogenesis, and function of S-layers. In: König H, Claus H, Varma A, editors. *Prokaryotic cell wall compounds-structure and biochemistry.* Berlin: Springer-Verlag; 2010. pp. 53–109.

28. Sleytr UB, Messner P, Pum D, Sára M. *Crystalline bacterial cell surface proteins. Molecular biology intelligence unit.* Austin: R.G. Landes Comp., and San Diego: Academic Press; 1996.

29. Sleytr UB, Messner P, Pum D, Sára M. Crystalline surface layers on eubacteria and archaeobacteria. In: Sleytr UB, Messner P, Pum D, Sára M, editors. *Crystalline bacterial cell surface proteins.* Austin, Texas, USA: R.G. Landes Company and Academic Press Inc.; 1996. pp. 211–25.

30. Baumeister W, Lembcke G. Structural features of archaebacterial cell envelopes. *J Bioenerg Biomembr* 1992;**24**:567–75.

31. Baumeister W, Wildhaber I, Phipps BM. Principles of organization in eubacterial and archaebacterial surface proteins. *Can J Microbiol* 1989;**35**:215–27.

32. Hovmöller S, Sjogren A, Wang DN. The structure of crystalline bacterial surface layers. *Prog Biophys Mol Biol* 1988;**51**:131–63.

33. Sleytr UB. Heterologous reattachment of regular arrays of glycoproteins on bacterial surfaces. *Nature* 1975;**257**:400–2.

34. Sleytr UB, Beveridge TJ. Bacterial S-layers. *Trends Microbiol* 1999;**7**:253–60.

35. Sleytr UB, Glauert AM. Bacterial cell walls and membranes. In: Harris JR, editor. *Electron microscopy of proteins* Vol. 3. London: Academic Press Inc.; 1982. pp. 41–76.

36. Sleytr UB, Messner P. Crystalline surface layers on bacteria. *Annu Rev Microbiol* 1983;**37**:311–39.

37. Sleytr UB, Messner P. Self-assembly of crystalline bacterial cell surface layers (S-layers). In: Plattner H, editor. *Electron microscopy of subcellular dynamics.* Boca Raton, Florida: CRC Press; 1989. pp. 13–31.

38. Sleytr UB, Messner P, Pum D. Analysis of crystalline bacterial surface layers by freezeteching, metal shadowing, negative staining and ultrathin sectioning. *Methods Microbiol* 1988;**20**:29–60.

39. Sleytr UB, Glauert AM. Analysis of regular arrays of subunits on bacterial surfaces: evidence for a dynamic process of assembly. *J Ultrastruct Res* 1975;**50**:103–16.

40. Beveridge TJ. Bacterial S-layers. *Curr Opin Struct Biol* 1994;**4**:204–12.

41. Hovmöller S. Crystallographic image processing applications for S-layers. In: Beveridge TJ, Koval S, editors. *Advances in paracrystalline bacterial surface layers. NATO ASI series, Series A: life sciences,* Vol. 252. New York: Plenum Press; 1993. pp. 13–21.

42. Müller DJ, Baumeister W, Engel A. Conformational change of the hexagonally packed intermediate layer of *Deinococcus radiodurans* monitored by atomic force microscopy. *J Bacteriol* 1996;**178**:3025–30.

43. Pum D, Tang J, Hinterdorfer P, Tocca Herrera JL, Sletyr UB. S-layer protein lattices studied by scanning force microscopy. In: Kumar CSSR, editor. *Biomimetic and bioinspired nanomaterials.* Weinheim, Germany: Wiley-VCH Verlag; 2010. pp. 459–510.

44. Engelhardt H. Are S-layers exoskeletons? The basic function of protein surface layers revisited. *J Struct Biol* 2007;**160**:115–24.

45. Veith A, Klingl A, Zolghadr B, Lauber K, Mentele R, Lottspeich F, Rachel R, Albers SV, Kletzin A. *Acidianus, Sulfolobus* and *Metallosphaera* surface layers: structure, composition and gene expression. *Mol Microbiol* 2009;**73**:58–72.

46. Bergey DH, Holt JG, Krieg NR, Sneath PHA. *Bergey's manual of determinative bacteriology.* Baltimore: Lippincott Williams & Wilkins; 1994.

47. Gram HC. Über die isolierte Färbung der *Schizomyceten* in Schnitt- und Trockenpräparaten. *Fortschr Med* 1884;**2**:185–9.

48. Beveridge TJ. Use of the gram stain in microbiology. *Biotech Histochem* 2001;**76**:111–8.

49. Sára M, Sleytr UB. S-layer proteins. *J Bacteriol* 2000;**182**:859–68.
50. Sleytr U, Messner P. Crystalline bacterial cell surface layers (S-layers). In: Lederberg J, editor. *Encyclopedia of microbiology*, Vol. 2. San Diego: Academic Press Inc.; 2000. pp. 899–906.
51. Sleytr UB, Sára M, Pum D, Schuster B. Characterization and use of crystalline bacterial cell surface layers. *Prog Surf Sci* 2001;**68**:231–78.
52. Engelhardt H. Mechanism of osmoprotection by archaeal S-layers: a theoretical study. *J Struct Biol* 2007;**160**:190–9.
53. Lortal S, van Heijenoort J, Gruber K, Sleytr UB. S-layer of *Lactobacillus helveticus* ATCC 12046: isolation, chemical characterization and re-formation after extraction with lithium chloride. *J Gen Microbiol* 1992;**138**:611–8.
54. Thompson SA, Shedd OL, Ray KC, Beins MH, Jorgensen JP, Blaser MJ. *Campylobacter fetus* surface layer proteins are transported by a type I secretion system. *J Bacteriol* 1998;**180**:6450–8.
55. Messner P, Sleytr UB. Separation and purification of S-layers from gram-positive and gram-negative bacteria. In: Hancock IC, Poxton IR, editors. *Bacterial cell surface techniques*. Chichester: John Wiley & Sons; 1988. pp. 97–104.
56. Sleytr UB, Glauert AM. Ultrastructure of the cell walls of two closely related clostridia that possess different regular arrays of surface subunits. *J Bacteriol* 1976;**126**:869–82.
57. Bröckl G, Behr M, Fabry S, Hensel R, Kaudewitz H, Biendl E, Konig H. Analysis and nucleotide sequence of the genes encoding the surface-layer glycoproteins of the hyperther-mophilic methanogens *Methanothermus fervidus* and *Methanothermus sociabilis*. *Eur J Biochem* 1991;**199**:147–52.
58. Kosma P, Wugeditsch T, Christian R, Zayni S, Messner P. Glycan structure of a heptose-containing S-layer glycoprotein of *Bacillus thermoaerophilus*. *Glycobiology* 1995;**5**:791–6.
59. Pei Z, Ellison 3rd RT, Lewis RV, Blaser MJ. Purification and characterization of a family of high molecular weight surface-array proteins from *Campylobacter fetus*. *J Biol Chem* 1988;**263**:6416–20.
60. Lechner J, Sumper M. The primary structure of a procaryotic glycoprotein. Cloning and sequencing of the cell surface glycoprotein gene of halobacteria. *J Biol Chem* 1987;**262**:9724–9.
61. Trachtenberg S, Pinnick B, Kessel M. The cell surface glycoprotein layer of the extreme halophile *Halobacterium salinarum* and its relation to *Haloferax volcanii*: cryo-electron tomography of freeze-substituted cells and projection studies of negatively stained envelopes. *J Struct Biol* 2000;**130**:10–26.
62. Sleytr UB, Messner P, Pum D, Sara M. Crystalline bacterial cell surface layers. *Mol Microbiol* 1993;**10**:911–6.
63. Sumper M, Wieland FT. Bacterial glycoproteins. In: Montreuil J, Vliegenthart JFG, Schachter H, editors. *Glycoproteins*. Amsterdam: Elsevier; 1995. pp. 455–73.
64. Messner P, Sleytr UB. Crystalline bacterial cell-surface layers. In: Rose AH, editor. *Advances in microbial physiology* Vol. 33. London: Academic Press Inc.; 1992. pp. 213–75.
65. Sára M, Sleytr UB. Crystalline bacterial cell surface layers (S-layers): from cell structure to biomimetics. *Prog Biophys Mol Biol* 1996;**65**:83–111.
66. Kuen B, Lubitz W, Sleytr UB, Messner P, Pum D, Sára M. Analysis of S-layer proteins and genes. In: Sleytr U, Messner P, Pum D, Sára M, editors. *Crystalline bacterial cell surface layer proteins*. Austin, TX: R.G. Landes Comp./Academic Press Inc.; 1996. pp. 77–102.
67. Messner P, Egelseer EM, Sleytr UB, Schäffer C. Bacterial surface layer glycoproteins and "non-classical" secondary cell wall polymers. In: Moran A, Holst O, Brennan PJ, von Itzstein M, editors. *Microbial glycobiology: structures, relevance and applications*. San Diego: Elsevier; 2009. pp. 109–28.

68. Messner P, Steiner K, Zarschler K, Schäffer C. S-layer nanoglycobiology of bacteria. *Carbohydr Res* 2008;**343**:1934–51.

69. Akca E, Claus H, Schultz N, Karbach G, Schlott B, Debaerdemaeker T, Declercq JP, König H. Genes and derived amino acid sequences of S-layer proteins from mesophilic, thermophilic, and extremely thermophilic methanococci. *Extremophiles* 2002;**6**:351–8.

70. Avall-Jääskeläinen S, Palva A. *Lactobacillus* surface layers and their applications. *FEMS Microbiol Rev* 2005;**29**:511–29.

71. Eidhin DN, Ryan AW, Doyle RM, Walsh JB, Kelleher D. Sequence and phylogenetic analysis of the gene for surface layer protein, *slpA*, from 14 PCR ribotypes of *Clostridium difficile*. *J Med Microbiol* 2006;**55**:69–83.

72. Kato H, Yokoyama T, Arakawa Y. Typing by sequencing the *slpA* gene of *Clostridium difficile* strains causing multiple outbreaks in Japan. *J Med Microbiol* 2005;**54**:167–71.

73. Poilane I, Humeniuk-Ainouz C, Durand I, Janoir C, Cruaud P, Delmee M, Popoff MR, Collignon A. Molecular characterization of *Clostridium difficile* clinical isolates in a geriatric hospital. *J Med Microbiol* 2007;**56**:386–90.

74. Radnedge L, Agron PG, Hill KK, Jackson PJ, Ticknor LO, Keim P, Andersen GL. Genome differences that distinguish *Bacillus anthracis* from *Bacillus cereus* and *Bacillus thuringiensis*. *Appl Environ Microbiol* 2003;**69**:2755–64.

75. Tu ZC, Zeitlin G, Gagner JP, Keo T, Hanna BA, Blaser MJ. *Campylobacter fetus* of reptile origin as a human pathogen. *J Clin Microbiol* 2004;**42**:4405–7.

76. Huber C, Ilk N, Rünzler D, Egelseer EM, Weigert S, Sleytr UB, Sára M. The three S-layer-like homology motifs of the S-layer protein SbpA of *Bacillus sphaericus* CCM 2177 are not sufficient for binding to the pyruvylated secondary cell wall polymer. *Mol Microbiol* 2005;**55**:197–205.

77. Ilk N, Völlenkle C, Egelseer EM, Breitwieser A, Sleytr UB, Sára M. Molecular characterization of the S-layer gene, *sbpA*, of *Bacillus sphaericus* CCM 2177 and production of a functional S-layer fusion protein with the ability to recrystallize in a defined orientation while presenting the fused allergen. *Appl Environ Microbiol* 2002;**68**:3251–60.

78. Jarosch M, Egelseer EM, Huber C, Moll D, Mattanovich D, Sleytr UB, Sára M. Analysis of the structure-function relationship of the S-layer protein SbsC of *Bacillus stearothermophilus* ATCC 12980 by producing truncated forms. *Microbiology* 2001;**147**:1353–63.

79. Kroutil M, Pavkov T, Birner-Gruenberger R, Tesarz M, Sleytr UB, Egelseer EM, Keller W. Towards the structure of the C-terminal part of the S-layer protein SbsC. *Acta Crystallogr Sect F Struct Biol Cryst Commun* 2009;**65**:1042–7.

80. Pavkov T, Egelseer EM, Tesarz M, Svergun DI, Sleytr UB, Keller W. The structure and binding behavior of the bacterial cell surface layer protein SbsC. *Structure* 2008;**16**:1226–37.

81. Horejs C, Pum D, Sleytr UB, Tscheliessnig R. Structure prediction of an S-layer protein by the mean force method. *J Chem Phys* 2008;**128**:1–11 065106.

82. Kinns H, Badelt-Lichtblau H, Egelseer EM, Sleytr UB, Howorka S. Identifying assembly-inhibiting and assembly-tolerant sites in the SbsB S-layer protein from *Geobacillus stearothermophilus*. *J Mol Biol* 2010;**395**:742–53.

83. Howorka S, Sára M, Wang Y, Kuen B, Sleytr UB, Lubitz W, Bayley H. Surface-accessible residues in the monomeric and assembled forms of a bacterial surface layer protein. *J Biol Chem* 2000;**275**:37876–86.

84. Kinns H, Howorka S. The surface location of individual residues in a bacterial S-layer protein. *J Mol Biol* 2008;**377**:589–604.

85. Vilen H, Hynonen U, Badelt-Lichtblau H, Ilk N, Jaaskelainen P, Torkkeli M, Palva A. Surface location of individual residues of SlpA provides insight into the *Lactobacillus brevis* S-layer. *J Bacteriol* 2009;**191**:3339–49.

86. Lupas A, Engelhardt H, Peters J, Santarius U, Volker S, Baumeister W. Domain structure of the *Acetogenium kivui* surface layer revealed by electron crystallography and sequence analysis. *J Bacteriol* 1994;**176**:1224–33.

87. Cava F, de Pedro MA, Schwarz H, Henne A, Berenguer J. Binding to pyruvylated compounds as an ancestral mechanism to anchor the outer envelope in primitive bacteria. *Mol Microbiol* 2004;**52**:677–90.

88. Chauvaux S, Matuschek M, Beguin P. Distinct affinity of binding sites for S-layer homologous domains in *Clostridium thermocellum* and *Bacillus anthracis* cell envelopes. *J Bacteriol* 1999;**181**:2455–8.

89. Ilk N, Kosma P, Puchberger M, Egelseer EM, Mayer HF, Sleytr UB, Sára M. Structural and functional analyses of the secondary cell wall polymer of *Bacillus sphaericus* CCM 2177 that serves as an S-layer-specific anchor. *J Bacteriol* 1999;**181**:7643–6.

90. Lemaire M, Miras I, Gounon P, Beguin P. Identification of a region responsible for binding to the cell wall within the S-layer protein of *Clostridium thermocellum*. *Microbiology* 1998;**144**:211–7.

91. Mader C, Huber C, Moll D, Sleytr UB, Sára M. Interaction of the crystalline bacterial cell surface layer protein SbsB and the secondary cell wall polymer of *Geobacillus stearothermophilus* PV72 assessed by real-time surface plasmon resonance biosensor technology. *J Bacteriol* 2004;**186**:1758–68.

92. Mesnage S, Fontaine T, Mignot T, Delepierre M, Mock M, Fouet A. Bacterial SLH domain proteins are non-covalently anchored to the cell surface via a conserved mechanism involving wall polysaccharide pyruvylation. *EMBO J* 2000;**19**:4473–84.

93. Mesnage S, Tosi-Couture E, Mock M, Fouet A. The S-layer homology domain as a means for anchoring heterologous proteins on the cell surface of *Bacillus anthracis*. *J Appl Microbiol* 1999;**87**:256–60.

94. Ries W, Hotzy C, Schocher I, Sleytr UB, Sára M. Evidence that the N-terminal part of the S-layer protein from *Bacillus stearothermophilus* PV72/p2 recognizes a secondary cell wall polymer. *J Bacteriol* 1997;**179**:3892–8.

95. Rünzler D, Huber C, Moll D, Kohler G, Sára M. Biophysical characterization of the entire bacterial surface layer protein SbsB and its two distinct functional domains. *J Biol Chem* 2004;**279**:5207–15.

96. May A, Pusztahelyi T, Hoffmann N, Fischer RJ, Bahl H. Mutagenesis of conserved charged amino acids in SLH domains of *Thermoanaerobacterium thermosulfurigenes* EM1 affects attachment to cell wall sacculi. *Arch Microbiol* 2006;**185**:263–9.

97. Petersen BO, Sára M, Mader C, Mayer HF, Sleytr UB, Pabst M, Puchberger M, Krause E, Hofinger A, Duus JO, Kosma P. Structural characterization of the acid-degraded secondary cell wall polymer of *Geobacillus stearothermophilus* PV72/p2. *Carbohydr Res* 2008;**343**:1346–58.

98. Moll D, Huber C, Schlegel B, Pum D, Sleytr UB, Sára M. S-layer-streptavidin fusion proteins as template for nanopatterned molecular arrays. *Proc Natl Acad Sci USA* 2002;**99**:14646–51.

99. Egelseer EM, Leitner K, Jarosch M, Hotzy C, Zayni S, Sleytr UB, Sára M. The S-layer proteins of two *Bacillus stearothermophilus* wild-type strains are bound via their N-terminal region to a secondary cell wall polymer of identical chemical composition. *J Bacteriol* 1998;**180**:1488–95.

100. Jarosch M, Egelseer EM, Mattanovich D, Sleytr UB, Sára M. S-layer gene *sbsC* of *Bacillus stearothermophilus* ATCC 12980: molecular characterization and heterologous expression in *Escherichia coli*. *Microbiology* 2000;**146**(Pt. 2):273–81.

101. Schäffer C, Kählig H, Christian R, Schulz G, Zayni S, Messner P. The diacetamidodideoxyuronic-acid-containing glycan chain of *Bacillus stearothermophilus* NRS 2004/3a represents

the secondary cell-wall polymer of wild-type *B. stearothermophilus* strains. *Microbiology* 1999;**145**:1575–83.

102. Schäffer C, Wugeditsch T, Kählig H, Scheberl A, Zayni S, Messner P. The surface layer (S-layer) glycoprotein of *Geobacillus stearothermophilus* NRS 2004/3a. Analysis of its glycosylation. *J Biol Chem* 2002;**277**:6230–9.

103. Egelseer EM, Danhorn T, Pleschberger M, Hotzy C, Sleytr UB, Sára M. Characterization of an S-layer glycoprotein produced in the course of S-layer variation of *Bacillus stearothermophilus* ATCC 12980 and sequencing and cloning of the sbsD gene encoding the protein moiety. *Arch Microbiol* 2001;**177**:70–80.

104. Weis WI. Cell-surface carbohydrate recognition by animal and viral lectins. *Curr Opin Struct Biol* 1997;**7**:624–30.

105. Ferner-Ortner J, Mader C, Ilk N, Sleytr UB, Egelseer EM. High-affinity interaction between the S-layer protein SbsC and the secondary cell wall polymer of *Geobacillus stearothermophilus* ATCC 12980 determined by surface plasmon resonance technology. *J Bacteriol* 2007;**189**:7154–8.

106. Antikainen J, Anton L, Sillanpaa J, Korhonen TK. Domains in the S-layer protein CbsA of *Lactobacillus crispatus* involved in adherence to collagens, laminin and lipoteichoic acids and in self-assembly. *Mol Microbiol* 2002;**46**:381–94.

107. Smit E, Oling F, Demel R, Martinez B, Pouwels PH. The S-layer protein of *Lactobacillus acidophilus* ATCC 4356: identification and characterisation of domains responsible for S-protein assembly and cell wall binding. *J Mol Biol* 2001;**305**:245–57.

108. Masuda K, Kawata T. Reassembly of the regularly arranged subunits in the cell wall os *Lactobacillus brevis* and their reattachment to cell walls. *Microbiol Immunol* 1980;**24**:299–308.

109. Masuda K, Kawata T. Reassembly of a regularly arranged protein in the cell wall of *Lactobacillus buchneri* and its reattachment to cell walls: chemical modification studies. *Microbiol Immunol* 1985;**29**:927–38.

110. Doig P, McCubbin WD, Kay CM, Trust TJ. Distribution of surface-exposed and non-accessible amino acid sequences among the two major structural domains of the S-layer protein of *Aeromonas salmonicida*. *J Mol Biol* 1993;**233**:753–65.

111. Dworkin J, Tummuru MK, Blaser MJ. A lipopolysaccharide-binding domain of the *Campylobacter fetus* S-layer protein resides within the conserved N terminus of a family of silent and divergent homologs. *J Bacteriol* 1995;**177**:1734–41.

112. Yang LY, Pei ZH, Fujimoto S, Blaser MJ. Reattachment of surface array proteins to *Campylobacter fetus* cells. *J Bacteriol* 1992;**174**:1258–67.

113. Bingle WH, Nomellini JF, Smit J. Linker mutagenesis of the *Caulobacter crescentus* S-layer protein: toward a definition of an N-terminal anchoring region and a C-terminal secretion signal and the potential for heterologous protein secretion. *J Bacteriol* 1997;**179**:601–11.

114. Bingle WH, Nomellini JF, Smit J. Secretion of the *Caulobacter crescentus* S-layer protein: further localization of the C-terminal secretion signal and its use for secretion of recombinant proteins. *J Bacteriol* 2000;**182**:3298–301.

115. Ford MJ, Nomellini JF, Smit J. S-layer anchoring and localization of an S-layer-associated protease in *Caulobacter crescentus*. *J Bacteriol* 2007;**189**:2226–37.

116. Boot HJ, Kolen CP, Pouwels PH. Identification, cloning, and nucleotide sequence of a silent S-layer protein gene of *Lactobacillus acidophilus* ATCC 4356 which has extensive similarity with the S-layer protein gene of this species. *J Bacteriol* 1995;**177**:7222–30.

117. Boot HJ, Kolen CP, Pouwels PH. Interchange of the active and silent S-layer protein genes of *Lactobacillus acidophilus* by inversion of the chromosomal slp segment. *Mol Microbiol* 1996;**21**:799–809.

118. Calabi E, Ward S, Wren B, Paxton T, Panico M, Morris H, Dell A, Dougan G, Fairweather N. Molecular characterization of the surface layer proteins from *Clostridium difficile*. *Mol Microbiol* 2001;**40**:1187–99.
119. Cerquetti M, Molinari A, Sebastianelli A, Diociaiuti M, Petruzzelli R, Capo C, Mastrantonio P. Characterization of surface layer proteins from different *Clostridium difficile* clinical isolates. *Microb Pathog* 2000;**28**:363–72.
120. Dworkin J, Blaser MJ. Molecular mechanisms of *Campylobacter fetus* surface layer protein expression. *Mol Microbiol* 1997;**26**:433–40.
121. Jakava-Viljanen M, Avall-Jääskeläinen S, Messner P, Sleytr UB, Palva A. Isolation of three new surface layer protein genes (*slp*) from *Lactobacillus brevis* ATCC 14869 and characterization of the change in their expression under aerated and anaerobic conditions. *J Bacteriol* 2002;**184**:6786–95.
122. Karjalainen T, Waligora-Dupriet AJ, Cerquetti M, Spigaglia P, Maggioni A, Mauri P, Mastrantonio P. Molecular and genomic analysis of genes encoding surface-anchored proteins from *Clostridium difficile*. *Infect Immun* 2001;**69**:3442–6.
123. Mignot T, Mesnage S, Couture-Tosi E, Mock M, Fouet A. Developmental switch of S-layer protein synthesis in *Bacillus anthracis*. *Mol Microbiol* 2002;**43**:1615–27.
124. Sára M, Kuen B, Mayer HF, Mandl F, Schuster KC, Sleytr UB. Dynamics in oxygen-induced changes in S-layer protein synthesis from *Bacillus stearothermophilus* PV72 and the S-layer-deficient variant T5 in continuous culture and studies of the cell wall composition. *J Bacteriol* 1996;**178**:2108–17.
125. Sára M, Sleytr UB. Comparative studies of S-layer proteins from *Bacillus stearothermophilus* strains expressed during growth in continuous culture under oxygen-limited and non-oxygen-limited conditions. *J Bacteriol* 1994;**176**:7182–9.
126. Thompson SA, Blaser MJ. Pathogenesis of *Campylobacter fetus* infections. In: Nachamkin I, Blaser MJ, editors. *Campylobacter*. 2nd ed. Washington, DC: ASM Press; 2000. pp. 321–47.
127. Avall-Jääskeläinen S, Kylä-Nikkilä K, Kahala M, Miikkulainen-Lahti T, Palva A. Surface display of foreign epitopes on the *Lactobacillus brevis* S-layer. *Appl Environ Microbiol* 2002;**68**:5943–51.
128. Scholz HC, Riedmann E, Witte A, Lubitz W, Kuen B. S-layer variation in *Bacillus stearothermophilus* PV72 is based on DNA rearrangements between the chromosome and the naturally occurring megaplasmids. *J Bacteriol* 2001;**183**:1672–9.
129. Egelseer EM, Idris R, Jarosch M, Danhorn T, Sleytr UB, Sara M. ISBst12, a novel type of insertion-sequence element causing loss of S-layer-gene expression in *Bacillus stearothermophilus* ATCC 12980. *Microbiology* 2000;**146**(Pt. 9):2175–83.
130. Gustafson CE, Chu S, Trust TJ. Mutagenesis of the paracrystalline surface protein array of *Aeromonas salmonicida* by endogenous insertion elements. *J Mol Biol* 1994;**237**:452–63.
131. Sleytr UB. Basic and applied S layer research: an overview. *FEMS Microbiol Rev* 1997;**20**:5–12.
132. Luckevich MD, Beveridge TJ. Characterization of a dynamic S layer on Bacillus thuringiensis. *J Bacteriol* 1989;**171**:6656–67.
133. Tsukagoshi N, Tabata R, Takemura T, Yamagata H, Udaka S. Molecular cloning of a major cell wall protein gene from protein-producing *Bacillus brevis* 47 and its expression in *Escherichia coli* and *Bacillus subtilis*. *J Bacteriol* 1984;**158**:1054–60.
134. Breitwieser A, Gruber K, Sleytr UB. Evidence for an S-layer protein pool in the peptidoglycan of *Bacillus stearothermophilus*. *J Bacteriol* 1992;**174**:8008–15.
135. Boot HJ, Kolen CP, Andreadaki FJ, Leer RJ, Pouwels PH. The *Lactobacillus acidophilus* S-layer protein gene expression site comprises two consensus promoter sequences, one of which directs transcription of stable mRNA. *J Bacteriol* 1996;**178**:5388–94.

136. Chu S, Gustafson CE, Feutrier J, Cavaignac S, Trust TJ. Transcriptional analysis of the *Aeromonas salmonicida* S-layer protein gene *vapA*. *J Bacteriol* 1993;**175**:7968–75.
137. Fisher JA, Smit J, Agabian N. Transcriptional analysis of the major surface array gene of *Caulobacter crescentus*. *J Bacteriol* 1988;**170**:4706–13.
138. Boot HJ, Pouwels PH. Expression, secretion and antigenic variation of bacterial S-layer proteins. *Mol Microbiol* 1996;**21**:1117–23.
139. Peters J, Peters M, Lottspeich F, Schafer W, Baumeister W. Nucleotide sequence analysis of the gene encoding the *Deinococcus radiodurans* surface protein, derived amino acid sequence, and complementary protein chemical studies. *J Bacteriol* 1987;**169**:5216–23.
140. Faraldo MM, de Pedro MA, Berenguer J. Sequence of the S-layer gene of *Thermus thermophilus* HB8 and functionality of its promoter in *Escherichia coli*. *J Bacteriol* 1992;**174**:7458–62.
141. Carl M, Dobson ME, Ching WM, Dasch GA. Characterization of the gene encoding the protective paracrystalline-surface-layer protein of *Rickettsia prowazekii*: presence of a truncated identical homolog in *Rickettsia typhi*. *Proc Natl Acad Sci USA* 1990;**87**:8237–41.
142. Noonan B, Trust TJ. Molecular analysis of an A-protein secretion mutant of *Aeromonas salmonicida* reveals a surface layer-specific protein secretion pathway. *J Mol Biol* 1995;**248**:316–27.
143. Awram P, Smit J. The *Caulobacter crescentus* paracrystalline S-layer protein is secreted by an ABC transporter (type I) secretion apparatus. *J Bacteriol* 1998;**180**:3062–9.
144. Kawai E, Akatsuka H, Idei A, Shibatani T, Omori K. *Serratia marcescens* S-layer protein is secreted extracellularly via an ATP-binding cassette exporter, the Lip system. *Mol Microbiol* 1998;**27**:941–52.
145. Blight MA, Holland IB. Heterologous protein secretion and the versatile *Escherichia coli* haemolysin translocator. *Trends Biotechnol* 1994;**12**:450–5.
146. Wandersman C. Secretion across the bacterial outer membrane. *Trends Genet* 1992;**8**:317–22.
147. Sleytr UB, Plohberger R. The dynamic process of assembly of two-dimensional arrays of macromolecules on bacterial cell walls. In: Baumeister W, Vogell W, editors. *Electron microscopy at molecular dimensions*. Berlin, Heidelberg, New York: Springer-Verlag; 1980. pp. 36–47.
148. Gruber K, Sleytr UB. Localized insertion of new S-layer during growth of *Bacillus stearothermophilus* strains. *Arch Microbiol* 1988;**149**:485–91.
149. Smit J, Agabian N. Cell surface patterning and morphogenesis: biogenesis of a periodic surface array during *Caulobacter* development. *J Cell Biol* 1982;**95**:41–9.
150. Harris WF, Scriven LE. Function of dislocations in cell walls and membranes. *Nature* 1970;**228**:827–9.
151. Messner P, Pum D, Sára M, Stetter KO, Sleytr UB. Ultrastructure of the cell envelope of the archaebacteria *Thermoproteus tenax* and *Thermoproteus neutrophilus*. *J Bacteriol* 1986;**166**:1046–54.
152. Pum D, Messner P, Sleytr UB. Role of the S layer in morphogenesis and cell division of the archaebacterium *Methanocorpusculum sinense*. *J Bacteriol* 1991;**173**:6865–73.
153. Györvary ES, Stein O, Pum D, Sleytr UB. Self-assembly and recrystallization of bacterial S-layer proteins at silicon supports imaged in real time by atomic force microscopy. *J Microsc* 2003;**212**:300–6.
154. Pum D, Sleytr UB. Large-scale reconstruction of crystalline bacterial surface layer proteins at the air-water interface and on lipids. *Thin Solid Films* 1994;**244**:882–6.
155. Pum D, Sleytr UB. Anisotropic crystal growth of the S-layer of *Bacillus sphaericus* CCM 2177 at the air/water interface. *Colloids Surf A Physicochem Eng Aspects* 1995;**102**:99–104.
156. Pum D, Sleytr UB. Monomolecular reassembly of a crystalline bacterial cell surface layer (S layer) on untreated and modified silicon surfaces. *Supramol Sci* 1995;**2**:193–7.

157. Messner P, Pum D, Sleytr UB. Characterization of the ultrastructure and the self-assembly of the surface layer of *Bacillus stearothermophilus* strain NRS 2004/3a. *J Ultrastruct Mol Struct Res* 1986;**97**:73–88.

158. Sleytr UB. Self-assembly of the hexagonally and tetragonally arranged subunits of bacterial surface layers and their reattachment to cell walls. *J Ultrastruct Res* 1976;**55**:360–77.

159. Jaenicke R, Welsch R, Sára M, Sleytr UB. Stability and self-assembly of the S-layer protein of the cell wall of *Bacillus stearothermophilus*. *Biol Chem Hoppe Seyler* 1985;**366**:663–70.

160. Pum D, Weinhandl M, Hödl C, Sleytr UB. Large-scale recrystallization of the S-layer of *Bacillus coagulans* E38-66 at the air/water interface and on lipid films. *J Bacteriol* 1993;**175**:2762–6.

161. Györvary E, O'Riordan A, Quinn A, Redmond G, Pum D, Sleytr UB. Biomimetic nanostructure fabrication: non-lithographic lateral patterning and self-assembly of functional bacterial S-layers at silicon supports. *Nano Lett* 2003;**3**:315–9.

162. Lopez AE, Moreno-Flores S, Pum D, Sleytr UB, Toca-Herrera JL. Surface dependence of protein nanocrystal formation. *Small* 2009;**6**:396–403.

163. Moreno-Flores S, Kasry A, Butt HJ, Vavilala C, Schmittel M, Pum D, Sleytr UB, Toca-Herrera JL. From native to non-native two-dimensional protein lattices through underlying hydrophilic/hydrophobic nanoprotrusions. *Angew Chem Int Ed Engl* 2008;**47**:4707–10.

164. Ulman A. Formation and structure of self-assembled monolayers. *Chem Rev* 1996;**96**:1533–54.

165. Györvary E, Schroedter A, Talapin DV, Weller H, Pum D, Sleytr UB. Formation of nanoparticle arrays on S-layer protein lattices. *J Nanosci Nanotechnol* 2004;**4**:115–20.

166. Schuster B, Sleytr UB. S-layer-supported lipid membranes. *J Biotechnol* 2000;**74**:233–54.

167. Schuster B, Sleytr UB. 2D-protein crystals (S-layers) as support for lipid membranes. In: Tien TH, Ottova A, editors. *Advances in planar lipid bilayers and liposomes*, Vol. 1. Amsterdam, The Netherlands: Elsevier Science; 2005. pp. 247–93.

168. Küpcü S, Sára M, Sleytr UB. Liposomes coated with crystalline bacterial cells surface protein (S-layer) as immobilization structures for macromolecules. *Biochim Biophys Acta* 1995;**1235**:263–9.

169. Mader C, Küpcü S, Sára M, Sleytr UB. Stabilizing effect of an S-layer on liposomes towards thermal or mechanical stress. *Biochim Biophys Acta* 1999;**1418**:106–16.

170. Mader C, Küpcü S, Sleytr UB, Sára M. S-layer-coated liposomes as a versatile system for entrapping and binding target molecules. *Biochim Biophys Acta* 2000;**1463**:142–50.

171. Toca-Herrera JL, Krastev R, Bosio V, Küpcü S, Pum D, Fery A, Sára M, Sleytr UB. Recrystallization of bacterial S-layers on flat polyelectrolyte surfaces and hollow polyelectrolyte capsules. *Small* 2005;**1**:339–48.

172. Pum D, Stangl G, Sponer C, Fallmann W, Sleytr UB. Deep UV patterning of monolayers of crystalline S layer protein on silicon surfaces. *Colloids Surf B* 1997;**8**:157–62.

173. Chung S, Shin SH, Bertozzi CR, De Yoreo JJ. Self-catalyzed growth of S layers via an amorphous-to-crystalline transition limited by folding kinetics. *Proc Natl Acad Sci USA* 2010;**107**:16536–41.

174. Fagan RP, Albesa-Jove D, Qazi O, Svergun DI, Brown KA, Fairweather NF. Structural insights into the molecular organization of the S-layer from *Clostridium difficile*. *Mol Microbiol* 2009;**71**:1308–22.

175. Norville JE, Kelly DF, Knight Jr. TF, Belcher AM, Walz T. 7A projection map of the S-layer protein sbpA obtained with trehalose-embedded monolayer crystals. *J Struct Biol* 2007;**160**:313–23.

176. Pavkov T, Oberer M, Egelseer EM, Sara M, Sleytr UB, Keller W. Crystallization and preliminary structure determination of the C-terminal truncated domain of the S-layer protein SbsC. *Acta Crystallogr D Biol Crystallogr* 2003;**59**:1466–8.

177. Whitelam S. Control of pathways and yields of protein crystallization through the interplay of nonspecific and specific attractions. *Phys Rev Lett* 2010;**105**: 088102.
178. Horejs C, Pum D, Sleytr UB, Peterlik H, Jungbauer A, Tscheliessnig R. Surface layer protein characterization by small angle x-ray scattering and a fractal mean force concept: from protein structure to nanodisk assemblies. *J Chem Phys* 2010;**133**: 175102.
179. Horejs C, Gollner H, Pum D, Sleytr UB, Peterlik H, Jungbauer A, Tscheliessnig R. Atomistic structure of monomolecular surface layer self-assemblies: towards functionalized nanostructures. *ACS Nano* 2011;**5**:2288–97.
180. Horejs C, Mitra MK, Pum D, Sleytr UB, Muthukumar M. Monte Carlo study of the molecular mechanisms of S-layer self-assembly. *J Chem Phys* 2011;**134**:125103.
181. Baumeister W, Engelhardt H. Three-dimensional structure of bacterial surface layers. In: Harris JR, Horne RW, editors. *Electron microscopy of proteins*, Vol. 6. London: Academic Press; 1987. pp. 109–54.
182. Sára M, Sleytr UB. Molecular sieving through S layers of *Bacillus stearothermophilus* strains. *J Bacteriol* 1987;**169**:4092–8.
183. Scherrer R, Gerhardt P. Molecular sieving by the *Bacillus megaterium* cell wall and protoplast. *J Bacteriol* 1971;**107**:718–35.
184. Sára M, Pum D, Sleytr UB. Permeability and charge-dependent adsorption properties of the S-layer lattice from *Bacillus coagulans* E38-66. *J Bacteriol* 1992;**174**:3487–93.
185. Sára M, Manigley C, Wolf G, Sleytr UB. Isoporous ultrafiltration membranes from bacterial cell envelope layers. *J Membr Sci* 1988;**36**:179–86.
186. Sára M, Sleytr UB. Membrane biotechnology: two-dimensional protein crystals for ultrafiltration purposes. In: Rehm HJ, editor. *Biotechnology*. Weinheim: VCH; 1988. pp. 615–36.
187. Sára M, Sleytr UB. Production and characteristics of ultrafiltration membranes with uniform pores from two-dimensional arrays of proteins. *J Membr Sci* 1987;**33**:27–49.
188. Sleytr UB, Sára M. *Structure with membrane having continuous pores*. U.S. Pat. 4,752,395.
189. Küpcü S, Sára M, Sleytr UB. Chemical modification of crystalline ultrafiltration membranes and immobilization of macromolecules. *J Membr Sci* 1991;**61**:167–75.
190. Sára M, Küpcü S, Sleytr UB. Crystalline bacterial cell surface layers used as ultrafiltration membranes and immobilization matrix. *Genet Eng Biotechnol* 1990;**10**:10–3.
191. Sára M, Küpcü S, Weiner C, Weigert S, Sleytr UB. Crystalline protein layers as isoporous molecular sieves and immobilisation and affinity matrices. In: Sleytr UB, Messner P, Pum D, Sára M, editors. *Immobilised macromolecules: application potentials*. London, UK: Springer-Verlag; 1993. pp. 71–86.
192. Sára M, Küpcü S, Weiner C, Weigert S, Sleytr UB. S-layers as immobilization and affinity matrices. In: Beveridge TJ, Koval SF, editors. *Advances in bacterial paracrystalline surface layers*. New York & London: Plenum Press; 1993.
193. Sleytr UB, Messner P, Pum D, Sára M. *Immobilised macromolecules: application potentials*. London, UK: Springer-Verlag; 1993.
194. Küpcü S, Sára M, Sleytr UB. Influence of covalent attachment of low molecular weight substances on the rejection and adsorption properties of crystalline proteinaceous ultra filtration membranes. *Desalination* 1993;**90**:65–76.
195. Weigert S, Sára M. Ultrafiltration membranes prepared from crystalline bacterial cell surface layers as model systems for studying the influence of surface properties on protein adsorption. *J Membr Sci* 1996;**121**:185–96.
196. Breitwieser A, Küpcü S, Howorka S, Weigert S, Langer C, Hoffmann-Sommergruber K, Scheiner O, Sleytr UB, Sára M. 2-D protein crystals as an immobilization matrix for producing reaction zones in dipstick-style immunoassays. *Biotechniques* 1996;**21**:918–25.

197. Küpcü S, Sleytr UB, Sara M. Two-dimensional paracrystalline glycoprotein S-layers as a novel matrix for the immobilization of human IgG and their use as microparticles in immunoassays. *J Immunol Methods* 1996;**196**:73–84.

198. Sára M, Sleytr UB. Use of regularly structured bacterial cell envelope layers as matrix for the immobilization of macromolecules. *Appl Microbiol Biotechnol* 1989;**30**:184–9.

199. Weiner C, Sára M, Sleytr UB. Novel protein A affinity matrix prepared from two-dimensional protein crystals. *Biotechnol Bioeng* 1994;**43**:321–30.

200. Crowther RA, Sleytr UB. An analysis of the fine structure of the surface layers from two strains of *Clostridia*, including correction for distorted images. *J Ultrastruct Res* 1977;**58**:41–9.

201. Sleytr UB, Sára M, Küpcü Z, Messner P. Structural and chemical characterization of S-layers of selected strains of *Bacillus stearothermophilus* and *Desulfotomaculum nigrificans*. *Arch Microbiol* 1986;**146**:19–24.

202. Bock K, Schuster-Kolbe J, Altman E, Allmaier G, Stahl B, Christian R, Sleytr UB, Messner P. Primary structure of the O-glycosidically linked glycan chain of the crystalline surface layer glycoprotein of *Thermoanaerobacter thermohydrosulfuricus* L111-69. Galactosyl tyrosine as a novel linkage unit. *J Biol Chem* 1994;**269**:7137–44.

203. Christian R, Messner P, Weiner C, Sleytr UB, Schulz G. Structure of a glycan from the surface-layer glycoprotein of *Clostridium thermohydrosulfuricum* strain L111-69. *Carbohydr Res* 1988;**176**:160–3.

204. Pum D, Sára M, Sleyt UB. Two-dimensional (glyco)protein crystals as patterning elements and immobilisation matrices for the development of biosensors. In: Sleytr UB, Messner P, Pum D, Sára M, editors. *Immobilised macromolecules: application potentials*. London, UK: Springer-Verlag; 1993. pp. 141–60.

205. Küpcü S, Mader C, Sára M. The crystalline cell surface layer of *Thermoanaerobacter thermohydrosulfuricus* L111-69 as an immobilization matrix: influence of the morphological properties and the pore size of the matrix on activity loss of covalently bound enzymes. *Biotechnol Appl Biochem* 1995;**21**:275–86.

206. Weber V, Weigert S, Sára M, Sleytr UB, Falkenhagen D. Development of affinity microparticles for extracorporeal blood purification based on crystalline bacterial cell surface proteins. *Ther Apher* 2001;**5**:433–8.

207. Sára M, Küpcü S, Sleytr UB. Biotechnological applications of S-layers. In: Sleytr UB, Messner P, Pum D, Sára M, editors. *Crystalline bacterial cell surface proteins*. Austin, Texas, USA: R.G. Landes Company and Academic Press Inc.; 1996. pp. 133–59.

208. Breitwieser A, Mader C, Schocher I, Hoffmann-Sommergruber K, Aberer W, Scheiner O, Sleytr UB, Sára M. A novel dipstick developed for rapid Bet v 1-specific IgE detection: recombinant allergen immobilized via a monoclonal antibody to crystalline bacterial cell-surface layers. *Allergy* 1998;**53**:786–93.

209. Sleytr UB, Pum D, Sára M, Schuster B. Molecular nanotechnology and nanobiotechnology with 2-D protein crystals. In: Nalwa HS, editor. *Encylopedia of nanoscience and nanotechnology* Vol. 5. San Diego: Academic Press; 2004. pp. 693–702.

210. Sleytr UB, Sára M. Bacterial and archaeal S-layer proteins: structure-function relationships and their biotechnological applications. *Trends Biotechnol* 1997;**15**:20–6.

211. Völkel D, Zimmermann K, Breitwieser A, Pable S, Glatzel M, Scheiflinger F, Schwarz HP, Sára M, Sleytr UB, Dorner F. Immunochemical detection of prion protein on dipsticks prepared with crystalline bacterial cell-surface layers. *Transfusion* 2003;**43**:1677–82.

212. Neubauer A, Pum D, Sleytr UB. An amperometric glucose sensor based on isoporous crystalline protein membranes as immobilization matrix. *Anal Lett* 1993;**26**:1347–60.

213. Neubauer A, Hödl C, Pum D, Sleytr UB. A multistep enzyme sensor for sucrose based on S-layer microparticles as immobilization matrix. *Anal Lett* 1994;**27**:849–65.

214. Schäffer C, Novotny R, Kupcu S, Zayni S, Scheberl A, Friedmann J, Sleytr UB, Messner P. Novel biocatalysts based on S-layer self-assembly of *Geobacillus stearothermophilus* NRS 2004/3a: a nanobiotechnological approach. *Small* 2007;**3**:1549–59.

215. Tschiggerl H, Breitwieser A, de Roo G, Verwoerd T, Schäffer C, Sleytr UB. Exploitation of the S-layer self-assembly system for site directed immobilization of enzymes demonstrated for an extremophilic laminarinase from *Pyrococcus furiosus*. *J Biotechnol* 2008;**133**:403–11.

216. Douglas K, Clark NA, Rothschild KJ. Nanometer molecular lithography. *Appl Phys Lett* 1986;**48**:676–8.

217. Douglas S, Beveridge TJ. Mineral formation by bacteria in natural microbial communities. *FEMS Microbiol Ecol* 1998;**26**:79–88.

218. Bergkvist M, Mark SS, Yang X, Angert ER, Batt CA. Bionanofabrication of ordered nanoparticle arrays: effect of particle properties and adsorption conditions. *J Phys Chem B* 2004;**108**:8241–8.

219. Dieluweit S, Pum D, Sleytr UB, Kautek W. Monodisperse gold nanoparticles formed on bacterial crystalline surface layers (S-layers) by electroless deposition. *Mater Sci Eng C* 2005;**25**:727–32.

220. Mertig M, Kirsch R, Pompe W, Engelhardt H. Fabrication of highly orineted nanocluster arrays by biomolecular templating. *Eur Phys J D* 1999;**9**:45–8.

221. Mertig M, Wahl R, Lehmann M, Simon P, Pompe W. Formation and manipulation of regular metallic nanoparticle arrays on bacterial surface layers: an advanced TEM study. *Eur Phys J D* 2001;**16**:317–20.

222. Shenton W, Pum D, Sleytr UB, Mann S. Biocrystal templating of CdS superlattices using self-assembled bacterial S-layers. *Nature* 1997;**389**:585–7.

223. Wahl R, Mertig M, Raff J, Selenska-Pobell S, Pompe W. Electron-beam induced formation of highly ordered palladium and platinum nanoparticle arrays on the S-layer of *Bacillus sphaericus* NCTC 9602. *Adv Mater Sci Technol* 2001;**13**:736–40.

224. Hall SR, Shenton W, Engelhardt H, Mann S. Site-specific organization of gold nanoparticles by biomolecular templating. *Chemphyschem* 2001;**3**:184–6.

225. Mark SS, Bergkvist M, Yang X, Angert ER, Batt CA. Self-assembly of dendrimer-encapsulated nanoparticle arrays using 2-D microbial S-layer protein biotemplates. *Biomacromolecules* 2006;**7**:1884–97.

226. Mark SS, Bergkvist M, Yang X, Teixeira LM, Bhatnagar P, Angert ER, Batt CA. Bionanofabrication of metallic and semiconductor nanoparticle arrays using S-layer protein lattices with different lateral spacings and geometries. *Langmuir* 2006;**22**:3763–74.

227. Badelt-Lichtblau H, Kainz B, Völlenkle C, Egelseer EM, Sleytr UB, Pum D, Ilk N. Genetic engineering of the S-layer protein SbpA of *Lysinibacillus sphaericus* CCM 2177 for the generation of functionalized nanoarrays. *Bioconjug Chem* 2009;**20**:895–903.

228. Sleytr UB, Huber C, Ilk N, Pum D, Schuster B, Egelseer E. S-Layers as a tool kit for nanobiotechnological applications. *FEMS Microbiol Lett* 2007;**267**:131–44.

229. Huber C, Liu J, Egelseer EM, Moll D, Knoll W, Sleytr UB, Sára M. Heterotetramers formed by an S-layer-streptavidin fusion protein and core-streptavidin as nanoarrayed template for biochip development. *Small* 2006;**2**:142–50.

230. Breitwieser A, Egelseer EM, Moll D, Ilk N, Hotzy C, Bohle B, Ebner C, Sleytr UB, Sára M. A recombinant bacterial cell surface (S-layer)-major birch pollen allergen-fusion protein (rSbsC/Bet v1) maintains the ability to self-assemble into regularly structured monomolecular lattices and the functionality of the allergen. *Protein Eng* 2002;**15**:243–9.

231. Völlenkle C, Weigert S, Ilk N, Egelseer E, Weber V, Loth F, Falkenhagen D, Sleytr UB, Sára M. Construction of a functional S-layer fusion protein comprising an immunoglobulin G-binding domain for development of specific adsorbents for extracorporeal blood

purification. *Appl Environ Microbiol,* 2004;**70**:1514–21, Highlight in Nature Rewiews Microbiology 2(5), 353.

232. Ilk N, Küpcü S, Moncayo G, Klimt S, Ecker RC, Hofer-Warbinek R, Egelseer EM, Sleytr UB, Sára M. A functional chimaeric S-layer-enhanced green fluorescent protein to follow the uptake of S-layer-coated liposomes into eukaryotic cells. *Biochem J* 2004;**379**:441–8.

233. Pleschberger M, Saerens D, Weigert S, Sleytr UB, Muyldermans S, Sára M, Egelseer EM. An S-layer heavy chain camel antibody fusion protein for generation of a nanopatterned sensing layer to detect the prostate-specific antigen by surface plasmon resonance technology. *Bioconjug Chem* 2004;**15**:664–71.

234. Tschiggerl H, Casey JL, Parisi K, Foley M, Sleytr UB. Display of a peptide mimotope on a crystalline bacterial cell surface layer (S-layer) lattice for diagnosis of Epstein-Barr virus infection. *Bioconjug Chem* 2008;**19**:860–5.

235. Kainz B, Steiner K, Möller M, Pum D, Schäffer C, Sleytr UB, Toca-Herrera JL. Absorption, steady-state fluorescence, fluorescence lifetime, and 2D self-assembly properties of engineered fluorescent S-layer fusion proteins of *Geobacillus stearothermophilus* NRS 2004/3a. *Biomacromolecules* 2010;**11**:207–14.

236. Kainz B, Steiner K, Sleytr UB, Pum D, Toca-Herrera JL. Fluorescence energy transfer in the bi-fluorescent S-layer tandem fusion protein ECFP-SgsE-YFP. *J Struct Biol* 2010;**172**:276–83.

237. Riedmann EM, Kyd JM, Smith AM, Gomez-Gallego S, Jalava K, Cripps AW, Lubitz W. Construction of recombinant S-layer proteins (rSbsA) and their expression in bacterial ghosts–a delivery system for the nontypeable Haemophilus influenzae antigen Omp26. *FEMS Immunol Med Microbiol* 2003;**37**:185–92.

238. Mesnage S, Tosi-Couture E, Fouet A. Production and cell surface anchoring of functional fusions between the SLH motifs of the *Bacillus anthracis* S-layer proteins and the *Bacillus subtilis* levansucrase. *Mol Microbiol* 1999;**31**:927–36.

239. Mesnage S, Weber-Levy M, Haustant M, Mock M, Fouet A. Cell surface-exposed tetanus toxin fragment C produced by recombinant *Bacillus anthracis* protects against tetanus toxin. *Infect Immun* 1999;**67**:4847–50.

240. Bingle WH, Nomellini JF, Smit J. Cell-surface display of a *Pseudomonas aeruginosa* strain K pilin peptide within the paracrystalline S-layer of *Caulobacter crescentus*. *Mol Microbiol* 1997;**26**:277–88.

241. Simon B, Nomellini JF, Chiou P, Binale W, Thornton J, Smit J, Leong JA. Recombinant vaccines against infectious hematopoietic necrosis virus: production line by the *Caulobacter crescentus* S-layer protein secretion system and evaluation in laboratory trials. *Dis Aquat Organ* 2001;**44**:400–2.

242. Duncan G, Tarling CA, Bingle WH, Nomellini JF, Yamage M, Dorocicz IR, Withers SG, Smit J. Evaluation of a new system for developing particulate enzymes based on the surface (S)-layer protein (RsaA) of *Caulobacter crescentus*: fusion with the beta-1,4-glycanase (Cex) from the cellulolytic bacterium *Cellulomonas fimi* yields a robust, catalytically active product. *Appl Biochem Biotechnol* 2005;**127**:95–110.

243. Nomellini JF, Duncan G, Dorocicz IR, Smit J. S-layer-mediated display of the immunoglobulin G-binding domain of streptococcal protein G on the surface of *Caulobacter crescentus*: development of an immunoactive reagent. *Appl Environ Microbiol* 2007;**73**:3245–53.

244. Nomellini JF, Li C, Lavallee D, Shanina I, Cavacini LA, Horwitz MS, Smit J. Development of an HIV-1 specific microbicide using *Caulobacter crescentus* S-layer mediated display of CD4 and MIP1alpha. *PLoS One* 2010;**5**:e10366.

245. Patel J, Zhang Q, McKay RM, Vincent R, Xu Z. Genetic engineering of *Caulobacter crescentus* for removal of cadmium from water. *Appl Biochem Biotechnol* 2010;**160**:232–43.

246. Huber C, Egelseer EM, Ilk N, Sleytr UB, Sára M. S-layer-streptavidin fusion proteins and S-layer-specific heteropolysaccharides as part of a biomolecular construction kit for application in nanobiotechnology. *Microelectron Eng* 2006;**83**:1589–93.

247. Pleschberger M, Neubauer A, Egelseer EM, Weigert S, Lindner B, Sleytr UB, Muyldermans S, Sára M. Generation of a functional monomolecular protein lattice consisting of an S-layer fusion protein comprising the variable domain of a camel heavy chain antibody. *Bioconjug Chem* 2003;**14**:440–8.

248. Bohle B, Breitwieser A, Zwölfer B, Jahn-Schmid B, Sára M, Sleytr UB, Ebner C. A novel approach to specific allergy treatment: the recombinant fusion protein of a bacterial cell surface (S-layer) protein and the major birch pollen allergen Bet v 1 (rSbsC-Bet v 1) combines reduced allergenicity with immunomodulating capacity. *J Immunol* 2004;**172**:6642–8.

249. Gerstmayr M, Ilk N, Schabussova I, Jahn-Schmid B, Egelseer EM, Sleytr UB, Ebner C, Bohle B. A novel approach to specific allergy treatment: the recombinant allergen-S-layer fusion protein rSbsC-Bet v 1 matures dendritic cells that prime Th0/Th1 and IL-10-producing regulatory T cells. *J Immunol* 2007;**179**:7270–5.

250. Ferner-Ortner-Beckmann, J., Tesarz, M., Egelseer, E. M., Sleytr, U. B. & Ilk, N. (submitted). S-layer-based biocatalysts carrying multimeric extremozymes. *Small*.

251. Kuen B, Sleytr UB, Lubitz W. Sequence analysis of the *sbsA* gene encoding the 130-kDa surface-layer protein of *Bacillus stearothermophilus* strain PV72. *Gene* 1994;**145**:115–20.

252. Fouet A, Mesnage S. *Bacillus anthracis* cell envelope components. *Curr Top Microbiol Immunol* 2002;**271**:87–113.

253. Smit J, Engelhardt H, Volker S, Smith SH, Baumeister W. The S-layer of *Caulobacter crescentus:* three-dimensional image reconstruction and structure analysis by electron microscopy. *J Bacteriol* 1992;**174**:6527–38.

254. Smit J, Grano DA, Glaeser RM, Agabian N. Periodic surface array in *Caulobacter crescentus*: fine structure and chemical analysis. *J Bacteriol* 1981;**146**:1135–50.

255. Gilchrist A, Fisher JA, Smit J. Nucleotide sequence analysis of the gene encoding the *Caulobacter crescentus* paracrystalline surface layer protein. *Can J Microbiol* 1992;**38**:193–202.

256. Walker SG, Smith SH, Smit J. Isolation and comparison of the paracrystalline surface layer proteins of freshwater caulobacters. *J Bacteriol* 1992;**174**:1783–92.

257. Apweiler R, Hermjakob H, Sharon N. On the frequency of protein glycosylation, as deduced from analysis of the SWISS-PROT database. *Biochim Biophys Acta* 1999;**1473**:4–8.

258. Spiro RG. Protein glycosylation: nature, distribution, enzymatic formation, and disease implications of glycopeptide bonds. *Glycobiology* 2002;**12**:43R–NaN.

259. Samuelson P, Gunneriusson E, Nygren PA, Stahl S. Display of proteins on bacteria. *J Biotechnol* 2002;**96**:129–54.

260. Zarschler K, Janesch B, Kainz B, Ristl R, Messner P, Schäffer C. Cell surface display of chimeric glycoproteins via the S-layer of *Paenibacillus alvei*. *Carbohydr Res* 2010;**345**:1422–31.

261. Kern JW, Schneewind O. BslA, a pXO1-encoded adhesin of *Bacillus anthracis*. *Mol Microbiol* 2008;**68**:504–15.

262. Kern J, Schneewind O. BslA, the S-layer adhesin of *B. anthracis*, is a virulence factor for anthrax pathogenesis. *Mol Microbiol* 2010;**75**:324–32.

263. Ní Eidhin DB, O'Brien JB, McCabe MS, Athié-Morales V, Kelleher DP. Active immunization of hamsters against *Clostridium difficile* infection using surface-layer protein. *FEMS Immunol Med Microbiol* 2008;**52**:207–18.

264. Ford LA, Thune RL. Immunization of channel catfish with crude, acid-extracted preparation of motile aeromonad S-layer protein. *Biomed Lett* 1992;**47**:355–62.

265. Poobalane S, Thompson KD, Ardo L, Verjan N, Han HJ, Jeney G, Hirono I, Aoki T, Adams A. Production and efficacy of an *Aeromonas hydrophila* recombinant S-layer protein vaccine for fish. *Vaccine* 2010;**28**:3540–7.

266. Malcolm AJ, Messner P, Sleytr UB, Smith RH, Unger FM. Crystalline bacterial cell surface layers (S-layers) as combined carrier/adjuvants for conjugated vaccines. In: Sleytr UB, Messner P, Pum D, Sára M, editors. *Immobilized macromolecules: application potentials*. London, UK: Springer-Verlag; 1993. pp. 195–207.

267. Smith RH, Messner P, Lamontagne LR, Sleytr UB, Unger FM. Induction of T-cell immunity to oligosaccharide antigens immobilized on crystalline bacterial surface layers (S-layers). *Vaccine* 1993;**11**:919–24.

268. Jahn-Schmid B, Siemann U, Zenker A, Bohle B, Messner P, Unger FM, Sleytr UB, Scheiner O, Kraft D, Ebner C. Bet v 1, the major birch pollen allergen, conjugated to crystalline bacterial cell surface proteins, expands allergen-specific T cells of the Th1/Th0 phenotype in vitro by induction of IL-12. *Int Immunol* 1997;**9**:1867–74.

269. Jahn-Schmid B, Graninger M, Glozik M, Küpcü S, Ebner C, Unger FM, Sleytr UB, Messner P. Immunoreactivity of allergen (Bet v 1) conjugated to crystalline bacterial cell surface layers (S-layers). *Immunotechnology* 1996;**2**:103–13.

270. Jahn-Schmid B, Messner P, Unger FM, Sleytr UB, Scheiner O, Kraft D. Toward selective elicitation of TH1-controlled vaccination responses: vaccine applications of bacterial surface layer proteins. *J Biotechnol* 1996;**44**:225–31.

271. Gerstmayr M, Ilk N, Jahn-Schmid B, Sleytr UB, Bohle B. Natural self-assembly of allergen-S-layer fusion proteins is no prerequisite for reduced allergenicity and T cell stimulatory capacity. *Int Arch Allergy Immunol* 2009;**149**:231–8.

272. Umelo-Njaka E, Nomellini JF, Bingle WH, Glasier LG, Irvin RT, Smit J. Expression and testing of *Pseudomonas aeruginosa* vaccine candidate proteins prepared with the *Caulobacter crescentus* S-layer protein expression system. *Vaccine* 2001;**19**:1406–15.

273. Mock M, Fouet A. Anthrax. *Annu Rev Microbiol* 2001;**55**:647–71.

274. Ausiello CM, Cerquetti M, Fedele G, Spensieri F, Palazzo R, Nasso M, Frezza S, Mastrantonio P. Surface layer proteins from *Clostridium difficile* induce inflammatory and regulatory cytokines in human monocytes and dendritic cells. *Microbes Infect* 2006;**8**:2640–6.

275. Brun P, Scarpa M, Grillo A, Palu G, Mengoli C, Zecconi A, Spigaglia P, Mastrantonio P, Castagliuolo I. *Clostridium difficile* TxAC314 and SLP-36kDa enhance the immune response toward a co-administered antigen. *J Med Microbiol* 2008;**57**:725–31.

276. O'Brien JB, McCabe MS, Athié-Morales V, McDonald GS, Ní Eidhin DB, Kelleher DP. Passive immunization of hamsters against *Clostridium difficile* infection using antibodies to surface layer proteins. *FEMS Microbiol Lett* 2005;**246**:199–205.

277. Rahman MH, Kawai K. Outer membrane proteins of *Aeromonas hydrophila* induce protective immunity in goldfish. *Fish Shellfish Immunol* 2000;**10**:379–82.

278. Baba T, Imamura J, Izawa K, Ikeda K. Immune protection in carp. *Cyprinus carpido* L. after immunization with Aeromonas hydrophila crude lipopolysaccaride. *J Fish Dis* 1988;**11**:237–44.

279. Azad IS, Shankar KM, Mohan CV, Kalita B. Uptake and processing of biofilm and free-cell vaccines of *Aeromonas hydrophila* in indian major carps and common carp following oral vaccination–antigen localization by a monoclonal antibody. *Dis Aquat Organ* 2000;**43**:103–8.

280. Lamers CHJ, De Haas MJH, Van Muiswinkel WB. The reaction of the immune system of fish vaccination: development of immunological memory in carp. *Cyprinus carpio* L., following direct immersion in *Aeromonas hydrophila* bacterin. *J Fish Dis* 1985;**8**:253–62.

281. Leung KY, Wong LS, Low KW, Sin YM. Mini-Tn5 induced growth- and protease- deficient mutants of *Aeromonas hydrophila* as live vaccines for blue gourami, *Trichogaster trichopterus* (Pallas). *Aquaculture* 1997;**158**:11–22.

282. Messner P, Unger FM, Sleytr U. Vaccine development based on S-layer technology. In: Sleytr UB, Messner P, Pum D, Sára M, editors. *Crystalline bacterial cell surface proteins.* Austin, Texas, USA: Landes Company and Academic Press Inc.; 1996. pp. 161–73.

283. Sleytr UB, Mundt W, Messner P. *Immunogenic composition containing ordered carriers.* US Patent Nr. 5,043,158.

284. Brown FGD, Hoey EM, Martin S, Rima BK, Trudgett A. Vaccine design. In: James K, Morris A, editors. *Molecular medical science series.* England, UK: John Wiley & Sons; 1993.

285. Powell MF, Newman MJ. Vaccine design: the subunit and adjuvant approach. In: Powell MF, Newman MJ, editors. *Pharmaceutical biotechnology.* New York, USA: Plenum Press; 1995.

286. Malcolm AJ, Best MW, Szarka RJ, Mosleh Z, Unger FM, Messner P, Sleytr UB. Surface layers of *Bacillus alvei* as a carrier for a *Streptococcus pneunoniae* conjugate vaccine. In: Beveridge J, Koval SF, editors. *Bacterial paracrystalline surface layers.* New York, USA: Plenum Press; 1993. pp. 219–33.

287. Messner P. Chemical composition and biosynthesis of S-layer. In: Sleytr UB, Messner P, Pum D, Sára M, editors. *Crystalline bacterial cell surface proteins.* Austin, Texas, USA: R.G. Landes Company and Academic Press Inc.; 1996. pp. 35–76.

288. Messner P, Unger FM, Sleytr UB. Vaccine development based on S-layer technology. In: Sleytr UB, Messner P, Pum D, Sara M, editors. *Crystalline bacterial cell surface proteins.* Austin, TX: R.G. Landes and Academic Press; 1996. pp. 161–73.

289. Hianik T, Küpcü S, Sleytr UB, Rybar P, Krivanek R, Kaatze U. Interaction of crystalline bacterial cell surface proteins with lipid bilayers in liposomes. *Colloids Surf A* 1999;**147**:331–9.

290. Küpcü S, Lohner K, Mader C, Sleytr UB. Microcalorimetric study on the phase behaviour of S-layer coated liposomes. *Mol Membr Biol* 1998;**15**:69–74.

291. Krivanek R, Rybar P, Küpcü S, Sleytr UB, Hianik T. Affinity interactions on a liposome surface detected by ultrasound velocimetry. *Bioelectrochemistry* 2002;**55**:57–9.

292. Bhatnagar PK, Awasthi A, Nomellini JF, Smit J, Suresh MR. Anti-tumor effects of the bacterium *Caulobacter crescentus* in murine tumor models. *Cancer Biol Ther* 2006;**5**:485–91.

293. Georgiou G, Stathopoulos C, Daugherty PS, Nayak AR, Iverson BL, Curtiss 3rd R. Display of heterologous proteins on the surface of microorganisms: from the screening of combinatorial libraries to live recombinant vaccines. *Nat Biotechnol* 1997;**15**:29–34.

294. Nomellini JF, Toporowski MC, Smit J. Secrtetion or presentation of recombinant proteins and peptides mediated by the S-layer of *Caulobacter crescentus.* In: Banex F, editor. *Expression technologies: current status and future trends.* Norfolk, UK: Horizon Scientific Press; 2004. pp. 477–524.

295. Ratner BD, Bryant SJ. Biomaterials: where we have been and where we are going. *Annu Rev Biomed Eng* 2004;Vol. 6. 41–75.

296. Zhang Z, Zhang M, Chen S, Horbett TA, Ratner BD, Jiang S. Blood compatibility of surfaces with superlow protein adsorption. *Biomaterials* 2008;**29**:4285–91.

297. Goodsell DS. *Bionanotechnology, Lessons from nature.* Hoboken, NJ: Wiley-Liss; 2004.

298. Whitesides GM, Mathias JP, Seto CT. Molecular selfassembly and nanochemistry: a chemical strategy for the synthesis of nanostructures. *Science* 1991;**254**:1312–9.

299. Galdiero S, Falanga A, Vitiello M, Raiola L, Russo L, Pedone C, Isernia C, Galdiero M. The presence of a single N-terminal histidine residue enhances the fusogenic properties of a membranotropic peptide derived from herpes simplex virus type 1 glycoprotein H. *J Biol Chem* 2010;**285**:17123–36.

300. Galdiero S, Galdiero M, Pedone C. Beta-barrel membrane bacterial proteins: structure, function, assembly and interaction with lipids. *Curr Protein Pept Sci* 2007;**8**:63–82.

301. Gerstein M, Hegyi H. Comparing genomes in terms of protein structure: surveys of a finite parts list. *FEMS Microbiol Rev* 1998;**22**:277–304.

302. Sleytr UB, Sára M, Pum D. Crystalline bacterial cell surface layers (S-layers): a versatile self-assembly system. In: Ciferri A, editor. *Supramolecular polymers*. New York, Basel: Marcel Dekker Inc.; 2000. pp. 177–213.
303. Viviani B, Gardoni F, Marinovich M. Cytokines and neuronal ion channels in health and disease. *Int Rev Neurobiol* 2007;**82**:247–63.
304. Ellis C, Smith A. Highlighting the pitfalls and possibilities of drug research. *Nat Rev Drug Discov* 2004;**3**:238–78.
305. De Rosa M. Archaeal lipids: structural features and supramolecular organization. *Thin Solid Films* 1996;**284–285**:13–7.
306. Hanford MJ, Peeples TL. Archaeal tetraether lipids. *Appl Biochem Biotechnol* 2002;**97**:45–62.
307. Stetter KO. Extremophiles and their adaptation to hot environments. *FEBS Lett* 1999;**452**:22–5.
308. Schuster B, Sleytr UB. Tailor-made crystalline structures of truncated S-layer proteins on heteropolysaccharides. *Soft Matter* 2009;**5**:334–41.
309. Schuster B, Pum D, Sleytr UB. S-layer stabilized lipid membranes (Review). *Biointerphases* 2008;**3**:FA3–FA11.
310. Diederich A, Sponer C, Pum D, Sleytr UB, Lösche M. Reciprocal influence between the protein and lipid components of a lipid-protein membrane model. *Colloids Surf B* 1996;**6**:335–46.
311. Wetzer B, Pum D, Sleytr UB. S-layer stabilized solid supported lipid bilayers. *J Struct Biol* 1997;**119**:123–8.
312. Weygand M, Kjaer K, Howes PB, Wetzer B, Pum D, Sleytr UB, Lösche M. Structural reorganization of phospholipid headgroups upon recrystallization of an S-layer lattice. *J Phys Chem B* 2002;**106**:5793–9.
313. Weygand M, Schalke M, Howes PB, Kjaer K, Friedmann J, Wetzer B, Pum D, Sleytr UB, Lösche M. Coupling of protein sheet crystals (S-layers) to phospholipid monolayers. *J Mater Chem* 2000;**10**:141–8.
314. Weygand M, Wetzer B, Pum D, Sleytr UB, Cuvillier N, Kjaer K, Howes PB, Losche M. Bacterial S-layer protein coupling to lipids: x-ray reflectivity and grazing incidence diffraction studies. *Biophys J* 1999;**76**:458–68.
315. Wetzer B, Pfandler A, Györvary E, Pum D, Lösche M, Sleytr UB. S-layer reconstitution at phospholipid monolayers. *Langmuir* 1998;**14**:6899–906.
316. Schuster B, Sleytr UB, Diederich A, Bahr G, Winterhalter M. Probing the stability of S-layer-supported planar lipid membranes. *Eur Biophys J* 1999;**28**:583–90.
317. Hirn R, Schuster B, Sleytr UB, Bayerl TM. The effect of S-layer protein adsorption and crystallization on the collective motion of a planar lipid bilayer studied by dynamic light scattering. *Biophys J* 1999;**77**:2066–74.
318. Schuster B. Biomimetic design of nanopatterned membranes. *NanoBiotechnology* 2005;**1**:153–64.
319. Schuster B, Pum D, Sára M, Sleytr UB. S-Layer proteins as key components of a versatile molecular construction kit for biomedical Nanotechnology. *Mini Rev Med Chem* 2006;**6**:909–20.
320. Lee BW, Faller R, Sum AK, Vattulainen I, Patra M, Karttunen M. Structural effects of small molecules on phospholipid bilayers investigated. *Fluid Phase Equilibria* 2004;**225**:63–8.
321. Györvary E, Wetzer B, Sleytr UB, Sinner A, Offenhäuser A, Knoll W. Lateral diffusion of lipids in silane-, dextrane- and S-layer protein-supported mono- and bilayers. *Langmuir* 1999;**15**:1337–47.
322. Schuster B, Pum D, Braha O, Bayley H, Sleytr UB. Self-assembled alpha-hemolysin pores in an S-layer-supported lipid bilayer. *Biochim Biophys Acta* 1998;**1370**:280–8.

323. Schuster B, Pum D, Sleytr UB. Voltage clamp studies on S-layer-supported tetraether lipid membranes. *Biochim Biophys Acta* 1998;**1369**:51–60.
324. Schuster B, Gufler PC, Pum D, Sleytr UB. Interplay of phospholipase A2 with S-layer-supported lipid monolayers. *Langmuir* 2003;**19**:3393–7.
325. Gufler PC, Pum D, Sleytr UB, Schuster B. Highly robust lipid membranes on crystalline S-layer supports investigated by electrochemical impedance spectroscopy. *Biochim Biophys Acta* 2004;**1661**:154–65.
326. Schuster B, Gufler PC, Pum D, Sleytr UB. S-layer proteins as supporting scaffoldings for functional lipid membranes. *IEEE Trans Nanobioscience* 2004;**3**:16–21.
327. Schuster B, Sleytr UB. The effect of hydrostatic pressure on S-layer-supported lipid membranes. *Biochim Biophys Acta* 2002;**1563**:29–34.
328. Bayley H, Cremer PS. Stochastic sensors inspired by biology. *Nature* 2001;**413**:226–30.
329. Bhakdi S, Tranum-Jensen J. Alpha-toxin of *Staphylococcus aureus*. *Microbiol Rev* 1991;**55**:733–51.
330. Revah F, Bertrand D, Galzi JL, Devillers-Thiery A, Mulle C, Hussy N, Bertrand S, Ballivet M, Changeux JP. Mutations in the channel domain alter desensitization of a neuronal nicotinic receptor. *Nature* 1991;**353**:846–9.
331. Schuster B, Sleytr UB. Single channel recordings of alpha-hemolysin reconstituted in S-layer-supported lipid bilayers. *Bioelectrochemistry* 2002;**55**:5–7.
332. Keizer HM, Andersson M, Chase C, Laratta WP, Proemsey JB, Tabb J, Long JR, Duran RS. Prolonged stochastic single ion channel recordings in S-layer protein stabilized lipid bilayer membranes. *Colloids Surf B Biointerfaces* 2008;**65**:178–85.
333. Castellana ET, Cremer PS. Solid supported lipid bilayers: from biophysical studies to sensor design. *Surf Sci Rep* 2006;**61**:429–44.
334. Chan YH, Boxer SG. Model membrane systems and their applications. *Curr Opin Chem Biol* 2007;**11**:581–7.
335. Katsaras J, Kucerka N, Nieh MP. Structure from substrate supported lipid bilayers. *Biointerphases* 2008;**3**:FB55–63.
336. Knoll W, Naumann R, Friedrich M, Robertson JWF, Lösche M, Heinrich F, McGillivray DJ, Schuster B, Gufler PC, Pum D, Sleytr UB. Solid supported lipid membranes: new concepts for the biomimetic functionalization of solid surfaces. *Biointerphases* 2008;**3**:FA125–35.
337. Reimhult E, Kumar K. Membrane biosensor platforms using nano- and microporous supports. *Trends Biotechnol* 2008;**26**:82–9.
338. Steinem C, Janshoff A. Multicomponent membranes on solid substrates: interfaces for protein binding. *Curr Opin Colloid Interface Sci* 2010;**15**:479–88.
339. Schuster B, Pum D, Sára M, Braha O, Bayley H, Sleytr UB. S-layer ultrafiltration membranes: a new support for stabilizing functionalized lipid membranes. *Langmuir* 2001;**17**:499–503.
340. Schuster B, Weigert S, Pum D, Sára M, Sleytr UB. New method for generating tetraether lipid membranes on porous supports. *Langmuir* 2003;**19**:2392–7.
341. Bangham AD, Standish MM, Watkins JC. Diffusion of univalent ions across the lamellae of swollen phospholipids. *J Mol Biol* 1965;**13**:238–52.
342. Cui HF, Ye JS, Leitmannova Lui A, Tien HT. Lipid microvesicles: on the four decades of liposome research. In: Leitmannova Lui A, Tien HT, editors. *Advances in planar lipid bilayers and liposomes*Vol. 4. Amsterdam: Elsevier; 2006. pp. 1–48.
343. Tien HT, Ottova-Leitmannova A. *Membrane biophysics: as viewed from experimental bilayer lipid membranes*. Elsevier, Amsterdam: Planar lipid bilayers and spherical liposomes; 2000.
344. Amselem AS, Yogev A, Zawoznik E, Friedman D. Emulsomes, a novel drug delivery technology. *Proc Int Symp Control Release Bioactive Mater* 1994;**21**:1369.
345. Vyas SP, Subhedar R, Jain S. Development and characterization of emulsomes for sustained and targeted delivery of an antiviral agent to liver. *J Pharm Pharmacol* 2006;**58**:321–6.

346. Andresen TL, Jensen SS, Jorgensen K. Advanced strategies in liposomal cancer therapy: problems and prospects of active and tumor specific drug release. *Prog Lipid Res* 2005;**44**:68–97.

347. Hollmann A, Delfederico L, Glikmann G, De Antoni G, Semorile L, Disalvo EA. Characterization of liposomes coated with S-layer proteins from *lactobacilli*. *Biochim Biophys Acta* 2007;**1768**:393–400.

348. Pum D, Sára M, Schuster B, Sleytr UB, Chen J, Jonoska N, Rozenberg G. Bacterial surface layer proteins: a simple but versatile biological self-assembly system in nature. In: Chen J, Jonoska N, Rosenberg G, editors. *Nanotechnology: science and computation*. Berlin: Springer; 2006. pp. 277–90.

349. Schuster B, Kepplinger C, Sleytr UB. Biomimetic S-layer stabilized lipid membranes. In: Toca-Herrera JL, editor. *Biomimetics in biophysics: model systems, experimental techniques and computation*. Kerala, India: Research Signpost; 2010. pp. 1–12.

350. Dieluweit S, Pum D, Sleytr UB. Formation of a gold superlattice on an S-layer with square lattice symmetry. *Supramol Sci* 1998;**5**:15–9.

351. Wahl R, Engelhardt H, Pompe W, Mertig M. Multivariate statistical analysis of two-dimensional metal cluster arrays grown in vitro on a bacterial surface layer. *Chem Mater* 2005;**17**:1887–94.

352. Maslyuk VV, Mertig I, Bredow T, Mertig M, Vyalikh DV, Molodtsov SL. Electronic structure of bacterial surface protein layers. *Phys Rev B* 2008;**77**:45419.

353. Vyalikh DV, Danzenbächer S, Mertig M, Kirchner A, Pompe W, Dedkov YS, Molodtsov SL. Electronic structure of regular bacterial surface layers. *Phys Rev Lett* 2004;**93**: 238103-1-238103-4.

354. Vyalikh DV, Kummer K, Kade A, Blüher A, Katzschner B, Mertig M, Molodtsov SL. Site-specific electronic structure of bacterial surface protein layers. *Appl Phys A* 2009;**94**:455–9.

355. Göbel C, Schuster B, Baurecht D, Sleytr UB, Pum D. S-layer templated bioinspired synthesis of silica. *Colloids Surf B Biointerfaces* 2009;**75**:565–72.

356. Kröger N, Lorenz S, Brunner E, Sumper M. Self-Assembly of highly phosphorylated silaffins and their function in biosilica morphoGenesis. *Science* 2002;**298**:584–6.

Viral Capsids as Self-Assembling Templates for New Materials

MICHEL T. DEDEO,[*,†]
DANIEL T. FINLEY,[*,†] AND
MATTHEW B. FRANCIS[*,†]

[*]*Department of Chemistry, University of California, Berkeley, California, USA*
[†]*Materials Sciences Division, Lawrence Berkeley National Laboratories, Berkeley, California, USA*

The self-assembling protein shells of viruses have provided convenient scaffolds for the construction of many new materials with well-defined nanoscale architectures. In some cases, the native amino acid functional groups have served as nucleation sites for the deposition of metals and semiconductors, leading to organic–inorganic composites with interesting electronic, magnetic, optical, and catalytic properties. Other approaches have involved the covalent

Progress in Molecular Biology
and Translational Science, Vol. 103
DOI: 10.1016/B978-0-12-415906-8.00002-9

353

modification of the protein monomers, typically with the goal of generating targeting delivery vehicles for drug and imaging cargo. Covalently modified capsid proteins have also been used to generate periodic arrays of chromophores for use in light harvesting and photocatalytic applications. All of these research areas have taken advantage of the low polydispersity, high chemical stability, and intrinsically multivalent properties that are uniquely offered by these biological building blocks.

I. Introduction

Many of the most exciting ideas and opportunities in materials science require the construction of complex structures with nanoscale features. The realization of these designs, however, presents a substantial synthetic challenge as this size scale lies between those that are traditionally addressed using organic synthesis and lithographic patterning techniques. Although much progress is being made on both sides of this "spatial divide," a growing number of laboratories have turned to self-assembly as an effective means to generate integrated multicomponent materials with nanoscale architectures.[1] This approach can generate both discrete and extended structures with high efficiency as individual components come together to satisfy intermolecular recognition forces. Despite the many recent advances in the *de novo* creation of self-assembling materials, the most complex and functional examples of this concept are still found in biological systems. Through the assembly of proteins, nucleic acids, lipids, and carbohydrates that have evolved to have just the right shapes, flexibilities, solvation, and electrostatic properties, nature can construct elaborate nanoscale "devices" that are capable of high-level functions. Compelling examples include structures for energy transduction, complex molecule synthesis and selective degradation, information processing, and even self-replication. Inspired by these examples, a growing number of research laboratories have taken up the task of using biomolecular self-assembly to make artificial analogs that begin to mimic some of these functions.

As one possibility, many examples have shown that new materials can be generated using modified peptides, capitalizing on the ease of functional group introduction and the rich assembly behavior exhibited even by simple sequences.[2–5] Nucleic acids have also been used to generate elaborate nanoscale structures that are difficult to create using other methods.[6,7] A particular advantage of this approach is the ability to design completely new structures using computational methods. However, the bulk of the functional materials in nature involve self-assembling proteins, which typically involve defined three-dimensional shapes and complex interfaces that come together through large surface interactions. As a result of these properties, protein-based materials can possess a rigidity over length scales ranging from nanometers to microns. This property is often challenging to achieve using other building blocks, and is important for the generation of materials that can propagate nanoscale features into bulk materials with much larger dimensions. Proteins also offer the opportunity for synthetic tailoring, as site-directed mutagenesis can be used to introduce new amino acids chosen from the natural pool or the rapidly growing list of artificial groups that can be incorporated.[8,9] This opens the door to the use of often sophisticated organic synthesis methods that can introduce new functional groups, from polymers to catalysts to nanoparticles, on the surfaces of the proteins either before or after the assembly occurs. As an additional advantage, many proteins can be produced on large scale through recombinant expression in bacteria or propagation in plants.

The protein shells of viruses, or "capsids," provide the most well-studied class of self-assembling proteins that has been used for materials science.[10–19] These structures serve to protect the genomic nucleic acids from degradation and are often active participants in the cell-specific delivery of the genes required for propagation.[20] In the general sense, viral capsids possess a high degree of symmetry and are assembled from many copies of a small number of distinct protein monomers. These features have resulted in the atomic-level characterization of many capsids using X-ray crystallography and a growing number of detailed morphological analyses using cryo-electron microscopy. Many viral capsids exist as hollow spherical structures with icosahedral symmetry, providing a high volume-to-surface area ratio for efficient genome encapsulation. Capsids with this geometry exhibit a characteristic "T" number, which is related to the total number of protein monomers ($T \times 60$) that comprise their structure.[20] As viral capsids with varying T numbers are known, it is possible to choose structures with different sizes and complexities to meet the synthetic challenge at hand. A number of viruses also exist as high-aspect-ratio rods, housing their nucleic acids in extended helices or circular loops. In some cases, purified viral capsid proteins can be used to generate more than one self-assembled state: tobacco mosaic virus (TMV) monomers can assemble into both disks or rods,[21,22] the brome mosaic virus (BMV) can form spherical capsids with either $T = 1$ or $T = 3$ symmetry,[23] and the length of bacteriophage M13 can be altered by altering its genome size.[24] A structural comparison of several viral capsids used for materials science applications is shown in Fig. 1, and a table listing many of their key properties is given in Fig. 2.

Taken together, viral capsids offer many different structural starting points for materials construction. A number of approaches detailed in the following sections use the intact particles without removing the nucleic acids inside, while some have shown that the genome can be removed to provide new spaces for chemical modification. In many examples, wild-type capsids can serve as scaffolds for the nucleation and growth of inorganic materials, while capsids

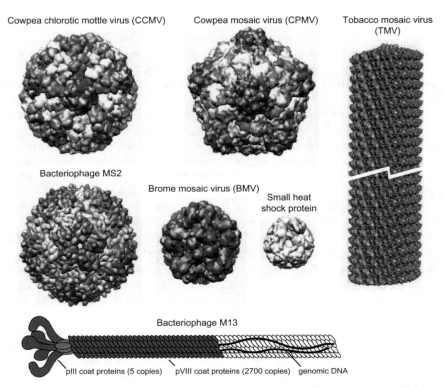

Cowpea chlorotic mottle virus (CCMV)   Cowpea mosaic virus (CPMV)   Tobacco mosaic virus (TMV)

Bacteriophage MS2

Brome mosaic virus (BMV)

Small heat shock protein

Bacteriophage M13

pIII coat proteins (5 copies)   pVIII coat proteins (2700 copies)   genomic DNA

FIG. 1. Structural comparison of viral capsids used for materials applications. Additional structural details are provided in the table in Fig. 2. (See Color Insert.)

bearing point mutations in the coat protein monomers are often used in biomedical and catalytic applications. The latter case allows the capsids to be modified using the rapidly growing set of chemoselective bioconjugation reactions,[39,40] providing multivalent platforms for enhanced target recognition *in vivo* or "amplified" imaging agents that can be detected with very high sensitivity. A summary of the chemical strategies that are commonly used for viral capsid modification appears in Fig. 3.

Based on these considerations, this chapter reviews the progress that has been made in the use of self-assembling viral capsids to make new materials. The examples have been grouped by the application area, including sections on carriers for biomedical imaging and drug delivery, templates for nanoparticle synthesis and the creation of inorganic structures, and positional scaffolds for complex materials with optical and catalytic behavior. Within each section, the material has been organized by the viral particle that has been modified, as

Virus	Size	Symmetry	Monomers/capsid	Expression system (ref)	Genome	PDB ID code
Cowpea chlorotic mottle virus	28 nm	T=3	180	Plant, P. pastoris[25]	ssRNA: 3171, 3100, 2173 bases	1CWP
Cowpea mosaic virus	30 nm	Pseudo T=3	120	Plant[26], S. cerevisiae	ssRNA: 6.6, 3.8 kb	1NY7
Bacteriophage MS2	27 nm	T=3	180	E. coli (infectious[27] and recombinant[28])	ssRNA: 3569 bases	2MS2
Bacteriophage Qβ	28 nm	T=3	180	E. coli (recombinant)[29]	ssRNA: 4217 bases	1QBE
Canine parvovirus	26 nm	T=1	60	Insect and mammalian cells (infectious)[30]	ssDNA: 5 kb	1P5Y
Brome mosaic virus	28 nm	T=3	180	Plant[31]	ssRNA: 3.2, 2.8, 2.1 kb	1JS9
Brome mosaic virus	19 nm	T=1	60	Plant[31]	–	1YC6
Turnip yellow mosaic virus	30 nm	T=3	180	Plant[32]	ssRNA: 6318 bases	1AUY
Red clover necrotic mosaic virus	36 nm	T=3	180	Plant[34]	ssRNA: 3889, 1448 bases	none
Small heat shock protein*	12 nm	Octahedral	24	E. coli (recombinant)[34]	–	1SHS
Tobacco mosaic virus	17 × 300 nm	Helical rod	2130	Plant[35], E. coli (recombinant)[36]	ssRNA: 6390 bases	2TMV
Chilo iridescent virus	140 nm	Icosahedral	Thousands	Whole insects[37], cultured insect cells[38]	Single linear dsDNA: 212,482 bp	none
Bacteriophage M13	6.6 × 880 nm	Helical rod	2700 (pVIII)	E. coli (infectious)	ssDNA: 6.4 kb	none

* From *Methanocaldococcus jannaschii*.

FIG. 2. Table of key properties for viral capsids commonly used to build new materials.

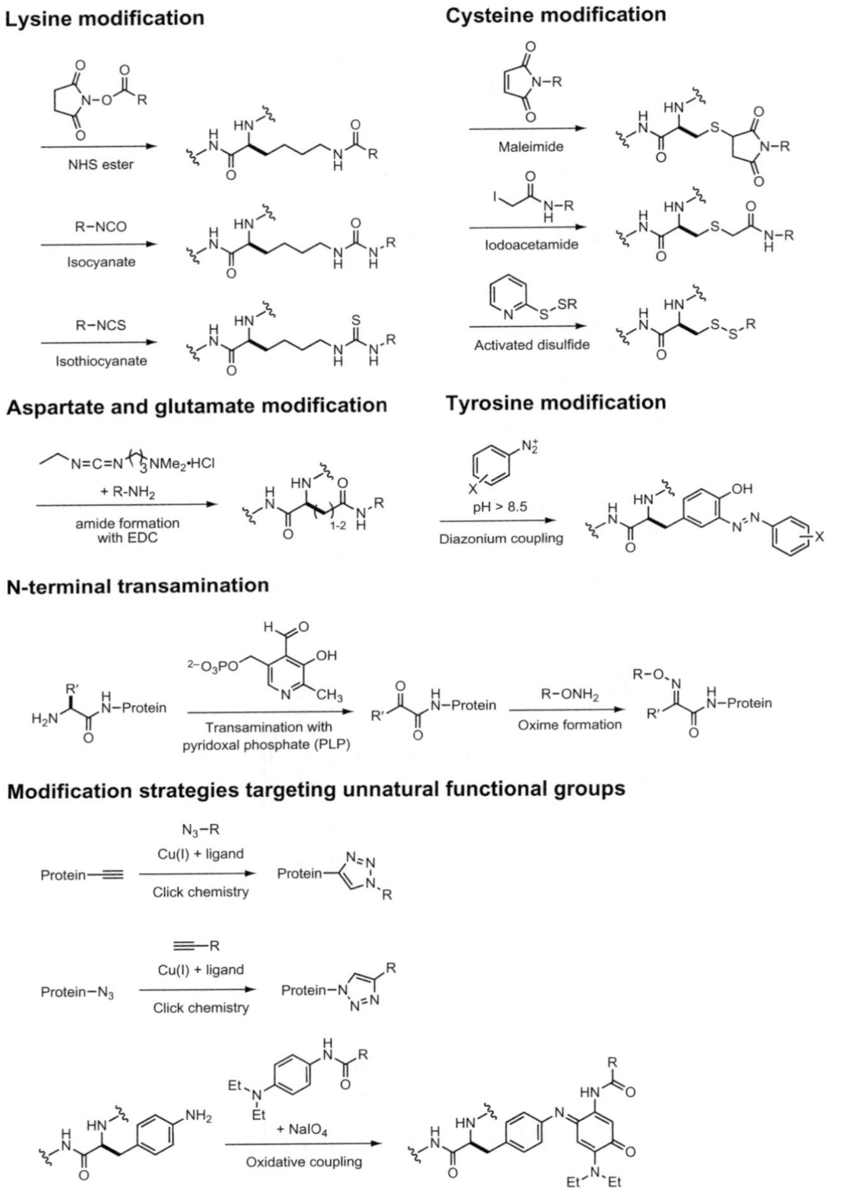
Lysine modification
NHS ester
R−NCO
Isocyanate
R−NCS
Isothiocyanate
Cysteine modification
Maleimide
Iodoacetamide
Activated disulfide
Aspartate and glutamate modification
+ R-NH2
amide formation with EDC
Tyrosine modification
pH > 8.5
Diazonium coupling
N-terminal transamination
N−Protein
Transamination with pyridoxal phosphate (PLP)
R−ONH2
Oxime formation
Modification strategies targeting unnatural functional groups
Cu(I) + ligand
Click chemistry
Protein
+ NaIO4
Oxidative coupling

FIG. 3. Summary of the common chemical strategies that are used to attach new functional groups to viral capsid surfaces. EDC = *N*-ethyl-*N*',*N*',-dimethylaminopropylcarbodiimide.

most research groups currently publish a series of studies using one or a limited number of different capsids. Although the chapter does not include related work on the use of ferritin[10,41–43] and heat shock protein cages,[10,34,44–46] the reader is encouraged to refer to the referenced works for examples of the use of these proteins for many of the same applications.

II. Covalent Modification of Viral Capsids for Applications in Medical Imaging and Drug Delivery

Perhaps the most intuitive applications for virus-based materials involve the delivery of therapeutic or imagable cargo to specific locations in living organisms. As carriers, viral capsids offer the advantages of small particle size (10–100 nm, which is ideal tumor accumulation),[47] low overall toxicity, and the ability to encapsulate and protect large numbers of cargo molecules within their interior volumes. They also provide a readily functionalizable external surface that can display multiple copies of a tissue-targeting group. These features have inspired several research groups to develop a host of chemical methods that can add the desired functionality to specific amino acid side chains introduced into the capsids before assembly. The strategies used to access these materials, the applications for which they are intended, and the rapidly advancing *in vivo* studies of viral capsid biodistribution and immunogenicity serve as the focus of this section. In addition to the applications that are discussed below, it should also be noted that filamentous viruses are commonly used to evolve peptides and small proteins with desired binding capabilities. As these applications lie outside of the scope of this chapter, the reader is instead directed to other reviews on this topic.[48,49] The use of viruses to deliver genetic cargo in the context of gene therapy[50,51] and the application of viruses as adjuvants for vaccines[52,53] have also been reviewed elsewhere, and so will not be covered here.

A. Cowpea Chlorotic Mottle Virus

The first report of using viral capsids for materials applications appeared in 1998, when Trevor Douglas and Mark Young showed that polymeric and metal ion cargo could be contained inside the protein shell of the cowpea chlorotic mottle virus (CCMV).[54] This early report relied on the ability of this capsid to "swell" when placed under specific pH conditions, allowing the access of exogenous molecules to the interior. They quickly followed up on this work with a series of publications involving the use of CCMV for the deposition of inorganic materials (described in the Section II.B of this chapter) as well as the use of CCMV in a number of therapeutic applications. Regarding the latter, they demonstrated in their early work that the capsids could be chemically modified through the targeting of native lysine and carboxylic acid residues.[55] This led to as many as 560 modifications on each capsid. They also introduced cysteine residues to allow the more controlled introduction of new functionality, such as 24 copies of a 24-amino acid peptide with antitumorogenesis properties. Although these high levels of modification caused the capsids to swell appreciably, they did not otherwise disrupt the morphology or lead to disassembly. In one study, they determined the capsid biodistribution by labeling them with fluorophores or radioactive iodine that could be tracked *in vivo*.[56] Their results showed that the viral capsids distributed throughout the animal, produced an immune response, and were significantly cleared after 24 h.

By capitalizing on the ability of wild-type CCMV capsids to bind metal cations, the Douglas and Young groups were also the first to report a virus-based MRI contrast agent.[57] By exchanging the Ca^{2+} ions required for capsid assembly *in vivo*, they were able to bind roughly 140 Gd^{3+} ions to each structure. These capsids were also determined to have both the highest total relaxivity per particle and the highest relaxivity per Gd^{3+} ion reported to date. However, the relatively weak affinity for the toxic Gd^{3+} ions ($K_d = 31$ μM) prevented the use of these capsids in a clinical setting. Not to be discouraged, these labs began to develop capsid delivery systems with significantly tighter binding for the metal ions. To do this, Douglas and Young turned to the introduction of the chelating agent DOTA, which has a K_d of 10^{-20} M for Gd^{3+}.[58] They attached it to lysines on the surface of CCMV capsids using NHS ester chemistry and then metallated it using $GdCl_3$. The resulting capsids exhibited a 10-fold increase in relaxivity for each capsid-bound Gd(III)–DOTA complex compared to unbound analogs, which they attributed to the dramatically slower rotation of the capsids.[59] The relaxivity of this second-generation system was less than that with the Gd^{3+} bound directly to the capsids, which could be explained by the lack of rigidity in the linker between the DOTA ligand and the protein backbone (thus undermining the slower rotation offered by the capsid). Alternately, the direct capsid binding motif could permit multiple water molecules to exchange at the same time, in contrast to free Gd(III)–DOTA complexes, which only bind a single water molecule at once. Nonetheless, this compromise still yielded capsids with very high relaxivity values and a clinically appropriate Gd(III) affinity. The structures of the Gd(III) chelating ligands used in this study and in others are shown in Fig. 4A.

In later studies, the Douglas and Young labs continued to pioneer the biomedical applications of viral capsids when they adapted CCMV for use in photodynamic therapy (PDT). To generate highly cytotoxic singlet oxygen in response to light, they attached a ruthenium bipyridine complex bound to an iodoacetamide-modified phenanthroline ligand (Fig. 5A) to cysteine residues introduced on the capsid surface.[46] Because the ruthenium bipyridine required

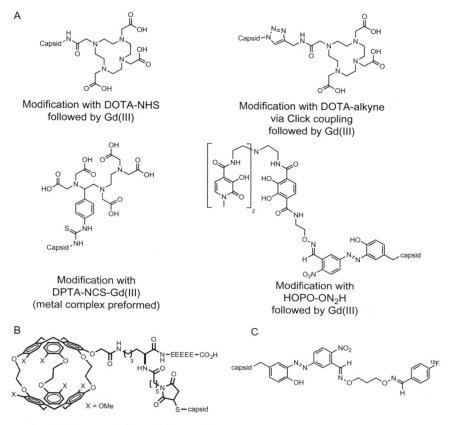

A

Modification with DOTA-NHS
followed by Gd(III)

Modification with DOTA-alkyne
via Click coupling
followed by Gd(III)

Modification with
DPTA-NCS-Gd(III)
(metal complex preformed)

Modification with
HOPO-ON₂H
followed by Gd(III)

B

C

X = OMe

FIG. 4. Summary of chemical functional groups attached to viral capsids for use in diagnostic imaging experiments. Applications of these conjugates include (A) proton MRI contrast enhancement, (B) ^{129}Xe-based MRI, and (C) positron emission tomography.

blue light (which does not penetrate tissue well), the authors suggested that their PDT system could be most appropriate for treating biofilm infections of the oral cavity. In particular, they sought to target multidrug-resistant bacteria, such as *Staphylococcus aureus*, since resistance to singlet oxygen would be difficult if not impossible to develop.

Because singlet oxygen is indiscriminate but very short-lived (the diffusion length is on the order of 50–100 nm[60]), the photosensitizer must be tightly and selectively associated with the target. To explore this strategy the authors compared two systems for targeting their CCMV–photosensitizer conjugates. The first approach was to coat the negatively charged capsids with poly-L-lysine, a cationic polymer, and then mix them with negatively charged bacteria.

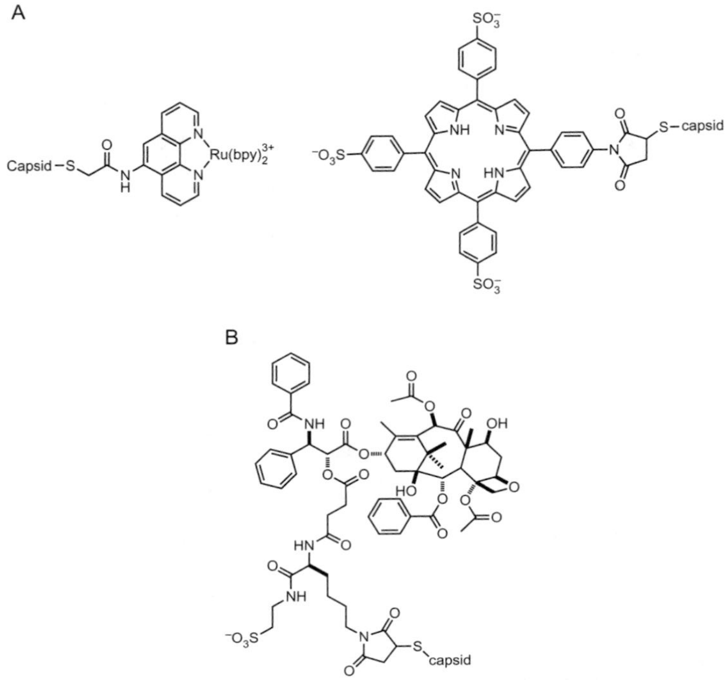

FIG. 5. Therapeutic cargo attached to viral capsids for the purpose of killing cancer cells and *S. aureus*. The compounds in (A) are potent generators of singlet oxygen for photodynamic therapy, while Taxol (B) disrupts the cell cycle through the inhibition of tubulin depolymerization.

The second approach involved appending biotin to the surface lysines of the capsid via the corresponding NHS ester, and then exposing the capsids to biotinylated antibodies coupled to streptavidin. The virtually irreversible association of biotin and streptavidin thus connected the antibodies to the photosensitizer-containing capsid. In this case, the antibody targeted *S. aureus*, but the broad utility of the system lies in the fact that *any* antibody raised to *any* biomolecular target could in principle be coupled using the same strategy. This second set of capsids led to significantly greater bacterial cell death upon illumination. These results were further substantiated by field emission scanning electron microscopy (FESEM) images, which showed sparse viral attachment with the electrostatic approach and very heavy binding mediated by the antibodies. This report highlighted the importance of using capsids to combine multiple copies of different functional groups, especially when receptors are nonabundant, and therefore a maximum of cargo must be delivered to each.

B. Cowpea Mosaic Virus and Bacteriophage Qβ

In studies that were contemporary with much of the CCMV work, the Finn and Johnson groups also provided many early demonstrations of the addition of new synthetic functionality to viral capsid surfaces.[61–63] In their initial reports, they developed a series of chemical methods that could target native lysine and introduced cysteine residues on the surface of the cowpea mosaic virus (CPMV). This gave them the ability to control both the positions and the overall coverage of the introduced groups, which they demonstrated using chromophores and other small molecules. In terms of adding biological targeting groups, they found in a later report that sugars derivatized with isothiocyanates could be attached to lysines on CPMV with varying surface density.[64] To test whether the protein-bound carbohydrates remained solution accessible, they mixed the CPMV conjugates with lectins, which are known to agglutinate natural glycoproteins. The observed aggregation confirmed that the sugars were displayed in a biologically recognizable manner.

Although carbohydrates can bind to some receptors *in vivo* and could lead to increased cellular uptake, the specific recognition of many cancer cell types is likely to require the addition of peptides and full-length proteins to the capsid carriers. Two of the first demonstrations of this capability were reported in a single publication by the Johnson lab in 2004.[65] As the longest genetic sequence that can be inserted directly into a CPMV coat protein is about 30 amino acids, they opted instead to express intact targeting proteins independently and then couple them to the fully assembled capsids. In one example of this approach, they first prepared a mutant of CPMV's large subunit to introduce 60 thiol groups to the capsid surface. They then expressed LRR-InlB, a 22-kDa protein shown to promote internalization of attached cargo, with its own free cysteine residue. To prepare the conjugate, a homobifunctional maleimide cross-linker was added to the CPMV capsids, followed by the LRR-InlB targeting group. The coupling of the proteins was confirmed using gel electrophoresis and electron microscopy, and could be carried out to the extent that Western blots indicated that all of the large subunits were linked to LRR-InlB.

To make a second type of targeted capsids, the Johnson group also coupled them to the Int8 module of Herstatin, which has been shown to promote uptake by cancer cells. Rather than introducing a cysteine into Int8, they converted the N-terminal amino group to a thiolate before using the homobifunctional cross-linking strategy described above. *In vitro* tests indeed showed the specific binding of the viral conjugate to cells expressing the HER2 receptor. Finally, a heterobifunctional NHS ester–maleimide cross-linker was used to couple a cysteine mutant of T4 lysozyme to the lysines of wild-type CPMV, taking advantage of up to five native lysines per subunit for maximal loading. Importantly, they demonstrated that the conjugated enzymes were still capable of hydrolyzing bacterial cell walls even after the coupling reaction.

To allow the attachment of an even wider set of synthetic groups, the viral capsid of CPMV has been explored extensively as a substrate for copper-catalyzed "Click" chemistry by the Finn group.[66] Based on the [3 + 2]Huisgen cycloaddition of azides to alkynes, this reaction provides a highly effective way to couple diverse sets of molecules with virtually complete functional group tolerance. The reaction is particularly well suited for bioconjugation because it is accelerated in aqueous media and can be used in conjunction with many other protein modification reactions.[67] To introduce the reactive groups onto CPMV, they modified native lysine and cysteine residues with either alkyne-substituted NHS esters or azides attached to bromoacetamides, respectively. Treatment of the modified capsids with copper(I) and a small molecule bearing the complementary functional group provided very highly levels of modification. With the later availability of improved ligands for the copper(I) ions, the Finn and Manchester groups demonstrated the installation of highly hindered molecules, including the large glycoprotein transferrin.[68] This protein was chosen because it is known to mediate cancer cell targeting and uptake. In a subsequent paper, the Finn group used a further optimized Click reaction to attach transferrin to the bacteriophage Qβ.[69] They showed that the conjugates were specifically internalized by cells expressing transferrin receptors. Although the rate of cell uptake was significantly improved with increasing numbers of transferrin ligands, separate binding experiments showed that this polyvalency did not improve affinity or avidity. Despite these surprising results, their report shows that Qβ has great potential as a drug delivery vehicle.

To begin the *in vivo* characterization of the rapidly growing collection of tissue-targeting CPMV conjugates, the Manchester and Finn groups sought to determine the fate of the virus during the first few days after its introduction.[70] By tracking both the viral RNA (which was packaged inside, serving as a model drug cargo) and externally attached fluorophores, they confirmed that both components were distributed to a wide range of tissues after either oral or intravenous administration. This suggested that the capsids remained intact and could indeed be used as successful delivery vehicles. To examine the biological uptake more closely, the Manchester and Stuhlmann labs published a subsequent paper describing the use of an AlexaFluor-labeled capsid to image chick embryos to a depth of up to 500 μm.[71] Using microscopy of intact and microtomed tissues, they found uptake of the capsids into the perinuclear compartments of vascular endothelial cells. The attachment of poly(ethylene glycol) (PEG) alongside the dye prevented this uptake, and also increased the circulation time. The superior resolution and brightness offered by the viral capsid carriers allowed the researchers to capture real-time videos that reported vascular changes associated with tumor formation.

To image deeper into tissue than can be achieved using microscopy, the Finn and Manchester groups have also functionalized CPMV and bacteriophage Qβ with MRI contrast enhancement agents.[72] In a two-step procedure, they first introduced azide groups on the exterior surface of the capsids using lysine/NHS ester chemistry, after which an alkyne-modified DOTA ligand (Fig. 4A) was installed using Cu(I) Click chemistry. The DOTA moiety was then complexed to Gd^{3+}, which is the most common metal ion used in MRI contrast agents. When compared to the commercially available Magnevist, the viral conjugates demonstrated a two- to threefold increase in relaxivity for each Gd(III) complex. Despite the large size of the noted above, they did not observe an increase over free Gd–DOTA complexes. This was attributed to a trade-off between the slower tumbling rate afforded by the capsid and a decreased access to water molecules. In a continuation of this collaboration, these groups demonstrated the use of the capsid conjugates for MRI contrast enhancement in mice, observing no *in vivo* toxicity.[73] For further reference, the Manchester group has published a review on the use of virus-based nanoparticles for diagnostic imaging.[74]

As a different type of cargo, the Manchester and Finn labs have reported the attachment of fullerenes, or "Buckyballs," to the exterior of CPMV and bacteriophage Qβ.[75] The material properties of fullerenes have created a significant interest in their use in a number of areas, including the generation of photovoltaic devices and as photosensitizers for photodynamic cancer therapy, but they are ordinarily very difficult to solubilize in water. Using Click chemistry, they were able to achieve high levels of fullerene attachment to bacteriophage Qβ capsids, compared to the levels that could be achieved using EDC/NHS coupling. These conjugates were observed to be taken up by cancer cells, suggesting promise for future applications in photoactivated tumor therapy.

A key consideration for the use of viral capsids for *in vivo* applications is the level to which they are recognized by the immune system. While the extent of this concern will vary according to the amounts of the capsid carriers that are introduced and the duration of the treatment, many applications will require the shielding of the capsid surfaces from antibody recognition at the very least. With a view toward this goal, the Finn, Johnson, and Manchester groups collaborated to attach PEG chains to the surface of CPMV capsids using NHS ester chemistry.[76] They found that this strategy was indeed successful in preventing the binding of monoclonal antibodies to the capsids.

C. Canine Parvovirus

The popularity of many of the viruses discussed herein is due in part to the fact that they infect plants or bacteria, rather than something resembling the scientists who are working with them. The downside of this is that the innate targeting capability of such viruses is irrelevant when placed inside a mammal, forcing researchers to go to great lengths to introduce their own targeting groups. Canine parvovirus (CPV), by contrast, has a natural affinity for transferrin receptors in dogs as well as in humans, which it can use to enter cells of either species. Certain tumors overexpress transferrin receptors, making CPV a promising scaffold for drug delivery to these sites. In the first report on its use as a potential drug delivery vehicle, the Manchester group modified up to 100 of the surface lysines with NHS ester dyes, which served as a model drug cargo.[77] This loading was not found to interfere with capsid uptake, which still colocalized with transferrin. Furthermore, the uptake remained dependent on expression of the transferrin receptor. This report thus demonstrated the possibility of generating a new class of cancer-targeting drug delivery vehicles from mammalian virus precursors.

D. Bacteriophage MS2

Another icosahedral virus that has been used extensively as a bioconjugation target is bacteriophage MS2, or simply MS2. Its capsid consists of a 27 nm hollow sphere, which, like many other $T = 3$ viruses, is formed from 180 sequence-identical monomers. As a somewhat unusual feature, the assembled MS2 capsid has 32 pores, each of which is about 2 nm in diameter. This allows small molecules to diffuse more readily into and out of the capsid. With a view toward the development of vaccines, early work by the Stockley group has shown that peptides can be inserted into the coat protein sequence for efficient display to the immune system.[78]

In terms of chemical modification, the Francis lab has focused extensively on the development of new bioconjugation strategies for both the interior and the exterior surfaces of the MS2 capsid. The first of these reports was in 2004, when they showed that it was possible to remove the ssRNA genome of the virus under alkaline conditions to produce stable, "empty" capsids.[27] These assemblies exhibited high stability, tolerating temperatures up to 60 °C and pH conditions ranging from 3 to 11. This treatment exposed a set of native tyrosine residues on the interior capsid surface, which could be modified chemically using diazonium salts at pH 8.5 or above. This resulted in the introduction of up to 180 copies of a new functional group in locations that would presumably not interfere with capsid biodistribution or be subject to enzymatic degradation *in vivo*. Subsequent reactions showed that the azo groups introduced using this strategy could be reduced using sodium dithionite, and that the resulting aminophenol could be used in further functionalization reactions.

More recent work used the diazonium coupling method to install fluorescent dyes on the inside of MS2 capsids while independently modifying external lysine groups with NHS ester PEG polymers. The PEG coating was shown to decrease the binding of polyclonal antibodies to the capsids by 90% using ELISA.[79] This study highlights the utility of these capsids for the creation of complex materials bearing different modifications on each surface. These structures thus also provided the first model of a virus-based delivery vector that could display targeting moieties on the external surface and over 100 synthetic cargo molecules within.

For use in imaging applications, the MS2 capsid has also been modified to display Gd(III)-based MRI contrast enhancement agents (Fig. 4A). As was the case for CCMV and CPMV (described above), these studies highlight the advantage of decreasing the tumbling rate of the metal complexes while simultaneously amplifying their signal by attaching multiple copies to the same delivery vehicle. The first MS2-based example of this concept came from the Kirschenbaum group,[80] which used the external lysines of MS2 to attach isothiocyanate-functionalized Gd–DTPA ligands. This installed over 500 Gd^{3+} atoms on each capsid and resulted in an overall relaxivity of 7200 mM^{-1} s^{-1} (over three orders of magnitude greater than that of a single Gd^{3+} ion). The Francis group also explored the use of MS2 as a scaffold to bind Gd^{3+},[81] but expanded the study to investigate the effects of internal versus external attachment. They installed aldehyde functional groups either on the interior or the exterior of the capsid by targeting tyrosine residues with diazonium salts and targeting lysines with NHS esters, respectively. A hydroxypyridonate (HOPO) ligand[82] functionalized with an aminooxy group was then attached to the aldehyde groups through oxime formation. They found that Gd–HOPO confined to the interior of the capsid exhibited higher overall relaxivities (up to 3900 mM^{-1} s^{-1}) than the externally bound ligands, as well as much better solubility for the modified capsids. A follow-up report[83] showed that both the rigid linker attaching the Gd–HOPO complex to the capsid and the fast diffusion of water in and out of the protein shell were both important to obtain the high relaxivities that were observed.

As an alternative to proton-based MRI, which directly observes the relaxation of water molecules occurring throughout the body, MRI imaging based on hyperpolarized xenon probes shows much promise for the highly sensitive detection of nonabundant biomarkers. This is due to the large chemical shift window of ^{129}Xe (which gives it the ability to report its specific chemical environment) and the virtually nonexistent level of background signal. To develop viral capsid-based materials that could be detected using this imaging method, the Pines, Wemmer, and Francis groups attached 120 cryptophane cages (Fig. 4B) to cysteine residues introduced inside genome-free MS2 capsids.[84] The cryptophane cages bound to xenon atoms in aqueous media, giving them a specific chemical shift that could be distinguished from unbound xenon atoms. In combination with chemical exchange saturation transfer (HYPER-CEST),[85] solutions of capsids could be detected with concentrations as low as 700 fM. The high sensitivity of this technique, along with the nontoxicity of xenon, makes this a very attractive future candidate for *in vivo* MRI imaging.

MS2 capsids have also been used to house radioactive species, such as ^{18}F, for detection using positron emission tomography (PET). This technique allows the imaging of positron-emitting radioisotopes with exceptionally high signal-to-noise ratios *in vivo*, and is widely used for cancer detection using [18F]-fluorodeoxyglucose. To generate capsids that could be similarly tracked *in vivo* using this technique, aldehyde groups were introduced inside MS2 capsids using the diazonium coupling strategy described above.[86] A bis(aminooxy) linker was then used to attach [18F]-fluorobenzaldehyde to these locations site selectively (Fig. 4C). By following the biodistribution of the capsids in rats using dynamic PET, the authors observed that the capsids exhibited somewhat prolonged circulation times relative to free [18F]-fluorobenzaldehyde. They also showed that additional small molecules attached to the inner surface of the capsid did not affect the biodistribution behavior.

In the previous examples of MS2 bioconjugation, the capsid exterior was only modified in a nonspecific fashion through the acylation of any of the three solvent-exposed lysine residues. This introduces many limitations for the installation of more complex targeting groups, as most of the cell-binding proteins and peptides that one would want to attach will also contain lysine residues. Desiring a more elegant means of site-specifically modify the external surfaces of MS2 capsids while retaining the ability to target cysteine residues on the interior, the Francis lab used the Schultz *in vivo* technique[8] to introduce a uniquely reactive amino acid side chain into the coat protein monomers.[28] They chose to introduce *p*-aminophenylalanine (*p*aF), producing 180 anilines on the external surface of each capsid. In the presence of sodium periodate (NaIO$_4$), the *p*aF residues were shown to participate in a chemoselective oxidative coupling reaction with *N,N*-dialkyl phenylenediamines that had been reported previously by their group.[87] This reaction was shown to be fast, high yielding, and selective for the *p*aF residues over all of the other natural functional groups on the protein.

With this new construct in hand, the Francis group went on to elaborate MS2 as a targeted drug delivery vehicle. In their first report, the external *p*aF residues were modified with *N,N*-dialkyl phenylenediamine-substituted DNA aptamers.[88] The specific nucleic acid sequences used in this study were identified by the Tan group[89] as having a high affinity for protein tyrosine kinase 7

(PTK7), which is overexpressed on many cancer cell types. The inside of the capsids were labeled with fluorescent dyes through the alkylation of introduced cysteine residues using maleimide chemistry. It was shown that these aptamer-functionalized capsids could effectively target Jurkat leukemia T cells with high specificity over other cell types, while capsids displaying an aptamer with a random sequence had no targeting capabilities.

A second report took this observation one step further by combining the capsids with external PTK7 targeting aptamers with porphyrin molecules, which act as photodynamic sensitizers upon illumination.[90] To introduce them, water-soluble porphyrins bearing maleimide groups were constructed and used to alkylate the interior cysteine residue. Using this strategy, up to 180 copies of the porphyrins (Fig. 5A) could be encapsulated within each capsid, and 20 copies of the DNA aptamers were displayed on the outside. As before, the dual-modified capsids were shown to bind to Jurkat cells effectively and showed low affinity for other cell lines. Once the modified viral particles had bound the receptors, illumination for 20 min caused the porphyrins to destroy the cancer cells through the generation of up to 300,000 molecules of singlet oxygen per capsid. Control experiments with cells lacking the PTK7 receptor displayed little to no capsid binding and subsequent cell death. Additionally, the authors mixed solutions of dual-modified Jurkat targeting MS2 capsids, Jurkat cells, and red blood cells. Jurkat cell death was observed, but the red blood cells were minimally perturbed. This illustrates the utility of PDT in selective destruction of targets in complex mixtures. The high spatial resolution and selectivity of this technique stems from the exceptionally high reactivity of the active species, namely, singlet oxygen. Singlet oxygen will undergo side reactions and become quenched before it can diffuse very far from where it is formed.

One final example of bacteriophage MS2 being used as a drug delivery platform also comes from the Francis lab. In this report, 120 copies of the potent but very hydrophobic anticancer drug Taxol (Fig. 5B) were incorporated into the interior of the capsids.[91] This was accomplished by producing a water-soluble linker that provided each Taxol molecule with a maleimide group for cysteine alkylation. These untargeted constructs displayed the ability to destroy MCF-7 breast cancer cells, presumably initiated by uptake of the capsids into the cells through pinocytosis. Endosomal and lysosomal compartments then disassembled the capsid, exposing the drugs and allowing hydrolysis of the linker, and releasing free Taxol.

E. Summary

Viral capsids have the potential to overcome a number of the challenges for the targeted delivery of therapeutics and diagnostic cargo. They offer mono-disperse, nontoxic structures that can solubilize hydrophobic drugs introduced through encapsulation or bioconjugation, facilitate chemotherapy by targeting the payload to cancer cells and minimizing exposure of healthy tissues, increase uptake into tumors through the EPR effect, and extend the half-life by preventing premature filtration by the kidneys. All of these roles require the addition of functionality to the capsid, performed using a rapidly growing set of bioconjugation reactions.

III. Viral Capsids as Templates for the Construction of Inorganic Materials

Many of the first examples of using viral capsids to make new materials involved their use as templates for the controlled deposition of metal ions and the precise positioning of inorganic nanoparticles. In general, these studies have been motivated by the monodisperse nanoscale sizes of viral capsids, as well as the powerful ability to introduce metal-binding amino acid side chains in specific locations on their surfaces. Often, the assembled capsid structures are subjected to combinations of metal ion salts and reducing agents, which cause metal to be deposited selectively on the protein surface. This can produce inorganic nanostructures with discrete shapes and sizes that are difficult to realize using other methods. Hollow spherical viral capsids can sequester nanoparticles within their cavities or be used as "nanoreactors" to grow nanoparticles of well-defined sizes. Filamentous and rod-shaped viruses can be used to template the growth of one-dimensional nanomaterials and can produce extended structures through liquid-crystal assembly properties.

A. Cowpea Chlorotic Mottle Virus

The earliest report of any viral capsid-based material came from the labs of Trevor Douglas and Mark Young in 1998.[54] In this work, they used 28-nm CCMV capsids that were devoid of genetic material as empty shells for the encapsulation of small molecule guests. Key to this work was the discovery that the CCMV capsid swelled at pH values above 6.5, leading to increased porosity and more facile diffusion of materials into the structure. This property, coupled with the natural positive charge lining the interior of the cavity stemming from 1620 basic residues, allowed the controlled sequestration of anionic species within the capsid. Initially, soluble tungstate anions (WO_4^{2-}) were introduced into the capsids at a high pH. The pH was then decreased, causing oligomerization of the tungstate to form insoluble paratungstate ($H_2W_{12}O_{42}^{10-}$) while simultaneously decreasing the porosity of the capsid and trapping the polyoxometalate within the cavity (Fig. 6A). This methodology was additionally used to mineralize decavanadate ($V_{10}O_{28}^{6-}$) inside CCMV. In this same report, the positive lining of the capsid was also used to capture the polyanionic polymer poly(anethol sulfonate) through electrostatic interactions. Expanding on this seminal work,

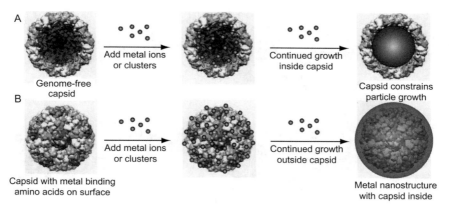

FIG. 6. General strategies for the templated growth of metal nanoparticles using viral capsids. (See Color Insert.)

Douglas, Young, and coworkers introduced genetic modifications into the viral coat protein in order to increase the range of materials that could be trapped inside.[92] This was driven by the desire to form iron oxide nanoparticles with controlled dimensions. To accomplish this, the positively charged interior lining was altered by mutating nine basic residues on each monomer to glutamic acids. The modified capsids were shown to bind ferrous ions, which then reacted with oxygen to form iron oxide particles with a size that was constrained by the interior diameter of the capsid. In separate work, they were able to use similar approaches to produce platinum nanoparticles within heat shock proteins and demonstrate their use for photocatalytic hydrogen production.[44]

As an alternative to this procedure, Dragnea and coworkers (along with Douglas and Young) reported the use of fully formed inorganic nanocrystals as templates for the assembly of CCMV coat proteins (Fig. 7A).[93] The motivation for this work was to investigate the assembly process of CCMV and to determine whether the nanoparticles could assist or even direct it. In this study, the authors first removed a large portion (34 amino acids) of the protein N-terminus, which resulted in a broadening of the resulting capsid size distribution. When assembled in the presence of gold nanocrystals, however, the capsids that encapsulated the particles once again formed a monodisperse population that was dependent on the nanocrystal size.

B. Chilo Iridescent Virus

The chilo iridescent virus (CIV) is not a simple icosahedral virus; instead, it exhibits a layered structure composed of a dsDNA–protein core that is surrounded by a lipid bilayer. This is in turn encased in a second protein shell, giving the entire ensemble a diameter of 140 nm. With the goal of making

A

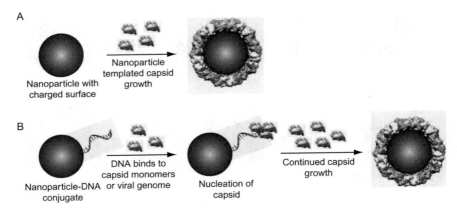

B

FIG. 7. General strategies for the templated growth of viral capsids on the surfaces of metal nanoparticles.

materials with desirable optical properties, Radloff and Vaia reported the use of CIV as a scaffold to position gold nanocrystals and subsequently form continuous nanoshells using electroless deposition techniques (Fig. 6B).[94] For the initial step, the authors relied on the native surface residues of CIV to bind negatively charged gold nanocrystals through electrostatic interactions. Then, reduction of gold(III) salts from solution afforded a thin shell of gold surrounding the entire virion. This approach of modifying the exterior of the capsid is complementary to the examples of CCMV, where the inner surface of the capsid was exploited for inorganic particle nucleation.

C. Cowpea Mosaic Virus

In addition to its applications in medical imaging and drug delivery, CPMV has also found use as a template for inorganic materials. In an intriguing report by Colvin, Johnson, and coworkers, body-centered cubic crystals of CPMV were formed to provide empty cavities and channels comprising about 50% of the overall crystal volume.[95] These repeating void spaces were then used for platinum and palladium deposition through the addition and subsequent reduction of metal salts. By using this approach, extended metal structures were formed with periodic nanoscale features that were templated by the regular packing arrangement established by the CPMV capsids.

D. Brome Mosaic Virus

The capsid of BMV is a simple icosahedral assembly with the ability to form both 28-nm ($T = 3$, 180 monomers) and 19-nm ($T = 1$, 60 monomers) particles. Similar to (and preceding) their work with CPMV, Dragnea and coworkers sought to assemble the BMV coat protein around preformed gold nanocrystals (Fig. 7A).[96] A critical step in doing so was determining that a tetraethylene glycol (TEG) coating on the gold nanoparticles was required for their effective encapsulation. They examined the effects of nanoparticle size on protein assembly, leading to the finding that the capsid size could be tuned by altering the diameter of the gold particle.[23] This result suggested that strong interactions took place between the encapsulated particle and the coat proteins during the assembly. A mechanism for this process was proposed, in which initial nonspecific protein adsorption first occurs on the nanocrystal surface. Once the nanoparticle has been covered with protein, the surface-bound capsomers can then fully assemble on the gold surface in response to a drop in pH. Through similar methodology, the Dragnea lab also showed that fluorescent quantum dots can also be encapsulated in BMV.[97]

E. Red Clover Necrotic Mosaic Virus

An additional example of an icosahedral virus that has been used for nanoparticle encapsulation is provided by the 35-nm capsid of the red clover necrotic mosaic virus (RCNMV). This virus, much like the CCMV, swells at pH values above 6.5. By increasing the pH even further, the capsid can be completely disassembled. Through the careful lowering of the pH, it is then possible to reassemble the intact capsid without the genomic RNA inside. Interestingly, the RNA-free capsid retains its native icosahedral geometry, but the diameter is reduced slightly to 30 nm. Franzen and coworkers used this controllable assembly property to encase gold nanoparticles using the RCNMV proteins, but took a different approach than previous examples by instead exploiting the natural RNA sequence to guide the assembly process.[33] To do this, they used gold nanoparticles functionalized with ssDNA sequences that were complementary to the RNA of RCNMV. Isolated RCNMV RNA hybridized to the nucleic acids on the nanoparticles, and the subsequent introduction of coat protein allowed the directed assembly of capsids onto the gold surfaces (Fig. 7B). It was shown that multiple sizes of gold nanoparticles could be encapsulated within the RCNMV, but when the size of the particle exceeded the size of the virus cavity, no complex structure formed. The authors then showed that both $CoFe_2O_4$ nanoparticles and quantum dots could be incorporated into RCNMV using the same chemistry,[98] displaying the versatility and promise of this technique. Fundamentally different from the examples above, in which the inorganic material was used as the scaffold to assemble the capsid, this method is unique in its use of the genomic RNA to template capsid formation around metal particles.

F. Bacteriophage M13

Unlike the spherical viruses described above, M13 is a filamentous phage with a large aspect ratio (880 nm in length, 6.6 nm in diameter) and a complex coat protein assembly. Its size and shape allows the virus to form liquid-crystalline domains on surfaces due to the high degree of interparticle attraction along the rod axis. Along the length of this large virus, an ssDNA genome is protected by 2700 copies of a pVIII major coat protein assembled in a helical pattern. One end of the virus displays five copies of a pIII minor coat protein, which is often fused to libraries of protein sequences for use in phage display techniques. Briefly, this process entails repeated cycles of mutation, selection of phage that displays pIII proteins with desired binding capabilities, and amplification. In this way, phage can be identified with the ability to bind to virtually any desired biological or even inorganic target. In-depth discussions of the phage display technique can be found in other references.[48,49]

The M13 phage display platform has been widely used by the Belcher lab to evolve peptides that can bind a variety of inorganic materials. In 2000, they showed that by generating large libraries of phage with differing peptides displayed on the pIII coat proteins, followed by extensive screening for binding against semiconductor crystals,[99] they were able to identify peptides that selectively bound to a specific semiconductor material. In this study, the single-crystal semiconductors GaAs(100), InP(100), and Si(100) could be selectively recognized by the evolved peptides, and the researchers were even able to show selectivity toward binding different faces of the same material, with the peptides able to differentiate between GaAs(100) and GaAs(111).

Through further evolution of peptides on the pIII protein, the researchers were able to generate M13 phage-bearing peptides capable of binding ZnS quantum dots.[100] They then used the liquid-crystalline behavior of the M13 to form ordered domains of the quantum dots that extended to millimeters in length. The application of a magnetic field was used to align the liquid-crystalline domains, providing an elegant means of ordering nanoscale objects over long distances.

Though most of the early work on inorganic material-binding peptides was done using the pIII protein, the pVIII major coat protein has also shown promise for the attachment of inorganic particles along the length of the phage. By evolving peptides on the pVIII protein in a manner that was analogous to the pIII protein evolution described above, the Belcher group was able to install 2700 sites on each M13 capsid for the binding of ZnS, CdS, CoPt, and FePt nanoparticles. This led to the construction of one-dimensional nanowires with high-aspect ratios.[101] Two years later, the Belcher lab used anionic tetraglutamate sequences displayed on the pVIII coat proteins to deposit continuous nanowires of cobalt oxide (Co_3O_4).[102] These structures were shown to be attractive candidates for anodes in lithium ion batteries due to their high surface area, ability to produce thin 2D arrays, and mechanical flexibility afforded by their core–shell structure. By also introducing a gold binding peptide motif alongside the tetraglutamate peptides, gold was incorporated into the Co nanowires to increase the battery capacity further. As an alternative approach to creating flexible, high-capacity electrodes for advanced lithium ion batteries, the Belcher lab has also used the pVIII protein to attach amorphous iron phosphate (α-FePO$_4$) while binding the pIII protein to carbon nanotubes.[103] Electrodes constructed with these nanowires surpassed the capacity and conductivity of bulk iron phosphate electrodes.

The Schaak laboratory used similar methodology for creating nanowires on M13, but instead exploited the native charge of the pVIII proteins to bind metal cations.[104] In this fashion, nanowires of Rh, Pd, or Ru nanoparticles could be created in a much less time- and energy-consuming process than through the use of phage display and directed evolution.

G. Tobacco Mosaic Virus

The TMV is another rod-shaped virus that has seen extensive exploration for the fabrication of new materials. The capsid of the native virus is formed from the assembly of 2130 copies of a single protein monomer around an ssRNA genome. This results in low-polydispersity helical rods that are 18 nm in diameter and 300 nm in length. These tube-like structures are hollow, with a 4-nm central pore that spans the entire length of the virus.

In an early report, the labs of Mann, Douglas, and Young used the naturally charged amino acids of TMV capsids to deposit a range of inorganic materials along their sides.[105] Various nanotube structures were created by passing H$_2$S gas through solutions containing TMV and CdCl$_2$ or Pb(NO$_3$)$_2$, leading to the nucleation of CdS or PbS on the viruses. Iron oxide mineralization and the sol–gel deposition of SiO$_2$ on the TMV exterior surfaces were also reported, displaying the versatility of TMV as the template.

Like the bacteriophage M13, TMV also will form liquid-crystalline domains because of its high-aspect ratio. The Mann group was able to use the liquid-crystal-forming behavior of TMV to their advantage as a template for the sol–gel condensation of SiO$_2$. Through this process, the void spaces between viruses were filled with silica, producing a porous silica meso-structure.[106] In a different report, Velev and coworkers were able to gain control of the long-range alignment of TMV liquid crystals by dragging the meniscus of a TMV solution between two glass plates.[107] This created a shear gradient in the solution, aligning the liquid-crystalline domains over centimeter distances. The introduction of NHS-modified gold nanoparticles to these liquid crystals resulted in the positioning of nanocrystals along the capsids. These sites were then used to direct the subsequent electroless deposition of silver to afford ordered nanowires aligned over distances greater than a centimeter.

In the case of metal deposition, the preceding example uses the strategy of first attaching preformed particles to the capsids, and then increasing their size through the addition of metal salts and reductants in solution. An interesting alternative approach was taken by Demir and Stowell, who initially attached the dye Cascade Blue to the exterior surface of TMV through NHS ester chemistry.[108] UV irradiation of the Cascade Blue molecules caused the photochemical reduction of Cu(I) ions in solution, leading to the formation of copper nanoparticles on the surface of TMV.

In addition to metals and metal oxides, conducting nanowires can also be templated by TMV using organic conducting polymers. Although the building blocks are different, the general concept is the same—the capsid serves as an initial nucleation site for the material, which grows to cover the protein surface. The Wang group displayed this methodology when they attached and polymerized pyrrole and aniline on the exterior surface of TMV to create conductive wires.[109] This serves as yet another example of the wide scope of materials that can be generated using capsid templates.

The central pore of TMV capsids has also served as a surface for controlled metal deposition. This was first demonstrated by the Culver and Mann laboratories in 2003, in which they demonstrated the ability to control nucleation of metal nanoparticles irrespective of whether they occurred on the interior or the exterior surface, by modifying the surface charge of the virus.[110] Later that year, Bittner and coworkers demonstrated the successful construction of continuous nickel and cobalt nanowires that were only a few atoms thick within the TMV pore.[111] Further work showed that this technique could also be applied to form copper nanowires in the same fashion.[112]

To improve metal deposition on the virus surface, the Culver lab turned to genetic manipulation of the coat protein. Cysteine residues were introduced in positions 2 and 3, which positioned them on the exterior surface and 3'-face of the rods. This promoted metal nucleation along the length due to thiol–metal interaction and led to a denser and more stable coverage of gold, silver, or palladium nanoparticles.[113] In a subsequent paper, a coat protein mutant was produced with a single cystine in position 3, which proved solvent-accessible on the 3'-face of TMV rods, but less reactive along its length.[114] This design allowed the Culver lab to attach TMV viruses to a gold surface in an "end-on" configuration through the formation of gold–sulfur covalent linkages. Because only one end of the virus displayed the thiol groups, directionally oriented 2D arrays of TMV capsids were formed perpendicular to the surface. Electroless plating of nickel or cobalt on these systems then produced metal nanotube arrays with a high surface area, displaying a twofold increase in battery capacity when used as an electrode in a nickel–zinc battery system.

Further versatility of TMV as an inorganic scaffold is illustrated in the work of Ozkan, Yang, and coworkers, who created components for digital memory devices by aligning platinum nanoparticles along the sides of TMV.[115] One final example comes from a closely related virus—the tomato mosaic virus (ToMV). In this report, Yamashita and coworkers were able to align magnetic nanoparticles within the pore of the rod-shaped ToMV.[116]

H. Summary

These examples highlight the utility of viral capsids for the construction of inorganic nanomaterials. When used as templates for metals and metal oxides, a number of spherical, core–shell, one-dimensional, and mesoscale architectures with nanometer-sized features can be created. Their well-defined sizes and shapes, coupled with the stability of these viral capsids to the often aggressive conditions that are required for metal ion deposition, have established them as promising building blocks for future electronic and optical materials.

IV. Capsid-Based Materials for Optical and Catalytic Applications

In order to capture the energy of sunlight, photosynthetic organisms arrange an impressively diverse collection of chromophores and photoprotective molecules around centralized electron-transfer groups (Fig. 8A).[120,121] This is achieved through the use of a protein scaffold that establishes the crucial spacings between and relative orientations of the components in order to maximize the efficiency of interchromophore Förster resonance energy transfer (FRET). This architecture has inspired a variety of successful artificial light harvesting systems based on polymers and dendrimers. However, these efforts are often hindered by the difficulties associated with the synthesis of organic frameworks containing multiple chromophores and the lack of available scaffolds that are rigid enough to maintain discrete interchromophore distances over large length scales. Viral capsids provide promising alternative scaffolds for the positioning of light harvesting molecules, as they readily assemble into regular, highly symmetric, rigid shapes with an ideal periodicity for FRET to occur between attached components. The following examples

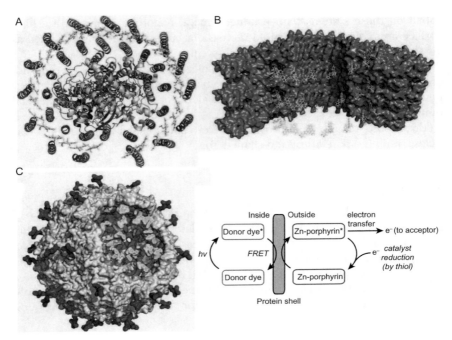

FIG. 8. Light harvesting and energy conversion systems templated by self-assembling proteins. (A) A structure is provided for the LH1 complex and the photosynthetic reaction center of *Rb. palustrus* (from Ref. 117), showing the highly optimized arrangement of bacteriochlorophylls (green) for light collection. Viral capsid-templated chromophore assemblies that mimic some aspects of natural light harvesting systems have been prepared using (B) TMV (Ref. 36), and (C) bacteriophage MS2 (Ref. 118) and M13 (Ref. 119, not shown). (See Color Insert.)

demonstrate the use of the capsid monomers to integrate multiple chromophores and electron-transfer groups, with the goal of creating complex functional assemblies with efficient energy collection and transduction capabilities.

A. Cowpea Mosaic Virus and Hepatitis B Virus

As noted above, CPMV was one of the first viral capsids to be covalently modified to create nanoscale materials.[61] In 2002, the Finn and Johnson groups began their work with virus-based materials by exploring the reactivity of the wild-type capsid with a maleimide fluorescein derivative. They observed that roughly 60 dyes could be attached to a capsid comprised two different monomers. They then installed an additional, much more reactive cysteine in a loop on the exterior surface of one of the monomers, which they found could be alkylated using fewer equivalents and shorter reaction times. This difference in reactivity allowed them to perform sequential reactions, selecting the introduced cysteine with moderate conditions, followed by more forcing conditions to modify the native cysteine. Using this differential reactivity, they first introduced fluorescein, followed by tetramethylrhodamine but did not analyze energy transfer. While these systems were not expressly built for optical or catalytic applications, they provided a proof of principle and inspiration for much of the work to follow.

In subsequent papers published the same year, the Finn and Johnson groups explored the reactivity of lysine residues and pushed the envelope for chromophore bioconjugation.[62] They showed that up to 240 chromophores could be installed using NHS ester chemistry. Lysines could also be targeted to introduce a biotin tag, which caused the formation of extended networks upon addition of avidin.[63] They also demonstrated the sequential and independent labeling of cysteine and lysine residues.

Four years later, Evans and coworkers took advantage of the high reactivity of CPMV's lysines to introduce 240 copies of NHS ester-modified ferrocene groups on each capsid.[122] This represented a new record for the number of redox-active groups on a single scaffold, surpassing previously reported dendrimer-based systems by almost a factor of four. Despite the crowding expected from placing four metal centers on each monomer, the ferrocene units were observed to behave independently using cyclic voltammetry. The authors proposed that the modified CPMV capsids could be used as electron storage reservoirs with the capability for multielectron transfer. Taking a different approach with another 30-nm icosahedral virus, the Finn group explored noncovalent catalyst attachment to the hepatitis B virus (HBV). To do this, they introduced a His_6 tag, commonly used for protein purification over immobilized nickel ions, and instead used it to bind iron protoporphyrin IX (or heme) groups from solution.[123] Despite the bulk of both the His_6 tag and the heme, they were able to attach an average of 80 copies to each capsid. As with the ferrocene case, the attached hemes were observed to behave as independent redox entities despite their proximity.

B. Tobacco Mosaic Virus

The coat protein of TMV has appeared in several reports as a means to arrange large arrays of synthetic chromophores. In contrast to the work done interfacing TMV with inorganic materials, light harvesting systems have been built from both the intact virus and the partially assembled coat proteins, either extracted from viruses or produced recombinantly. The self-assembly properties of TMV have been thoroughly studied, allowing researchers to access different metastable structures by changing the pH and ionic strength of the solution.[21] The most appealing structures for synthetic manipulation have been double disks, consisting of two stacked rings of 17 monomers each, and helical rods, which contract slightly to a periodicity of 16.33 monomers per turn and can range from tens to thousands of nanometers in length. The subunits in both structures are tightly packed so that attaching a chromophore to the same position on each protein yields a well-ordered array with interchromophore spacing well suited for energy transfer via FRET.

TMV has multiple distinct surfaces but few inherently reactive sites for covalent modification, which researchers have overcome in a number of creative ways. In the first example of using covalent chemistry to produce TMV-based materials, the Francis lab targeted native tyrosines on the exterior surface using a newly developed diazonium coupling reaction, and targeted native carboxylic acids in the pore of the rod to insert polymers, chromophores, and chelating groups.[35] Shortly thereafter, two reports were published introducing cystine mutants to facilitate the modification of different surfaces of the TMV rod. The groups of Harris and Culver introduced a cystine on the exterior, which they were able to functionalize with chromophores.[124] They also developed a clever surface immobilization strategy in which the end of the genomic RNA was revealed and hybridized to a complementary strand patterned on a gold surface. To modify the pore of TMV, the Majima group introduced a cysteine at position 99 or 100, which they were able to modify with a pyrene maleimide when the protein was disassembled into monomers.[125] Rather than interfering with the assembly, the pyrene seemed to encourage it by promoting the formation of longer rods than were formed from unmodified protein. Fluorescence measurements of the pyrene-containing rods indicated the formation of excimers, suggesting that the increased assembly may be due to favorable π-stacking in the pore.

The Francis group developed a system to collect light across a broad spectrum by creating a closely spaced array of mixed chromophores with spectral properties that allowed FRET down an energy gradient, analogous to natural photosynthetic systems.[36] Recombinantly derived TMV coat protein lacks genomic RNA, which usually winds through an internal channel in the assembled rods. Taking advantage of this newly available "surface" for covalent modification, the Francis group introduced a cysteine into position 123 to allow the straightforward modification of monomers with a variety of commercially available maleimide dyes. Subsequent assembly into rods or disks positioned the dyes to within 1.5 nm, which is well within the Förster radius for efficient energy transfer between identical dyes as well as donor–acceptor neighbors (Fig. 8B).

The modular nature of the system allowed measuring the energy transfer with a variety of chromophore ratios and combinations, in both the disk and rod assemblies. Studying systems of two spectrally distinct donor and acceptor dyes provided easier deconvolution and characterization, but limited efficiency (up to 47%). Much higher efficiencies, exceeding 90%, were observed following the incorporation of a third dye to bridge the gap between the donor and acceptor of the simpler systems. This highly efficient architect_ _ _..ws broad spectrum light to be collected and transferred over relatively large distances, permitting an acceptor to collect light from a maximal number of donors.

More detailed analysis of the energy transfer rates were performed using time resolved spectroscopic methods in a subsequent collaboration ʷith the Fleming group.[126] Additional studies performed with the Geissler ʳoup showed that energy transfer is possible not only to the nearest neighbor but to neighbors in all directions, providing many paths for energy to travel froᵤ the light-absorbing chromophore to the final acceptor.[127] This property is crucial to provide defect tolerance, for example, arising from the photobleaching of pigments.

In addition to the synthetic chromophores discussed above, the empty RNA channel has played host to molecules as large as porphyrins, as reported by the Majima group.[128] By introducing a cysteine at position 127, they created a similar system in which their donor zinc porphyrins and acceptor freebase porphyrins were conjugated to TMV via a maleimide linker. This held them in close proximity to one another upon monomer assembly into short rods. One advantage of using porphyrins is their very high extinction coefficient, though the large gap between excitation and emission prohibited significant donor-to-donor FRET, and therefore long-distance energy transfer.

Having established the possibility of collecting light with TMV-based system, the next challenges were converting the light into useful energy and collecting that energy on a large scale. To address the first challenge, the Francis group redesigned the TMV coat protein to introduce a new reactive site in the pore.[129] Their ultimate goal was to place an electron-transfer catalyst there to act as a final energy acceptor from the surrounding chromophores in the RNA channel. To do this, they first cut and reassembled the coat protein gene to link the preexisting termini on the exterior and introduce new ones in the pore—a process called circular permutation. The repositioned N-termini were then selectively converted into aldehydes using a PLP-mediated transamination reaction,[130] providing an orthogonal reactive handle in the central channel. Donor chromophores were placed in the RNA channel as before to serve as an "antenna," and an acceptor was coupled to the aldehydes just a few nanometers away, allowing it to collect light from the surrounding donors either in a rod or disk assembly. In the latter arrangement, the acceptor mimicked the reaction center of the LH1 complex, with the future goal of replacing it with an electron-transfer catalyst. This permuted TMV further differed from the known TMV coat protein by forming robust disk assemblies under a broader range of conditions, making them more promising candidates for integration into devices.

The second challenge involves interfacing TMV-based light harvesting systems with devices to collect the high-energy electrons produced, either in the form of chemical bonds or as a photocurrent. For the latter system, an electrically conductive component might be required to connect light harvesting components to an electrode. Carbon nanotubes have a number of properties that would make them ideal in this role, such as high conductivity and a demonstrated ability to interface with a variety of photocatalytic materials.[131,132] Since TMV and nanotubes do not associate on their own, the Francis group developed a method to pretreat both TMV rods and carbon nanotubes such that they would bind together. In parallel, an exterior tyrosine of TMV was modified to display aldehydes using a diazonium coupling method,[35] while the carbon nanotubes were solubilized with a surfactant-like pyrene–PEG–aminooxy bifunctional linker, which bound the nanotubes with the hydrophobic pyrene.[133] Upon mixing, the aminooxy and aldehydes reacted to form oximes, linking the TMV and nanotubes. In this proof-of-principle work, the PEG was chosen to solubilize the nanotubes despite is poor conductivity. In the future, it would need to be replaced with a more conductive polymer to transfer electrons from TMV to nanotubes.

C. Photocatalytic Systems Based on Bacteriophage MS2

Photosynthesis uses the energy of absorbed light to form high-energy electrons, which are ultimately used to generate chemical bonds. As an alternative to TMV-based photovoltaic systems emphasized above, artificial light harvesting systems can also be designed to generate high-energy electron carriers, for example, reducing $NADP^+$ to NADPH. The requirements for this system are essentially identical to those described above: the ability to array donor chromophores within FRET distance of both each other and a photocatalytic acceptor. The icosahedral protein shell of MS2 provides these capabilities in a distinctly different structure. In a report by the Francis group, an array of 180 donor chromophores was created by modifying a cysteine introduced to the interior of the empty capsid.[118] A zinc porphyrin was chosen as a suitable acceptor, and targeted to the unnatural amino acid *p*aF installed on the exterior surface through the Schultz *in vivo* method.[8,28] Because the protein shell is only about 1 nm thick, the separation between the interior donors and the exterior acceptors was well within FRET distance (Fig. 8C). The large number of exterior reactive sites (180) allowed the researchers to screen a variety of catalyst loadings (20–120 zinc porphyrins) to study the energy transfer process.

The flexibility of the system also allowed a comparison between donors designed to sensitize the porphyrin at either of two wavelengths (350 or 500 nm) where the porphyrin did not strongly absorb. Upon illuminating the donor, the energy transfer to the zinc porphyrin could be measured by the reduction of methyl viologen, which produced a colorimetric response.

The catalytic cycle was completed by addition of the sacrificial reductant, β-mercaptoethanol, to reduce the resulting porphyrin radical cations. When coupled to the chromophore array, the porphyrins were capable of photoreduction over a broader set of input wavelengths. Illumination at the peak donor absorption band produced an almost four-fold increase in the reduction of methyl viologen compared to a system with no donor.

D. Bacteriophage M13

Bacteriophage M13 is another helical virus that has been used to template arrays of dyes and photocatalysts. Its utility for light harvesting was first demonstrated by Scolaro and coworkers, who found that cationic porphyrins would spontaneously adsorb to its surface.[134] These porphyrins served as FRET acceptors for energy absorbed by a surface tryptophan introduced through mutation. However, the light harvesting efficiency was limited both by the low extinction coefficient of tryptophan and the poor overlap between donor emission and acceptor excitation.

The Belcher lab has created light harvesting catalytic systems with M13 as well, taking advantage of their extensive experience interfacing phage with inorganic particles to design complex structures. They began by attaching zinc porphyrins to the 2700 pVIII coat proteins that run the length of the virus.[135] These were introduced by targeting the N-terminus and a lysine on each monomer, providing distances of 1–2.4 nm between the reactive sites. This was close enough to allow significant amounts of electronic coupling. Using transient pump–probe spectroscopy, exciton delocalization was indeed observed in the porphyrin system as quenched fluorescence emission and decreased lifetime. Building on this work in a subsequent report, Belcher and coworkers used phage display to develop an M13 mutant that could also bind iridium oxide (IrO_2) clusters after porphyrin attachment.[119] These clusters could then nucleate the encapsulation of the virus–porphyrin assembly with an iridium oxide shell under mild conditions. The porphyrins then served as photosensitizers, oxidizing the iridium shell, which in turn oxidized water. The integration of the porphyrins and the catalysts with the viral template was found to improve the quantum yield relative to systems with no porphyrin or no M13 template, highlighting the potential of using self-assembling protein structures to arrange nanoscale components for optimal solar energy collection and conversion.

In a final example, the Belcher group used a similar approach to develop catalysts for conversion of ethanol into hydrogen and carbon dioxide, displaying the versatility of the M13 phage for templating nanoscale catalysts.[136] Through a process of biopanning, or selected evolution, the lab identified an M13 mutant capable of binding metal ions along its length. This binding is based on extra glutamate residues in the pVIII protein as discussed in the preceding section, and can be used to nucleate the formation of CeO$_2$ nanowires doped with nickel and rhodium. When tested in reactions and compared to nontemplated catalysts, these nanowires displayed greater stability, which was attributed to decreased phase separation, greater resistance to surface deactivation, and smaller catalyst pore size. These demonstrations of using M13 to template improved inorganic materials suggest that this technique has great potential to enhance other catalytic systems.

E. Summary

The past 10 years of research has provided strong demonstrations of the utility of viral capsids as scaffolds for light harvesting and catalysis. The ability to position orthogonal reactive groups with precision allows the integration of different interacting device components, and the self-assembly of repeating units permits large arrays to be constructed with relative ease. Artificial systems mimicking various aspects of photosynthesis have been developed, and should increase in complexity as additional reactive handles are introduced to attach components such as photoprotection groups. In the future, these viral components may be incorporated into a range of devices.

V. Summary and Future Challenges

The examples discussed in this chapter clearly indicate that viral capsids (and by extension, many other self-assembling proteins) have much to offer for the construction of new materials. Although this success has been based on many different capsids that have been modified using a highly creative list of synthetic approaches, the common theme is the use of the well-defined protein structure to template nanoscale architectures that are very difficult to achieve on this size scale. These examples also highlight many of the future challenges that must be addressed before viral capsids reach their full synthetic potential.

First among these is the development of new chemical methods that can be used to attach synthetic functionality to their surfaces. Although many of the above examples have made effective use of cysteine and lysine modifications, the generation of more complex materials with increased number of functional components will require additional chemistries that can be combined with tried-and-true bioconjugation methods. Artificial amino acids provide one promising way to implement these new reactive strategies.

It should also be noted that most viral capsids exhibit a very high level of symmetry that is difficult to break in a controlled fashion. While this situation is generally satisfactory for the creation of core–shell materials for drug delivery, many future applications in energy collection and conversion will require asymmetric arrays of chromophores and photocatalysts that can direct the energy flow. Perhaps the best way to achieve this will be to design new protein monomers that can mimic the main features of viral capsid assembly, but through the use of a richer set of sequences that come together with defined positional relationships in the final structures. These approaches will certainly benefit from the availability of emerging theoretical tools that can model and predict complex self-assembly behavior.

For optical and electronic materials, completely new methods are required to interface biologically templated materials with device surfaces. Our current understanding of the ways in which proteins interact with surfaces is limited at best, making the generation of well-oriented protein arrays a very difficult task. This is coupled with the fact that proteins have inherent insulating properties, and thus will most likely require embedded metals or arrays of well-positioned metal complexes to achieve efficient electron transfer. In general, proteins have good photostability, and many can withstand a significantly wide range of temperature and pH conditions without losing their three-dimensional structure. However, even proteins extracted from thermophilic organisms will place limitations on the range of temperatures that can be used for processing and fabrication. While not insurmountable, these considerations will provide many challenges as biomolecules are wed with inorganic materials for commercial applications.

In terms of biomedical applications, careful studies are needed to obtain a better picture of the fate of modified viral capsids that find their way into the bloodstream. While many of them are likely to elicit an immune response, it may be possible to attenuate this through shielding with polymer chains or altering the sequences of the protein monomers. In addition, there are still few if any examples of using viral capsids for the recognition of specific targets *in vivo*. While this concept is likely to be demonstrated in the coming years, many such studies will be required to understand the complex and currently nonintuitive influences of multivalency, overall charge, size, aspect ratio, and cargo loading on the biodistribution of these particles. Although drug delivery may be an end goal of these studies, *in vivo* imaging techniques are currently poised to provide much of this information.

ACKNOWLEDGMENTS

Our efforts to develop new bioconjugation strategies, capsid-based delivery agents, and protein–polymer hybrid materials have been generously supported by the NIH (GM072700), the DOD Breast Cancer Research Program (BC061995), the NSF (0449772), and the Director, Office of Science, Materials Sciences and Engineering Division, U.S. Department of Energy, under Contract No. DE-AC02-05CH11231.

REFERENCES

1. Whitesides GM, Mathias JP, Seto CT. Molecular self-assembly and nanochemistry: a chemical strategy for the synthesis of nanostructures. *Science* 1991;**29**:1312–9.
2. Cui H, Webber MJ, Stupp SI. Self-assembly of peptide amphiphiles: from molecules to nanostructures to biomaterials. *Pept Sci* 2010;**94**:1–18.
3. Rajagopal K, Schneider JP. Self-assembling peptides and proteins for nanotechnological applications. *Curr Opin Struct Biol* 2004;**14**:480–6.
4. Zhang S. Fabrication of novel biomaterials through molecular self-assembly. *Nat Biotechnol* 2003;**21**:1171–8.
5. Zhao X, Pan F, Xu H, Yaseen M, Shan H, Hauser CA, et al. Molecular self-assembly and applications of designer peptide amphiphiles. *Chem Soc Rev* 2010;**39**:3480–98.
6. Seeman NC. Nanomaterials based on DNA. *Annu Rev Biochem* 2010;**79**:65–87.
7. Nangreave J, Han D, Liu Y, Yan H. DNA origami: a history and current perspective. *Curr Opin Chem Biol* 2010;**14**:608–15.
8. Xie J, Schultz PG. A chemical toolkit for proteins—an expanded genetic code. *Nat Rev Mol Cell Biol* 2006;**7**:775–82.
9. Link AJ, Mock ML, Tirrell DA. Non-canonical amino acids in protein engineering. *Curr Opin Biotechnol* 2003;**14**:603–9.
10. Flenniken ML, Uchida M, Liepold LO, Kang S, Young MJ, Douglas T. Cage architectures as nanomaterials. *Curr Top Microbiol Immunol* 2009;**327**:71–93.
11. Douglas T, Young M. Viruses: making friends with old foes. *Science* 2006;**312**:873–5.
12. Evans DJ. The bionanoscience of plant viruses: templates and synthons for new materials. *J Mater Chem* 2008;**18**:3746–54.
13. Fischlechner M, Donath E. Viruses as building blocks for materials and devices. *Angew Chem Int Ed* 2007;**46**:3184–93.
14. Lin T. Structural genesis of the chemical addressability in a viral nano-block. *J Mater Chem* 2006;**16**:3673–81.
15. Niemeyer CM. Nanoparticles, proteins, and nucleic acids: biotechnology meets materials science. *Angew Chem Int Ed* 2001;**40**:4128–58.
16. Singh P, Gonzalez MJ, Manchester M. Viruses and their uses in nanotechnology. *Drug Dev Res* 2006;**67**:23–41.
17. Steinmetz NF, Evans DJ. Utilisation of plant viruses in bionanotechnology. *Org Biomol Chem* 2007;**5**:2891–902.
18. Uchida M, Klem M, Allen M, Suci P, Flenniken M, Gillitzer E, et al. Biological containers: protein cages as multifunctional nanoplatforms. *Adv Mater* 2007;**19**:1025–42.
19. Young M, Debbie W, Uchida M, Douglas T. Plant viruses as biotemplates for materials and their use in nanotechnology. *Annu Rev Phytopathol* 2008;**46**:361–84.
20. Harrison SC. Principles of virus structure. In: Fields BN, et al. editors. *Virology*. New York: Raven Press; 1985. p. 27–44.
21. Klug A. The tobacco mosaic virus particle: structure and assembly. *Philos Trans R Soc Lond B* 1999;**354**:531–5.
22. Butler PJG. Self-assembly of tobacco mosaic virus: the role of an intermediate aggregate in generating both specificity and speed. *Philos Trans R Soc Lond B* 1999;**354**:537–50.
23. Sun J, DuFort C, Daniel M, Murali A, Chen C, Gopinath K, et al. Core-controlled polymorphism in virus-like particles. *Proc Natl Acad Sci USA* 2007;**104**:1354–9.
24. Specthrie L, Bullitt E, Horiuchi K, Model P, Russel M, Makowski L. Construction of a microphage variant of filamentous bacteriophage° 1. *J Mol Biol* 1992;**228**:720–4.

25. Brumfield S, Willits D, Tang L, Johnson JE, Douglas T, Young M. Heterologous expression of the modified coat protein of *cowpea chlorotic mottle bromovirus* results in the assembly of protein cages with altered architectures and function. *J Gen Virol* 2004;**85**:1049–53.
26. Siler DJ, Babcock J, Bruening G. Electrophoretic mobility and enhanced infectivity of a mutant of cowpea mosaic virus. *Virology* 1976;**71**:560–7.
27. Hooker JM, Kovacs EW, Francis MB. Interior surface modification of bacteriophage MS2. *J Am Chem Soc* 2004;**126**:3718–9.
28. Carrico ZM, Romanini DW, Mehl RA, Francis MB. Oxidative coupling of peptides to a virus capsid containing unnatural amino acids. *Chem Commun* 2008;1205–7.
29. Kozlovska TM, Cielens I, Dreilina D, Dislers A, Baumanis V, Ose V, et al. Recombinant RNA phage Qβ capsid particles synthesized and self-assembled in *Escherichia coli*. *Gene* 1993;**137**:133–7.
30. Yuan W, Parrish CR. Canine parvovirus capsid assembly and differences in mammalian and insect cells. *Virology* 2001;**279**:546–57.
31. Sun JH, Adkins S, Faurote G, Kao CC. Initiation of (−) strand RNA synthesis catalyzed by the BMV RNA-dependent RNA polymerase: synthesis of oligoribonucleotides. *Virology* 1996;**225**:1–12.
32. Barnhill HN, Reuther R, Ferguson PL, Dreher T, Wang Q. Turnip yellow mosaic virus as a chemoaddressable bionanoparticle. *Bioconjug Chem* 2007;**18**:852–9.
33. Loo L, Guenther RH, Basnayake VR, Lommel SA, Franzen S. Controlled encapsidation of gold nanoparticles by a viral protein shell. *J Am Chem Soc* 2006;**128**:4502–3.
34. Flenniken ML, Willits DA, Brumfield S, Young M, Douglas T. The small heat shock protein cage from *Methanococcus jannaschii* is a versatile nanoscale platform for genetic and chemical modification. *Nano Lett* 2003;**3**:1573–6.
35. Schlick TL, Ding Z, Francis MB. Dual-surface modification of the tobacco mosaic virus. *J Am Chem Soc* 2005;**127**:3718–23.
36. Miller RA, Presley AD, Francis MB. Self-assembling light-harvesting systems from synthetically modified tobacco mosaic virus coat proteins. *J Am Chem Soc* 2007;**129**:3104–9.
37. Webby RJ, Kalmakoff J. Comparison of the major capsid protein genes, terminal redundancies, and DNA–DNA homologies of two New Zealand iridoviruses. *Virus Res* 1999;**59**:179–89.
38. Barray S, Devauchelle G. Protein synthesis in cells infected by Chilo iridescent virus (Iridovirus type 6). *Arch virol* 1985;**86**:315–26.
39. Hermanson GT. *Bioconjugate techniques*. 2nd ed. San Diego: Academic Press; 2008.
40. Tilley SD, Joshi NS, Francis MB. The chemistry and chemical reactivity of proteins. In: Begley T, editor. *The Wiley encyclopedia of chemical biology*. Weinheim: Wiley-VCH; 2008.
41. Allen M, Willits D, Mosolf J, Young M, Douglas T. Protein engineering of a viral cage for constrained nanomaterials synthesis. *Adv Mater* 2002;**14**:1562–5.
42. Allen M, Willits D, Young M, Douglas T. Constrained synthesis of cobalt oxide nanomaterials in the 12-subunit protein cage from *Listeria innocua*. *Inorg Chem* 2003;**42**:6300–5.
43. Ensign D, Young M, Douglas T. Photocatalytic synthesis of copper colloids from Cu (II) by the ferrihydrite core of ferritin. *Inorg Chem* 2004;**43**:3441–6.
44. Varpness Z, Peters JW, Young M, Douglas T. Biomimetic synthesis of a H_2 catalyst using a protein cage architecture. *Nano Lett* 2005;**5**:2306–9.
45. Abedin MJ, Liepold L, Suci P, Young M, Douglas T. Synthesis of a cross-linked branched polymer network in the interior of a protein cage. *J Am Chem Soc* 2009;**131**:4346–54.
46. Suci PA, Varpness Z, Gillitzer E, Douglas T, Young M. Targeting and photodynamic killing of a microbial pathogen using protein cage architectures functionalized with a photosensitizer. *Langmuir* 2007;**23**:12280–6.
47. Maeda H, Wu J, Sawa T, Matsumura Y, Hori K. Tumor vascular permeability and the EPR effect in macromolecular therapeutics: a review. *J Control Release* 2000;**65**:271–84.

48. Smith GP, Petrenko VA. Phage display. *Chem Rev* 1997;**97**:391–410.
49. Kehoe JW, Kay BK. Filamentous phage display in the new millennium. *Chem Rev* 2005;**105**:4056–72.
50. Verma IM, Weitzman MD. GENE THERAPY: twenty-first century medicine. *Annu Rev Biochem* 2005;**74**:711–38.
51. Schaffer DV, Koerber JT, Lim K. Molecular engineering of viral gene delivery vehicles. *Annu Rev Biomed Eng* 2008;**10**:169–94.
52. Garcea RL, Gissmann L. Virus-like particles as vaccines and vessels for the delivery of small molecules. *Curr Opin Biotechnol* 2004;**15**:513–7.
53. Gonzalez MJ, Plummer EM, Rae CS, Manchester M. Interaction of cowpea mosaic virus (CPMV) nanoparticles with antigen presenting cells *in vitro* and *in vivo*. *PLoS ONE* 2009;**4**:e7981.
54. Douglas T, Young M. Host–guest encapsulation of materials by assembled virus protein cages. *Nature* 1998;**393**:152–5.
55. Gillitzer E, Willits D, Young M, Douglas T. Chemical modification of a viral cage for multivalent presentation. *Chem Commun* 2002;2390–1.
56. Kaiser CR, Flenniken ML, Gillitzer E, Harmsen AL, Harmsen AG, Jutila MA, et al. Biodistribution studies of protein cage nanoparticles demonstrate broad tissue distribution and rapid clearance *in vivo*. *Int J Nanomed* 2007;**2**:715–33.
57. Allen M, Bulte JWM, Liepold L, Basu G, Zywicke HA, Frank JA, et al. Paramagnetic viral nanoparticles as potential high-relaxivity magnetic resonance contrast agents. *Magn Reson Med* 2005;**54**:807–12.
58. Liepold L, Anderson S, Willits D, Oltrogge L, Frank JA, Douglas T, et al. Viral capsids as MRI contrast agents. *Magn Reson Med* 2007;**58**:871–9.
59. Caravan P, Ellison JJ, McMurry TJ, Lauffer RB. Gadolinium (III) chelates as MRI contrast agents: structure, dynamics, and applications. *Chem Rev* 1999;**99**:2293–352.
60. Hatz S, Poulsen L, Ogilby PR. Time-resolved singlet oxygen phosphorescence measurements from photosensitized experiments in single cells: effects of oxygen diffusion and oxygen concentration. *Photochem Photobiol* 2008;**84**:1284–90.
61. Wang Q, Lin T, Tang L, Johnson JE, Finn MG. Icosahedral virus particles as addressable nanoscale building blocks. *Angew Chem Int Ed* 2002;**41**:459–62.
62. Wang Q, Kaltgrad E, Lin T, Johnson JE, Finn MG. Natural supramolecular building blocks: wild-type cowpea mosaic virus. *Chem Biol* 2002;**9**:805–11.
63. Wang Q, Lin T, Johnson JE, Finn MG. Natural supramolecular building blocks: cysteine-added mutants of cowpea mosaic virus. *Chem Biol* 2002;**9**:813–9.
64. Raja KS, Wang Q, Finn MG. Icosahedral virus particles as polyvalent carbohydrate display platforms. *Chembiochem* 2003;**4**:1348–51.
65. Chatterji A, Ochoa W, Shamieh L, Salakian SP, Wong SM, Clinton G, et al. Chemical conjugation of heterologous proteins on the surface of cowpea mosaic virus. *Bioconjug Chem* 2004;**15**:807–13.
66. Wang Q, Chan TR, Hilgraf R, Fokin VV, Sharpless KB, Finn MG. Bioconjugation by copper (I)-catalyzed azide-alkyne [3+2] cycloaddition. *J Am Chem Soc* 2003;**125**:3192–3.
67. Moses JE, Moorhouse AD. The growing applications of click chemistry. *Chem Soc Rev* 2007;**36**:1249–62.
68. Sen Gupta S, Kuzelka J, Singh P, Lewis WG, Manchester M, Finn MG. Accelerated bioorthogonal conjugation: a practical method for the ligation of diverse functional molecules to a polyvalent virus scaffold. *Bioconjug Chem* 2005;**16**:1572–9.
69. Banerjee D, Liu AP, Voss NR, Schmid SL, Finn MG. Multivalent display and receptor-mediated endocytosis of transferrin on virus-like particles. *Chembiochem* 2010;**11**:1273–9.

70. Rae CS, Wei Khor I, Wang Q, Destito G, Gonzalez MJ, Singh P, et al. Systemic trafficking of plant virus nanoparticles in mice via the oral route. *Virology* 2005;**343**:224–35.
71. Lewis JD, Destito G, Zijlstra A, Gonzalez MJ, Quigley JP, Manchester M, et al. Viral nanoparticles as tools for intravital vascular imaging. *Nat Med* 2006;**12**:354–60.
72. Prasuhn Jr. DE, Yeh RM, Obenaus A, Manchester M, Finn MG. Viral MRI contrast agents: coordination of Gd by native virions and attachment of Gd complexes by azide–alkyne cycloaddition. *Chem Commun* 2007;1269.
73. Singh P, Prasuhn D, Yeh RM, Destito G, Rae CS, Osborn K, et al. Bio-distribution, toxicity and pathology of cowpea mosaic virus nanoparticles *in vivo*. *J Control Release* 2007;**120**:41–50.
74. Manchester M, Singh P. Virus-based nanoparticles (VNPs): platform technologies for diagnostic imaging. *Adv Drug Deliv Rev* 2006;**58**:1505–22.
75. Steinmetz NF, Hong V, Spoerke ED, Lu P, Breitenkamp K, Finn MG, et al. Buckyballs meet viral nanoparticles: candidates for biomedicine. *J Am Chem Soc* 2009;**131**:17093–5.
76. Raja KS, Wang Q, Gonzalez MJ, Manchester M, Johnson JE, Finn MG. Hybrid virus-polymer materials. 1. Synthesis and properties of PEG-decorated cowpea mosaic virus. *Biomacromolecules* 2003;**4**:472–6.
77. Singh P, Destito G, Schneemann A, Manchester M. Canine parvovirus-like particles, a novel nanomaterial for tumor targeting. *J Nanobiotechnol* 2006;**4**:2.
78. Mastico RA, Talbot SJ, Stockley PG. Multiple presentation of foreign peptides on the surface of an RNA-free spherical bacteriophage capsid. *J Gen Virol* 1993;**74**:541–8.
79. Kovacs EW, Hooker JM, Romanini DW, Holder PG, Berry KE, Francis MB. Dual-surface-modified bacteriophage MS2 as an ideal scaffold for a viral capsid-based drug delivery system. *Bioconjug Chem* 2007;**18**:1140–7.
80. Anderson EA, Isaacman S, Peabody DS, Wang EY, Canary JW, Kirshenbaum K. Viral nanoparticles donning a paramagnetic coat: conjugation of MRI contrast agents to the MS2 capsid. *Nano Lett* 2006;**6**:1160–4.
81. Hooker JM, Datta A, Botta M, Raymond KN, Francis MB. Magnetic resonance contrast agents from viral capsid shells: a comparison of exterior and interior cargo strategies. *Nano Lett* 2007;**7**:2207–10.
82. Xu J, Franklin SJ, Whisenhunt DW, Raymond KN. Gadolinium complex of *tris*[(3-hydroxy-1-methyl- 2-oxo-1,2-didehydropyridine-4-carboxamido)ethyl]-amine: a new class of gadolinium magnetic resonance relaxation agents. *J Am Chem Soc* 1995;**117**:7245–6.
83. Datta A, Hooker JM, Botta M, Francis MB, Aime S, Raymond KN. High relaxivity gadolinium hydroxypyridonate-viral capsid conjugates: nanosized MRI contrast agents. *J Am Chem Soc* 2008;**130**:2546–52.
84. Meldrum T, Seim KL, Bajaj VS, Palaniappan K, Wu W, Francis MB, et al. A xenon-based molecular sensor assembled on an MS2 viral capsid scaffold. *J Am Chem Soc* 2010;**132**:5936–7.
85. Schröder L, Lowery T, Hilty C, Wemmer D, Pines A. Molecular imaging using a targeted magnetic resonance hyperpolarized biosensor. *Science* 2006;**314**:446–9.
86. Hooker J, O'Neil J, Romanini D, Taylor S, Francis MB. Genome-free viral capsids as carriers for positron emission tomography radiolabels. *Mol Imaging Biol* 2008;**10**:182–91.
87. Hooker JM, Esser-Kahn AP, Francis MB. Modification of aniline containing proteins using an oxidative coupling strategy. *J Am Chem Soc* 2006;**128**:15558–9.
88. Tong GJ, Hsiao SC, Carrico ZM, Francis MB. Viral capsid DNA aptamer conjugates as multivalent cell-targeting vehicles. *J Am Chem Soc* 2009;**131**:11174–8.
89. Shangguan D, Li Y, Tang Z, Cao ZC, Chen HW, Mallikaratchy P, et al. Aptamers evolved from live cells as effective molecular probes for cancer study. *Proc Natl Acad Sci USA* 2006;**103**:11838–43.

90. Stephanopoulos N, Tong GJ, Hsiao SC, Francis MB. Dual-surface modified virus capsids for targeted delivery of photodynamic agents to cancer cells. *ACS Nano* 2010;4:6014–20.

91. Wu W, Hsiao S, Carrico Z, Francis M. Genome-free viral capsids as multivalent carriers for taxol delivery. *Angew Chem Int Ed* 2009;48:9493–7.

92. Douglas T, Strable E, Willits D, Aitouchen A, Libera M, Young M. Protein engineering of a viral cage for constrained nanomaterials synthesis. *Adv Mater* 2002;14:415–8.

93. Aniagyei SE, Kennedy CJ, Stein B, Willits DA, Douglas T, Young M, et al. Synergistic effects of mutations and nanoparticle templating in the self-assembly of cowpea chlorotic mottle virus capsids. *Nano Lett* 2009;9:393–8.

94. Radloff C, Vaia RA, Brunton J, Bouwer GT, Ward VK. Metal nanoshell assembly on a virus bioscaffold. *Nano Lett* 2005;5:1187–91.

95. Falkner JC, Turner ME, Bosworth JK, Trentler TJ, Johnson JE, Lin T, et al. Virus crystals as nanocomposite scaffolds. *J Am Chem Soc* 2005;127:5274–5.

96. Chen C, Daniel M, Quinkert ZT, De M, Stein B, Bowman VD, et al. Nanoparticle-templated assembly of viral protein cages. *Nano Lett* 2006;6:611–5.

97. Dixit SK, Goicochea NL, Daniel M, Murali A, Bronstein L, De M, et al. Quantum dot encapsulation in viral capsids. *Nano Lett* 2006;6:1993–9.

98. Loo L, Guenther RH, Lommel SA, Franzen S. Encapsidation of nanoparticles by red clover necrotic mosaic virus. *J Am Chem Soc* 2007;129:11111–7.

99. Whaley SR, English DS, Hu EL, Barbara PF, Belcher AM. Selection of peptides with semiconductor binding specificity for directed nanocrystal assembly. *Nature* 2000;405:665–8.

100. Lee S, Mao C, Flynn CE, Belcher AM. Ordering of quantum dots using genetically engineered viruses. *Science* 2002;296:892–5.

101. Mao C, Solis DJ, Reiss BD, Kottmann ST, Sweeney RY, Hayhurst A, et al. Virus-based toolkit for the directed synthesis of magnetic and semiconducting nanowires. *Science* 2004;303:213–7.

102. Nam KT, Kim D, Yoo PJ, Chiang C, Meethong N, Hammond PT, et al. Virus-enabled synthesis and assembly of nanowires for lithium ion battery electrodes. *Science* 2006;312:885–8.

103. Lee YJ, Yi H, Kim W, Kang K, Yun DS, Strano MS, et al. Fabricating genetically engineered high-power lithium-ion batteries using multiple virus genes. *Science* 2009;324:1051–5.

104. Avery KN, Schaak JE, Schaak RE. M13 bacteriophage as a biological scaffold for magnetically-recoverable metal nanowire catalysts: combining specific and nonspecific interactions to design multifunctional nanocomposites. *Chem Mater* 2009;21:2176–8.

105. Shenton W, Douglas T, Young M, Stubbs G, Mann S. Inorganic–organic nanotube composites from template mineralization of tobacco mosaic virus. *Adv Mater* 1999;11:253–6.

106. Fowler CE, Shenton W, Stubbs G, Mann S. Tobacco mosaic virus liquid crystals as templates for the interior design of silica mesophases and nanoparticles. *Adv Mater* 2001;13:1266–9.

107. Kuncicky D, Naik R, Velev O. Rapid deposition and long-range alignment of nanocoatings and arrays of electrically conductive wires from tobacco mosaic virus. *Small* 2006;2:1462–6.

108. Demir M, Stowell MHB. A chemoselective biomolecular template for assembling diverse nanotubular materials. *Nanotechnology* 2002;13:541–4.

109. Niu Z, Liu J, Lee LA, Bruckman MA, Zhao D, Koley G, et al. Biological templated synthesis of water-soluble conductive polymeric nanowires. *Nano Lett* 2007;7:3729–33.

110. Dujardin E, Peet C, Stubbs G, Culver JN, Mann S. Organization of metallic nanoparticles using tobacco mosaic virus templates. *Nano Lett* 2003;3:413–7.

111. Knez M, Bittner AM, Boes F, Wege C, Jeske H, Maiβ E, et al. Biotemplate synthesis of 3-nm nickel and cobalt nanowires. *Nano Lett* 2003;3:1079–82.

112. Balci S, Bittner A, Hahn K, Scheu C, Knez M, Kadri A, et al. Copper nanowires within the central channel of tobacco mosaic virus particles. *Electrochim Acta* 2006;51:6251–7.

113. Lee S, Royston E, Culver JN, Harris MT. Improved metal cluster deposition on a genetically engineered tobacco mosaic virus template. *Nanotechnology* 2005;**16**:S435–41.

114. Royston E, Ghosh A, Kofinas P, Harris MT, Culver JN. Self-assembly of virus-structured high surface area nanomaterials and their application as battery electrodes. *Langmuir* 2008;**24**:906–12.

115. Tseng RJ, Tsai C, Ma L, Ouyang J, Ozkan CS, Yang Y. Digital memory device based on tobacco mosaic virus conjugated with nanoparticles. *Nat Nanotechnol* 2006;**1**:72–7.

116. Kobayashi M, Seki M, Tabata H, Watanabe Y, Yamashita I. Fabrication of aligned magnetic nanoparticles using tobamoviruses. *Nano Lett* 2010;**10**:773–6.

117. Roszak AW, Howard TD, Southall J, Gardiner AT, Law CJ, Isaacs NW, et al. Crystal structure of the RC–LH1 core complex from *Rhodopseudomonas palustris*. *Science* 2003;**302**:1969.

118. Stephanopoulos N, Carrico ZM, Francis MB. Nanoscale integration of sensitizing chromophores and porphyrins with bacteriophage MS2. *Angew Chem Int Ed* 2009;**48**:9498–502.

119. Nam YS, Magyar AP, Lee D, Kim J, Yun DS, Park H, et al. Biologically templated photocatalytic nanostructures for sustained light-driven water oxidation. *Nat Nanotechnol* 2010;**5**:340–4.

120. Nelson N, Ben-Shem A. The complex architecture of oxygenic photosynthesis. *Nat Rev Mol Cell Biol* 2004;**5**:971.

121. Freer A, Prince S, Sauer K, Papiz M, Lawless AH, McDermott G, et al. Pigment–pigment interactions and energy transfer in the antenna complex of the photosynthetic bacterium *Rhodopseudomonas acidophila*. *Structure* 1996;**4**:449–62.

122. Steinmetz N, Lomonossoff G, Evans D. Decoration of cowpea mosaic virus with multiple, redox-active, organometallic complexes. *Small* 2006;**2**:530–3.

123. Prashun Jr. D, Kuzelka J, Strable E, Udit AK, Cho S, Lander GC, et al. Polyvalent display of heme on hepatitis B virus capsid protein through coordination to hexahistidine tags. *Chem Biol* 2008;**15**:513–9.

124. Yi H, Nisar S, Lee SY, Powers MA, Bentley WE, Payne GF, et al. Patterned assembly of genetically modified viral nanotemplates via nucleic acid hybridization. *Nano Lett* 2005;**5**:1931–6.

125. Endo M, Wang H, Fujitsuka M, Majima T. Pyrene-stacked nanostructures constructed in the recombinant tobacco mosaic virus rod scaffold. *Chem Eur J* 2006;**12**:3735–40.

126. Ma Y, Miller RA, Fleming GR, Francis MB. Energy transfer dynamics in light-harvesting assemblies templated by the tobacco mosaic virus coat protein. *J Phys Chem B* 2008;**112**:6887–92.

127. Miller RA, Stephanopoulos N, McFarland JM, Rosko AS, Geissler PL, Francis MB. Impact of assembly state on the defect tolerance of TMV-based light harvesting arrays. *J Am Chem Soc* 2010;**132**:6068–74.

128. Endo M, Fujitsuka M, Majima T. Porphyrin light-harvesting arrays constructed in the recombinant tobacco mosaic virus scaffold. *Chem Eur J* 2007;**13**:8660–6.

129. Dedeo MT, Duderstadt KE, Berger JM, Francis MB. Nanoscale protein assemblies from a circular permutant of the tobacco mosaic virus. *Nano Lett* 2010;**10**:181–6.

130. Gilmore J, Scheck R, Esser-Kahn A, Joshi N, Francis M. N-terminal protein modification through a biomimetic transamination reaction. *Angew Chem Int Ed* 2008;**47**:7777–88.

131. Guldi DM, Rahman GMA, Zerbetto F, Prato M. Carbon nanotubes in electron donor–acceptor nanocomposites. *Acc Chem Res* 2005;**38**:871–8.

132. Tasis D, Tagmatarchis N, Bianco A, Prato M. Chemistry of carbon nanotubes. *Chem Rev* 2006;**106**:1105–36.

133. Holder P, Francis M. Integration of a self-assembling protein scaffold with water-soluble single-walled carbon nanotubes. *Angew Chem Int Ed* 2007;**46**:4370–3.

134. Scolaro LM, Castriciano MA, Romeo A, Micali N, Angelini N, Lo Passo C, et al. Supramolec-ular binding of cationic porphyrins on a filamentous bacteriophage template: toward a noncovalent antenna system. *J Am Chem Soc* 2006;**128**:7446–7.

135. Nam YS, Shin T, Park H, Magyar AP, Choi K, Fantner G, et al. Virus-templated assembly of porphyrins into light-harvesting nanoantennae. *J Am Chem Soc* 2010;**132**:1462–3.

136. Neltner B, Peddie B, Xu A, Doenlen W, Durand K, Yun DS, et al. Production of hydrogen using nanocrystalline protein-templated catalysts on M13 phage. *ACS Nano* 2010;**4**:3227–35.

Index